D1267419

On Gaia

ON GAIA

A CRITICAL INVESTIGATION OF THE

RELATIONSHIP BETWEEN LIFE AND EARTH

Toby Tyrrell

PRINCETON UNIVERSITY PRESS

PRINCETON AND OXFORD

Copyright © 2013 by Princeton University Press
Requests for permission to reproduce material from this work should be
sent to Permissions, Princeton University Press
Published by Princeton University Press, 41 William Street, Princeton,
New Jersey 08540
In the United Kingdom: Princeton University Press, 6 Oxford Street,
Woodstock, Oxfordshire OX20 1TW

press.princeton.edu

All Rights Reserved

ISBN (pbk.) 978-0-691-12158-1

British Library Cataloging-in-Publication Data is available

This book has been composed in Minion Pro

Printed on acid-free paper. ∞

Printed in the United States of America

1 3 5 7 9 10 8 6 4 2

TABLE OF CONTENTS

To Helen

PREFACE

In 1972 James Lovelock made an interesting proposal. Life is not solely a passenger on a fortuitously habitable planet, he suggested. Instead, life is at the controls of the planetary environment, and has been so down through the geological ages. Continued habitability of Earth, probably for more than three billion years, has not been a coincidence, he proposed, but rather life has kept it that way. In the thirty years or so since its inception, this Gaia hypothesis has variously inspired, infuriated, and intrigued a whole generation of environmental scientists. Today it is probably more widely credited than ever, although it by no means enjoys unanimous support. Articles about it appear in the pages of *Nature* and other major journals, prestigious international conferences have been devoted to it, and vast international research programs now address topics that Lovelock was one of the first to consider.

Many books have been written about the Gaia hypothesis, but this is the first to carry out a critical examination. Gaia is a grand idea, of awesome scope. But is it correct? In this book I "dissect" the hypothesis and examine in turn each of the separate lines of reasoning that have been put forward in its support. Each line of reasoning is individually subjected to close scrutiny. Although I have by now of course developed an opinion, at the start I was undecided and approached the hypothesis with an open mind. Here I lay out the results of my evaluation. The aim is to present a thorough and penetrating investigation in a cool and dispassionate manner, guided entirely by the available evidence and by careful logical reasoning based upon it. This book is therefore the opposite of a straightforward and uncritical paraphrasing of Lovelock's views. The emphasis instead is to be initially skeptical about everything, avoiding automatic unthinking acceptance of any viewpoints, whether for or against Gaia. Although the book reaches a clear conclusion, I set out the facts and arguments on both sides so that you can form your own opinion.

Dissecting a hypothesis as all-encompassing as Gaia requires this book to stretch across many different scientific specialisms. This book concerns itself with life, evolution, and ecology, as well as with geology and paleoclimatology. It also deals with all parts of the Earth's climate system, including the living parts, the ice sheets, the atmosphere, the oceans, and the land surface as well as the rocks beneath it. This wide-ranging analysis takes into account the results of many different scientific studies, including curious findings from little-known corners of the natural world. Within this book you will encounter hummingbirds in Caribbean islands and the similarity of their straight or curved beaks to the flowers they extract nectar from, the puzzle of occasional misplaced large stones in otherwise exclusively fine-grained marine sedimentary rocks, Walsby's square archaeon from the Dead Sea, the everlasting durability of the waste that corals generate, and differences in the way that Australian snakes bear young depending on climate (they don't always lay eggs). The relevance of all these and many more individual observations is made clear.

The critical examination of Gaia in this book involves consideration of many details such as these, but also involves stepping back to appreciate the bigger picture that emerges. In terms of that wider view, this analysis of Gaia engages with some fundamental issues. The book maintains a tight-beam focus on evaluating the Gaia hypothesis and includes only topics that contribute to a deeper understanding of its plausibility. Nevertheless, during the journey we find ourselves pondering some of the great questions about the nature of our planet, its history, and how it came to give rise to us. This analysis throws up interesting insights and sheds light from new directions on questions such as: "just how good a planet is Earth for life?" "what effect has life had on Earth habitability?" and "how is it that our planet has remained continuously hospitable for life over billions of years?"

Some exhilarating recent scientific developments are covered. We know a lot more today about climate, the Earth system, and even the mechanics of evolution than we did when Gaia was first proposed in the 1970s. Research activity in Earth system science has burgeoned, and numerous scientific journals exist today that didn't at the time Gaia was first proposed. Our understanding has been improved and changed by discoveries of environmental and climate instability as revealed by gas bubbles in ice cores, fluid inclusions in evaporite deposits, and magnetic mineral orientations and dropstones pointing to Snowball Earth events. Our knowledge of the limits to life (extremophile tolerances) and of the effects of temperature on life and evolution has developed rapidly over the last few decades. Given that more than thirty years have passed since Gaia was first proposed, it is now an appropriate time to take stock of the hypothesis in the light of all of this more recent, and fascinating, information. In the context of this subsequent work, I ask the question: "Does the Gaia hypothesis hold up in court?" I bring the hypothesis face to face with modern evidence and undertake a skeptical but hopefully fair-minded evaluation.

On Gaia

GAIA, THE GRAND IDEA

THIS FIRST CHAPTER introduces the Gaia hypothesis and two competing hypotheses.

1.1. A BRIEF HISTORY

Gaia, the idea that life moderates the global environment to make it more favorable for life, was first introduced in 1972 in an academic paper titled "Gaia as Seen through the Atmosphere" in the journal *Atmospheric Environment*, followed rapidly by two other papers both in 1974: "Atmospheric Homeostasis by and for the Biosphere" in the journal *Tellus*, and "Biological Modulation of the Earth's Atmosphere" in the journal *Icarus*.[1] James Lovelock was sole author of the first paper and coauthor with Lynn Margulis of the latter two. Both were already scientists of some note. Lovelock had already pursued a successful career inventing chemical instruments, including, most famously, the electron capture detector. This device, when coupled to a gas chromatograph, allows for the detection of trace chemical substances even at extremely low concentrations. Before that, Lovelock had worked for twenty years at the United Kingdom's National Institute for Medical Research in Mill Hill, London, carrying out research in biomedical science.

Use of the electron capture detector started to become widespread due to its great utility, and through his consultancy work with it Lovelock was invited to participate in a NASA project to work out how to ascertain if Mars contained life. The two Viking spacecraft, now revered in history as the first spacecraft ever to land on the surface of another planet, were just then being designed, and a major priority was to decide which instruments to put on board. Reflection on this problem of how to detect the presence of life stimulated Lovelock's first thoughts on the Gaia hypothesis.

Lynn Margulis was a groundbreaking microbiologist at Boston University. She had long been championing her own (separate) revolutionary idea, one that is now widely accepted. It proposed that in the evolutionary distant past one primitive cell managed to "enslave" another (engulf it without killing it) and in the process benefited from the new capabilities of the enslaved cell. She proposed that such "endosymbiosis" had occurred a number of times. Mitochondria, chloroplasts, and flagella are all part of the machinery of individual cells; they are subcomponents of many single-celled creatures and of individual cells in multicellular organisms. According to the endosymbiosis theory they are all suggested to be relics of long-ago-assimilated single cells.[2] Each cell of every animal and plant, including those making up human bodies, is from this perspective seen as an evolutionary amalgam of several different ancient lineages. This theory, initially treated with some considerable skepticism (an early Margulis paper on it was rejected by as many as fifteen different scientific journals before being accepted), is now the consensus view. Although she was not the first to conceive of the idea, Margulis was the first to support it with direct microbiological observations, and it was in large part thanks to her continued championing of it, against strong opposition, that it came to be widely accepted. It has considerable implications. For example, it requires some modification of the idea that evolution proceeds solely by selection among organisms, each of which is a slightly modified descendant of the previous generation. Among the unicellular microbes at least, evolution has at times created a radically new species in a single jump, as a novel intracellular symbiont has been acquired.

Although Margulis jointly authored some of the early papers and remained a champion of Gaia, the hypothesis has always been first and foremost the brainchild of James Lovelock. Following the Lovelock and Margulis papers and some other papers in academic journals, none of which generated large amounts of interest or attention, Lovelock brought out a book called *Gaia: A New Look at Life on Earth*.[3] When this book came out, in 1979, it brought Gaia to scientific prominence at last. The book stimulated a mixture of admiration and opposition among scientists. Many evolutionary biologists, in particular, were very critical, for reasons that will be explained in the next chapter. Some of the biologists' objections were subsequently countered by modifying the hypothesis and also by the production of a now-famous model, Daisyworld. This model demonstrated the theoretical possibility of stable regulation of planetary temperature by organisms that are still adhering to biologically plausible rules of behavior and reproduction.[4] The Daisyworld paper, by Andrew Watson and James Lovelock, came out in 1983 in the academic journal *Tellus*. In the ensuing years Lovelock produced two more books: *The Ages of Gaia: A Biography of Our Living Earth*, in 1988, and *Gaia: The Practical Science of Planetary Medicine*, in 1991.[5]

In the 1980s the Gaia hypothesis was considered both interesting and controversial and continued to attract a mixture of agreement, interest, doubt, and rejection. By the mid-1980s it was decided that there was sufficient interest to merit organizing an international conference. In 1988 a prestigious Chapman Conference of the American Geophysical Union brought together advocates and interested skeptics in a wide-ranging scientific discussion of the hypothesis.[6] Further international conferences on Gaia were convened at Oxford University in 1994, 1996, and 1999, with membership primarily by invitation. In 2000 a second open Scientists on Gaia Chapman Conference was held in Valencia, Spain.[7]

As Lovelock, now in his nineties, has become less active, others have taken up the torch. Tim Lenton, for instance, an Earth system scientist at the University of Exeter, has written many papers on Gaia, including a review article in *Nature* in 1998.[8] The Gaia hypothesis had achieved a degree of scientific respectability.[9] However, a brief review of its reception in books published since 2007 shows that while it is now accepted gladly by some, it also continues to stimulate intense debate.[10] This was also revealed in back-and-forth exchanges in the pages of the journal *Climatic Change* in 2002 and 2003.[11] Nevertheless, when interviewed for a biography published in 2009, Lovelock claimed that Gaia has made the transition from being just a hypothesis to being solid science.[12]

The degree to which Gaia has been accepted by a large part of the scientific community, including those in its higher echelons, was highlighted by the Amsterdam Declaration on Global Change.[13] This document is a synthesis of the work of four international research umbrella organizations, including the International Geosphere-Biosphere Programme (IGBP) and the World Climate Research Programme (WCRP) and was discussed at a conference attended by more than one thousand scientific delegates. The second paragraph of the declaration asserts: "Research carried out over the past decade under the auspices of the four programmes to address these concerns has shown that: The Earth System behaves as a single, self-regulating system comprised of physical, chemical, biological and human components." The wording could almost have been lifted from one of Lovelock's books. A *Nature* editor, reporting on the second Chapman Conference in 2000, judged that "James Lovelock's theory of the biotic regulation of Earth has now emerged with some respectability following close scrutiny by the biogeochemical community."[14]

Is the scientific respectability and the continuing prominence justified? Read the rest of this book if you want to find out.

1.2. The Hypothesis

The Gaia hypothesis is nothing if not daring and provocative. It proposes planetary regulation by and for the biota, where the "biota" is the collection of all life. It suggests that life has conspired in the regulation of the global environment so as to keep conditions comfortable. During the more than two (probably more than three) billion years that life has existed as a continuous presence on Earth,

Lovelock suggests that life has had a hand on the tiller of environmental control. And the intervention of life in the regulation of the planet has been such as to promote stability and keep conditions favorable for life.

That, in a nutshell, is the hypothesis. Providing a more precise definition is, however, made difficult by a couple of factors: (1) the hypothesis has not stayed constant but instead has been modified over time in response to criticisms; and (2) Lovelock's publications do not provide a completely clear definition, although others have tried subsequently to clarify it for him, as described below.

Gaia is not a hard-and-fast, well-defined concept. It is not a "set menu." Rather it is more like a loosely defined smörgåsbord, from which "diners" can take their pick from a collection of several related hypotheses, often couched in rather vague terms. The lack of clarity presents a problem for those of us who want to analyze and evaluate Gaia. It may even seem a poor basis for a book such as this one. However, fortunately, there are central components of Gaia that are fundamental to all definitions, and it is these that I examine in this book. These concepts are at the heart of the hypothesis and are present regardless of which variant is chosen:

A. Earth is a favorable habitat for life.
B. It has been so over geologic time as the environment has remained fairly stable.
C. This is partly due to life's role in shaping the environment. For instance, life has influenced the chemical composition of the atmosphere and the sea.

In Lovelock's own words, the hypothesis has been defined in various different ways over the years:

We have since defined Gaia as a complex entity involving the Earth's biosphere, atmosphere, oceans and soil; the totality constituting a feedback or cybernetic system which seeks an optimal physical and chemical environment for life on this planet. The maintenance of relatively constant conditions by active control may be conveniently described by the term "homeostasis." (Lovelock 1979)

The main part of the book . . . is about a new theory of evolution, one that does not deny Darwin's great vision but adds to it by observing that the evolution of the species of organisms is not independent of the evolution of their material environment. Indeed the species and their environment are tightly coupled and evolve as a single system. What I shall be describing is the evolution of the largest living organism, Gaia. (Lovelock 1988)

The concept that the Earth is actively maintained and regulated by life on its surface. (Ibid.)

Gaia theory predicts that the climate and chemical composition of the Earth are kept in homeostasis for long periods until some internal contradiction or external force causes a jump to a new stable state. (Ibid.)

Gaia is the Earth seen as a single physiological system, an entity that is alive at least to the extent that, like other living organisms, its chemistry and temperature are self-regulated at a state favourable for life. (Lovelock 1991)

The top-down view of the Earth as a single system, one that I call Gaia, is essentially physiological. It is concerned with the working of the whole system, not with the separated parts of a planet divided arbitrarily into the biosphere, the atmosphere, the lithosphere, and the hydrosphere. (Ibid.)

Organisms and their environment evolve as a single, self-regulating system. (Lovelock 2003b)

The hypothesis that living organisms regulate the atmosphere in their own interest. (Ibid.)

By the end of the 1980s there was sufficient evidence, models and mechanisms, to justify a provisional Gaia theory. Briefly, it states that organisms and their material environment evolve as a single coupled system, from which emerges the sustained self-regulation of climate and chemistry at a habitable state for whatever is the current biota. (Ibid.)

In Gaia theory, organisms change their material environment as well as adapt to it. (Ibid.)

And from Tim Lenton:

The Gaia theory proposes that organisms contribute to self-regulating feedback mechanisms that have kept the Earth's surface environment stable and habitable for life. (Lenton 1998)

Some changes have been made to the hypothesis over time. A first correction was to alter the proposed life effect on the environment, from one of making it optimal to one of making it comfortable:

The first edition of this book used the terms optimum and optimize too freely; Gaia does not optimize the environment for life. I should have said that it keeps the environment constant and close to a state comfortable for life. (Lovelock 1979, in the preface to a 1987 revised edition)

A second clarification was to renounce any ascribing of purpose or intent to the biota. It was made clear that any biotic regulation of the environment must be automatic and unconscious. The reason for their impacts on the environment is not because the organisms responsible consciously want to help out their brothers in life:

At first we explained the Gaia hypothesis in words such as "Life, or the biosphere, regulates or maintains the climate and the atmospheric composition at an optimum for itself." This definition was imprecise, it is true; but neither Lynn Margulis nor I have ever proposed that the planetary regulation is purposeful. (Lovelock 1991)

This topic is returned to in the next chapter.

An attempt to seek greater clarity of definition came from James Kirchner, an Earth scientist at the University of California, Berkeley. The first Scientists on Gaia conference, in 1988, was the first time that large numbers of both proponents and interested skeptics got together in open debate. As might be expected, there were some surprises. One novel contribution was a paper by James Kirchner titled "The Gaia Hypotheses: Are They Testable? Are They Useful?"[15] This paper revealed significant differences between the definitions of Gaia in the various papers and books and attempted to classify them. Kirchner proposed the following taxonomy of hypotheses, on a gradient from weakly to strongly controversial: (1) *influential Gaia*, which asserts only that biology affects the physical and chemical environment to some degree; (2) *coevolutionary Gaia*, which limits itself to stating that the biota and environment are somehow coupled; (3) *homeostatic Gaia*, which emphasizes the stabilizing effect of the biota; (4) *teleological Gaia*, which implies that the biosphere is a contrivance specifically arranged for the benefit of the biota; and (5) *optimizing Gaia*, which suggests that the biosphere is optimized in favor of the biota.

Kirchner's paper was perhaps unreasonable in one regard, in that the two weakest forms were weaker than adopted in any major Gaian publication. Nevertheless, the paper made a vital contribution by pointing to ambiguities in how Gaia was defined. It was an important influence on thinking at that time. Lovelock later retreated from the two stronger forms of Gaia. Even with regard to homeostatic Gaia, Kirchner called for greater clarity in terms of the definition: "What is stability? Does it mean resistance to change, resilience under change, or bounds on the magnitude of change?" Stability is dealt with in greater detail in chapter 8.

1.3. Supporting Evidence

It is one thing to form a hypothesis, but something else entirely to show that the hypothesis should be taken seriously. Anyone can formulate a hypothesis, in seconds. Demonstrating that it is a true picture of reality is quite another matter. Why should we believe the Gaia hypothesis? What reasons are there to suggest that it corresponds to nature, or, in other words, that it accords with the reality of how the Earth's environment has been regulated and maintained? In his books and papers, Lovelock has presented a wide variety of evidence. There are three main facts, or classes of facts, that he has advanced in support of the hypothesis. These three assertions are carefully examined in this book:

> *Assertion No. 1: the environment is very well suited to the organisms that inhabit it.*

In the first and subsequent Gaia books, the tolerances of organisms for temperature, pH, salinity, and other environmental parameters were described. The fact that the Earth environment satisfies these tolerances (is habitable) is put forward as strong evidence for Gaia. This assertion is critically analyzed in chapters 3, 4, and 5 of this book.

> *Assertion No. 2: the Earth's atmosphere is a biological construct whose composition is far from expectations of (abiotic) chemical equilibrium.*

Comparison of the atmospheres of Earth and of neighboring lifeless planets (Mars and Venus) shows that the Earth's atmosphere contains an unusually high percentage of oxygen and an unusually low percentage of carbon dioxide (CO_2). The extraordinarily high (21%) proportion of oxygen in Earth's atmosphere persists in spite of high annual production rates of methane (CH_4) and other substances that are highly reactive with oxygen, depleting atmospheric oxygen as they combine with it to form new molecules. The reason that the atmosphere has remained rich in oxygen in the face of ongoing large sinks for oxygen is that it is also continually replenished, by photosynthesis. It is argued that the atmosphere is, therefore, the result of a dynamic equilibrium between different biological fluxes. This assertion is considered in chapter 6.

> *Assertion No. 3: the Earth has been a stable environment over time, despite variable external forcings.*

The Earth has remained habitable for billions of years of geological time and, Lovelock argues, has been rather stable. And this is despite various assaults on planetary regulation that have taken place. Comets and asteroids have hit the Earth, one of which killed the dinosaurs at the end of the Cretaceous around sixty-five million years ago. But after each assault conditions have improved once again and life has eventually recovered from the perturbations. The standard model of how stars change with time strongly suggests that the Sun was ~30% fainter four billion years ago, and has become steadily and progressively brighter (hotter) over time, but the Earth's oceans have not boiled away as a result (this is known as the "faint young Sun paradox"). Lovelock, in tune with the understanding at the time the books were written, suggested that Earth's climate has remained fairly stable over time, and that if anything it has become cooler rather than warmer. The stability of the environment is assessed in chapter 8.

1.4. Alternative Hypotheses

1.4.1. The Geological Hypothesis

The Gaia hypothesis is not the only hypothesis concerning the relationship between life and environment on Earth. At the time that the Gaia hypothesis was first proposed, the dominant paradigm among geologists and others was that the nature of the Earth's environment is principally determined by a mixture of geological forces and astronomical processes.

Geological processes affecting the environment include: (1) changing continental configuration due to plate tectonics (continental drift); (2) variations over time in the rates of volcanic activity, including occasional enormous outpourings of lava (the "flood basalts" such as those making up the Deccan Traps in India and the Siberian Traps in Russia); and (3) fluctuations in mid-ocean ridge spreading rates.

Astronomical processes of relevance to Earth habitability include: (1) changes in solar heating due to increasing luminosity of the Sun over time; (2) periodic changes in Earth's orbit around the Sun, known as Milankovitch cycles; and (3) collisions of extraterrestrial bodies with the Earth.

According to this way of thinking life has been a passenger on Earth, helplessly buffeted around by externally driven changes in the environment. Life adapts to the changing environment but does not itself affect it. Throughout the rest of this book this hypothesis will be referred to as the geological hypothesis.

1.4.2. The Coevolutionary Hypothesis

A second competitor hypothesis to Gaia derives from the ecological/evolutionary concept of coevolution.[16] If you watch even a small proportion of the natural history programs on TV, you cannot fail to be impressed by the precise matches between certain organisms living in close dependence with each other. Take for instance flowers such as *Passiflora mixta* and sword-billed hummingbirds (*Ensifera ensifera*) in high Andean forests. The *Passiflora* hold their nectar at the inner end of long, pendent floral tubes. The length of the tubes makes their nectar almost inaccessible. However, sword-billed hummingbirds can feed from these flowers because they alone in these forests possess sufficiently long beaks—up to 10 centimeters in length, as long as the whole of the rest of their bodies (fig. 1.1). It is agreed that this strange and intricate coupling between the two organisms is a result of the process of coevolution. The plant depends on the bird for pollination, which it brings about by dusting the bird's head with

a b

Figure 1.1. Coevolved species. (*a*) sword-billed hummingbird (*Ensifera ensifera*) feeding from the long tubes of *Passiflora mixta*, (*b*) cuckoo chick ejecting one of the host's eggs from the nest; a host chick has previously been ejected, as shown at the bottom of the drawing. (*a*) Luis A. Mazariegos H. (*b*) Adapted from an original by David Quinn. In *Cuckoos, Cowbirds, and Other Cheats*, by N.B. Davies. T & AD Poyser, an imprint of Bloomsbury Publishing, Plc (2000).

pollen as it dips its beak into the flower. Therefore if the bird's beak evolves to get longer, so too must the length of the flower, because all short-flower variants will not exchange pollen and so will fail to contribute genes to the next generation. Likewise, the bird depends on the flower for food (nectar), and genetic mutations leading to beaks that are too short and unable to reach the nectar will quickly be culled from the population by natural selection. On the other hand, the bird is not much disadvantaged if its beak is too long, because it is still able to access the nectar. The only stable outcome to the evolutionary interaction between these two species is one in which the bird's beak is exactly the right size for the flower, even if this entails what seems to us to be a ridiculously long beak and flower. It should be said that there are many other codependent pollinators and flowering plants. In all cases there are close-fitting matches between flowers and beaks[17] (in some cases both are curved), but usually involving more economical shorter flowers and beaks.

Coevolution is also apparent in the interactions between predators and prey and between hosts and parasites or diseases. The human immune system, for instance, consists of a wonderfully complex array of systems evolved to combat infection by viruses and bacteria. Our bodies, when infected with an illness, develop specific antibodies that in time can clear us of the disease and prevent reinfection. However, viruses have evolved in turn so as to be better able to penetrate the massed defenses of the immune system. The ability to mutate rapidly into new forms that the antibodies no longer recognize allows the AIDS virus (and, slightly less rapidly, the influenza virus) to overcome human immune systems with such devastating effects.

The phenomenon of coevolution is also apparent in the interactions between cuckoos and their hosts. Birds susceptible to cuckoo deception pay a high price. All their own young (whether eggs or chicks) are mercilessly ejected from the nest by the growing cuckoo chick, which has evolved an appropriately shaped hollow in its back for the purpose (fig.1.1b). Host reproduction is thereby prevented for the season, leading to a strong evolutionary pressure for the host to respond with countermeasures. Recent research shows, intriguingly, that in some cases they do; the hosts are not always unwitting dupes. Many hosts have a first line of defense in that they are able to discern even small differences between their own and parasitic eggs. Cuckoos have responded in turn by evolving eggs that are effectively indistinguishable in appearance from those of their host. Recent research[18] has shown that one potential host, the Australian superb fairy-wren, has developed a second line of defense against those cuckoos that take advantage of it. Whereas hosts are usually rather bad at identifying the interlopers once they have hatched out, even when the interlopers are up to five times the size of bona fide chicks, in this case the fairy-wren has evolved to be able to distinguish its own chicks on the basis of the sound of their begging calls. The arms race has been escalated and taken to a new level. And the next stage? You may have guessed it already: Horsfield's bronze cuckoo chicks now mimic the call of fairy-wren chicks.

Another curious detail of a coevolutionary arms race involving cuckoos, this time concerning the use of identifying patterns on the eggs, will be described in chapter 3. But the main point here is not specific to cuckoos. Coevolutionary arms races of various types are apparent all over the natural world,[19] wherever two species subject to evolution are highly dependent on each other.

Climate scientist Stephen Schneider first proposed, in 1984 with Randi Londer,[20] that this concept of coevolution could be relevant to describing the interaction between climate and life:

> In a sense, climate and life grew up together, each exerting fundamental controlling influences on the other. Climate and life have *coevolved*, to borrow a term developed by population biologists Paul Ehrlich and Peter Raven. (Schneider and Londer 1984)
>
> Over the geologically long term, life contributes to the evolution of the environment, just as the changing environment shapes the formation or sustainability of organisms. . . . Life and climate can be said to have coevolved; their mutual influence is well-established over geological history. (Ibid.)

In contrast to the geological proposal, Schneider and Londer's proposal asserts that life has had an enormous impact on the planetary environment. The reciprocal is obviously true, that changes in the planetary environment have had an enormous impact on life. This coevolutionary hypothesis, unlike the geological hypothesis, assumes a two-way traffic. Not only does the nature of the environment shape the nature of life, but, according to this hypothesis, life also acts as a force that shapes the planetary environment.

There is one obvious difference, however, between the coevolution of climate and life and the coevolution of two life forms. Climate cannot evolve back, in the normal sense of the word. Whereas biological species have been shaped by mutation-prone reproduction followed by natural selection of the resulting variable descendants, climate has not evolved in this way. There is no equivalent cumulative process that builds better-adapted oceans or atmospheres over time. At any one time there is only one atmosphere and so there is no selection between variants. According to Schneider and Londer's coevolutionary hypothesis, the Earth's environment has undoubtedly been affected by the presence of life, and has changed over time, but not under the influence of the optimizing process of evolution by natural selection.

The main difference between the coevolutionary hypothesis and Gaia is that the former makes no claims about the wider outcome of the interaction. Whereas Gaia suggests that the outcome of the interaction has stabilized the planet and kept it favorable for life, coevolution is neutral about any such claims:

> Regardless of whether you believe in a collective biological purpose behind environmental and biological evolution . . . it is a profound realization that the physical, chemical, and biological subcomponents of the earth all interact and, whether by accident or design, mutually alter their collective destiny. (Schneider and Londer 1984)
>
> I saw an apt analogy, in that climate and life have coevolved. In other words, both life and the inorganic environment, including the meteorological elements, have followed evolutionary paths different from what otherwise would have happened over geological time had the other not been there. Coevolution does not require negative any more than positive feedbacks, just mutual interactions. (Schneider 1996)

Coevolution implies only that life and environment have both changed over time, and that changes in either have had effects on the other. There is no imputation that the result of the interaction should be beneficial for life. The coevolution hypothesis is free of any connotations that, once life had evolved and started to influence climate, the planet was bound to remain habitable thereafter.

TABLE 1.1.
Differences between the three hypotheses compared in this book

	HYPOTHESIS		
	Gaia	Geological	Coevolution
Has the environment played a major role in shaping the biota?	yes	yes	yes
Has the biota played a major role in shaping the environment?	yes	no	yes
Has the role of the biota been generally favorable in terms of keeping the environment comfortable for life?	yes	n/a	no opinion
Once started, was the continuation of life on Earth made more likely by the intervention of the biota?	yes	n/a	no opinion

1.5. WHY SHOULD WE CARE?

A major reason that Gaia has generated such interest, and continues to do so, is that it offers an explanation of how the Earth's climate and environment system works, and of how the Earth has been kept habitable throughout vast stretches of geologic time. Gaia is without doubt big-picture science; it is awe inspiring and majestic in its scope. It proffers answers to deep questions such as "What sort of place is our planet?" "Why has the Earth remained continuously habitable for so long?" "How does our planet work?" "How is a planet kept fit for life?" For all of these reasons it is of considerable interest to probe into whether Gaia is actually a correct or an incorrect view of our world and how it functions.

A second reason for interest does not concern how the Earth has come to be the way it is, but rather concerns how our planet will cope in the future as it continues to be buffeted by human impacts upon it. The last few decades have seen a revolution in our appreciation of the fragility of our planet. Whereas human activities used to be considered trivial (mere "drops in the ocean") in relation to the natural processes of the Earth, we are now rapidly learning otherwise. From the ozone hole to global warming to ocean acidification, our collective capacity to make great changes to our planet is becoming all too apparent. Our ability as a global society to steer a safe course through the next few centuries will depend in large part on our perspicacity in terms of understanding how our planet works and how it will respond to various anthropogenic perturbations.

The sagacity of our collective wisdom about planetary management will be affected by the correctness of our overall paradigm view of how the Earth's climate system works. In the last chapter of this book, I argue that acceptance or rejection of Gaia has the potential to color our view of how the Earth works as a system. In order to be successful stewards of our planet as a life-support system, our understanding of its natural dynamics must be based on a correct overarching view. We will then be better placed to understand how our perturbations will interact with the natural system, and to what degree they will disturb it. For this reason it is of considerable importance to determine whether or not Gaia is an accurate worldview.

1.6. Comparison of Hypotheses

As I review evidence for and against Gaia in the rest of this book, I will also relate it to the other two hypotheses just described. The feasibility of all three hypotheses will be evaluated in relation to the burgeoning knowledge about the Earth system. At the end of this book an assessment is made as to which of the three hypotheses appears to be most compatible with the large classes of facts and bodies of evidence that are introduced through the course of the book. Their explanatory power is compared. The task in this book is therefore not just to evaluate Gaia, but also to compare its performance against that of the other two hypotheses.

This first chapter has set the scene for the rest of the book: the Gaia hypothesis has been described, as have also the major assertions and lines of argument put forward in its support. It has been contrasted with two competing hypotheses. Now, in the following chapters, each assertion and line of reasoning is compared against the available evidence, in particular new evidence unearthed during the last thirty years. Together these chapters go to make up a careful and penetrating evaluation of whether the Gaia hypothesis is supported by current knowledge.

Chapter 2

GOOD CITIZENS OR SELFISH GENES?

MUCH OF THIS book involves examination of individual pieces of evidence relevant to the Gaia hypothesis. Later chapters will subject each of Lovelock's three assertions (those just described in chapter 1) to close examination. The degree of match between the three competing hypotheses (Gaia, coevolution, geological) and the evidence will also be compared. Before all this can be embarked on, however, let us first look at the relationship between Gaia and natural selection. This is a search to see if we can understand why Gaia should occur. We are interested in whether there is any reason to expect something like Gaia to appear as an emergent property from the rules of natural selection, in the same way for instance that air pressure is an emergent property from the velocities of individual gas molecules. If Gaia were to arise straightforwardly and predictably out of natural selection, then because of the large body of evidence behind natural selection Gaia could also easily be accepted. But it turns out there is no such straightforward link.

There are two themes to this chapter. Firstly, the history of Gaia's reception by biologists is described. There were long-running arguments between Gaians and evolutionary biologists. Richard Dawkins, W. Ford Doolittle, and others strongly criticized Lovelock for his naïveté about how evolution works. They questioned whether natural selection would ever be likely to give rise to a phenomenon such as Gaia. I describe the background to the arguments following the publication of the first Gaia book (and why its timing was particularly bad), the response by Gaians, and where this now leaves us. This is mostly history. However, some interesting modern developments are also introduced to the debate, including: (a) the recently advanced "niche construction" concept, (b) exciting new experimental tests of whether kin selection is the most usual explanation of evolved behaviors, and (c) recent hypotheses about how cooperation evolved in eusocial insects.

A second theme is to look at the natural world and consider the circumstances in which biology regulates environments in reality. I take advantage of numerous studies into many different branches of the tree of life and describe how, when taken en masse, they show that some family groups carry out cooperative manipulation of environments (thermoregulation, for instance) when the family members are closely interrelated, but not otherwise. Given that the global biota is, on average, much more distantly related than those family groups, this analysis does not explain why something like Gaia should occur at the global level. It does not reveal any obvious mechanistic basis for how Gaian-type regulation could arise predictably out of natural selection.

2.1. INTRODUCTION

Even a cursory review of animal behavior reveals that there is far more to nature than just symbiosis and cooperation, as this section will show. Charles Darwin was one of the first to make this clear. Darwin's theory of the origin of species was very much based on competition between individuals: Darwin noted the enormous overproduction of seeds, spawn, and eggs compared to what is needed simply to replace the dead. Therefore, if the world is not to become increasingly densely populated over time, the large majority of progeny must die before adulthood.[1]

However regrettable it may be for animal suffering, Darwin recognized that this intense competition, with few winners and many losers,[2] is how the natural world operates. It provides the basis for natural selection. Although chance also plays a role in which particular individuals win out in each generation, over time the intense competition[3] tends to weed out any stunted or faulty progeny, and favor progeny that, if they have mutated, have mutated in useful ways. The intense competition with such a high losers:winners ratio, in which only winners provide blueprints for future generations, acts as a gigantic sieve that over time tends to select for organisms that work well in the environments they live in.

As Darwin realized, with some dismay, this vision is one of "Nature red in tooth and claw." Because most die young in nature, succumbing, for instance, to starvation, disease, or predation, this is not a comfortable view of the natural world.[4] Although Darwin's picture of a struggle for survival and survival of the fittest is still accepted today, nearly 150 years later, there has also been an increasing recognition of cooperation. It has been increasingly appreciated that nature is not solely about combat and eat or be eaten; there are also many cases where teamwork and altruism dominate.[5] Termite mounds, beaver dams, and reciprocal grooming in chimpanzees are just some instances in which fierce competition is replaced by mutually beneficial communalism. There are many natural marvels, bees' nests to name but one, that are products of intricately cooperating communities. Many plants benefit from mycorrhizal associations with fungi in order to better extract mineral nutrients from the soil, with the fungi benefiting from carbohydrates supplied by the plants. Endosymbioses were described in chapter 1, and lichens are of course symbiotic associations between fungi and algae.

Nature is a mixture of apparent cruelty and kindness, of economy and waste,[6] of competition and cooperation. It is certainly not ubiquitously the happy cooperative that a love of nature can sometimes tempt us into believing. This was brought home to me personally during my teenage years, when I enjoyed some time as a ranch hand in the foothills of the Andes. It was an amazing experience. My strongest memory of that time, however, is of one of the ranch owner's dogs catching and eating an iguana in the garden. I watched in horror as the dog patiently ate the iguana alive. It started on the tip of the tail and slowly gnawed its way up to the hindquarters and then beyond. The iguana made occasional halfhearted (or so it seemed) attempts to escape, but mostly seemed to passively accept being slowly consumed. Unless you have had a similar experience you may find it hard to appreciate what it felt like to observe this, but perhaps if you can imagine having to watch a dog eat a cat, from the back end forward, while the cat is still conscious, you will get some idea.

The point here has nothing to do with this individual iguana or my personal experience—it is rather that such acts of terrible cruelty are ever present in nature; they are not at all unusual. We now know of many examples of animal practices that must inflict, most often unintentionally, extreme pain on the recipients. Innumerable such acts take place every day, far from human sight, somewhere out in the wild. Some species of birds (including representatives of the herons, pelicans, boobies, and eagles) practice obligate siblicide, or in other words one sibling invariably kills another. In one black eagle nest, an observer counted the larger chick of the two mercilessly peck its younger and smaller sibling 1,569 times during the seventy-two hours of the latter's harsh but thankfully fairly brief lifespan between its emergence from the shell and its death.[7] Wasps of the family Ichneumonidae (of which there are more than 50,000 species worldwide) parasitize many insect groups, including butterflies, moths, bees, ants, and beetles. The adult female typically injects her eggs inside a caterpillar, larva, or pupa. As the wasp eggs mature into larvae themselves, they have readily available food (living meals), which they consume from within, taking care of course to save the critical internal organs until last so as to prolong the freshness of the food. Wolves sometimes start eating the internal organs of their prey before having killed it, and big cats grab zebras by the throat in order to kill by suffocation. Most people will have seen cats playing with mice ("play" depends on which animal's perspective one takes), and killer whales have also been observed tossing baby seals in apparent play. Adult elephants are protected from predation by their size, but a not infrequent cause of death in old age is starvation due to worn out teeth,[8] presumably leading to a horribly drawn-out and painful decline as they lose the ability to eat sufficient food. A fairly familiar problem to many sheep farmers is that of crows pecking out lambs' eyeballs. The point of this litany is not to shock you, but rather to bring home the point that the real world in its natural state can be a savage place.

Darwin, having spent a lifetime studying natural history, was moved to comment: "What a book a Devil's Chaplain might write on the clumsy, wasteful, blundering, low and horribly cruel works of nature." The process of evolution unthinkingly favors what works, without justice or scruples or consideration of pain inflicted. Indeed pain itself is, obviously, just like pleasure, a product of the evolutionary process. Clearly, the natural world is not solely a happy cooperative, although there are numerous cases of cooperation and altruism within it.

The Jekyll-and-Hyde nature of Nature is encapsulated rather elegantly in the story of a pair of fish, the bluestreak cleaner wrasse (*Labroides dimidiatus*) and its mimic, the bluestriped fangblenny (*Plagiotremus rhinorhynches*). Larger fish on Indonesian reefs come to "cleaning stations" where they allow smaller fish such as the bluestreak cleaner wrasse to remove their external parasites, in a win-win mutualistic symbiosis. This is a satisfying example of the unforced cooperation and symbiosis that takes place day in, day out in nature, with the unrelated cleaner and recipient fish interacting in such a way as to give benefits to both sides. But the darker side of nature is shown in the behavior of the bluestriped fangblenny, which, when in the vicinity of the cleaners, changes its color so as to resemble the bluestreak cleaner wrasse. It uses its camouflage to ambush the larger fish and lunge at them, tearing away tissue and scales.

The point is also neatly encapsulated by the behavior of another parasite remover, the oxpecker birds of the family Buphagidae. Birds of this species ride astride large African herbivores such as rhinos and remove any parasitic ticks they find on their hides. So far so good, but the oxpeckers sully their reputation for providing a helpful service by their tendency to peck at any festering wounds on the host animal. The chance of a free and tasty morsel of flesh is not overlooked, despite the significant detriment to their hosts.

Evolution has produced both cooperation and competition between organisms; it has produced mutually beneficial symbioses as well as predation and parasitism; it has produced the capacity to experience both pleasure and pain. We know that evolution causes organisms to become better adapted to their environments, but other organisms almost always feature strongly in those environments, and part of becoming better adapted often involves the more efficient exploitation of other organisms, often as a food resource. Evolution evidently converges toward whatever works, heedless of suffering, happiness, or any concerns about whether it is efficient to continually produce such an incredible incessant flood of new organisms, most of which never even survive to adulthood.

The question of interest in this chapter is whether the Gaia hypothesis is consistent with this knowledge of the natural world. At the philosophical level, would we expect evolution, which allows caterpillars to be eaten alive from within, at the same time to produce a global biota that acts in a coordinated fashion to keep the planetary temperature comfortable? Evolution has in many instances produced cooperation, but it is not organized along purely benevolent lines—it has produced numerous species that inflict terrible pain on others. Is it at the same time plausible that it has produced a biota that works together smoothly to induce a comfortable global environment? In the rest of this chapter this question is investigated in greater detail.

2.2. The Selection Wars: Individual, Kin, or Group Selection?

In 1962 the book *Animal Dispersion in Relation to Social Behaviour*, by Vero Wynne-Edwards, was published.[9] This book was notable in explicitly suggesting that animal behavior could be for the good of the species or group. So, for instance, Wynne-Edwards suggested that low-ranking male grouse passed up opportunities to mate with female grouse in order to prevent excessive population growth (with consequences for exhaustion of resources), rather than for some more selfish reason such as fear of being caught and punished by the dominant male grouse. In the same vein, Wynne-Edwards also suggested that predators might show restraint in consuming prey in order to maintain foodstocks for future generations, and that the reason that animals performing long-distance migrations return to the exact same location is to enable them to perpetuate the success of their local group. This sort of thinking (behaviors for the good of the group or species) was not uncommon at this time, although not usually stated so explicitly, and became known by the term *group selection*. Other group-selectionist-type thinking was also apparent in the writings of Konrad Lorenz, who attributed ritual displays in animal conflicts as being evolved with the good of the species in mind.[10] He thought that the displays, which allow many disputes to be settled without physical damage to the disputants, evolved because of the wider benefits rather than simply because of the benefits to individuals.

Wynne-Edwards's book stirred considerable opposition, which is perhaps not surprising given the many known opposing instances of within-species killings. To cite just a couple of examples, Jane Goodall has documented tribal warfare in chimpanzees,[11] and male lions have been observed to kill cubs upon taking over a new pride. In the mid-1960s Bill Hamilton and George Williams, in particular, put forward a new and alternative vision[12] to that of Wynne-

Edwards. Their vision hypothesized that kin selection[13] (whereby the actions of an individual will be in the interests of itself and of its close kin, but not of unrelated members of the same species) is the dominant process. John Maynard Smith also published a paper[14] in 1964 called "Group Selection and Kin Selection," and together these works set the scene for an extended debate between proponents of kin and group selection. There was vehement discussion of this issue over the following decades in the fields of animal behavior and evolution. Scientists such as Richard Dawkins later joined the debate, arguing persuasively that kin selection arises because selection takes place on genes rather than on individuals per se: "if adaptations are to be treated as 'for the good of' something, that something is the gene."[15] This whole debate helped to clarify the nature of the parameter or quantity that evolution works to optimize: is it the transmission of genes, the fitness of individuals, the so-called inclusive fitness (the fitness of individuals plus the fitness of their relatives, that is, kin, weighted according to how closely related), the fitness of wider groups, or the fitness of species?

The debate still rumbles on today, but in the opinion of most experts it has been won in favor of kin selection. While the production of group-oriented behavior by evolution is theoretically possible,[16] it is not strongly supported by observations,[17] and most think it is a relatively infrequent occurrence in evolution. Even some behaviors that at first sight might seem clinching cases for group selection are now found to be explicable by kin selection. Coalitions of male turkeys in the wild cooperate to attract females, but only the most dominant male in the coalition ever mates. At first sight it seems there is no benefit for the subordinate males in helping the coalition if they then forego reproduction, unless they are doing it for the good of the group or species; genetic studies, however, are revealing that the coalition males are in fact close kin.[18] A similar explanation has been discovered for another otherwise puzzling phenomenon. Social amoebas (slime molds) are solitary for most of their lives but at certain times numerous individual cells come together to form fruiting bodies. Intriguingly, each single cell is for some of its life a free-living entity, at other times just one cell within a coordinated colonial body. Within the fruiting bodies, those cells that go to make up the stalk forego any chance of future reproduction (that is restricted to those cells that migrate to the top of the stalk and turn into spores). Again, investigators have found[19] when studying one species (*Dictyostelium purpureum*) that the amoebas choose to make fruiting bodies with close genetic relatives (they must, somehow, be able to distinguish them), and the stalk cells are therefore foregoing reproduction in favor of their close kin.

Kin selection is compatible not only with instances of group-oriented behavior (such as cooperation while hunting in groups or mutual grooming), when group members are closely related,[20] but also with intergroup conflict or mutual lack of concern, which should be most prevalent when group members are unrelated. Many groups are aggregations of relatively unrelated individuals, such as schools of fish and the large herds of wildebeest that migrate across the Serengeti. Whereas family groups of elephants execute joint defense against predators, within wildebeest herds it is each for him/herself (or each mother and child). It has been shown, in wonderful computer simulations ("boids": http://www.red3d.com/cwr/boids/), that the apparently choreographed and co-ordinated schooling behavior of fish or birds can in fact emerge out of only a few simple rules tying individual (selfish) behavior to those of the nearest neighbors in the flock or school. We can explain the schooling/flocking phenomenon by self-interest ("safety in numbers") without any need to invoke organisms being "good team players."

The overall principle of kin selection is relatively simple and has been formulated mathematically.[21] More informally, it says that organisms are prepared to make sacrifices for relatives, because relatives carry many of the same genes. Kin selection is able to explain[22] why, in colonies of some eusocial[23] insects, many or most individuals (members of the worker or soldier castes) are incapable of reproduction or forego it, and even sacrifice their lives in suicide attacks (honeybees leave their stings embedded in attackers of the communal nest, and the loss of their stinger is fatal to them), all in the cause of the common good. Kin selection also helps us understand why eusocial insects work together so smoothly and selflessly to make intricate constructions such as honeycombs, termite mounds, and ants' nests. These are easily understood via kin selection because members of eusocial insect colonies tend to all be close family members, as discussed below in section 2.6.3. Inbuilt sterility and suicidal nest defense are, on the other hand, completely confusing and in fact inexplicable from the standard view of natural selection, which assumes that individual organisms are optimized to each try and leave behind the maximum number of viable offspring; only from the viewpoint of kin selection can it be made sense of.

And it is not only insects; a bizarre species of mammals, the naked mole rat (*Heterocephalus glaber*), also fits the pattern. Reproduction in each underground colony is the exclusive activity of a single breeding female who bullies other females and breeds with only one or two males. The restriction to just two or three parents in a colony of up to 250 individuals leads to close family ties (a high degree of inbreeding), which in turn leads to extreme genetic similarity.[24] Nonbreeding colony members (the large majority) are observed to engage in group-helping behavior. Kin selection also allows us an insight into why many primates and other animals appear to be most cooperative with fellow members of their tribe (typically their closer relatives) but antagonistic to strangers (to whom they are less likely to be closely related). Kin/gene selection is thus immensely satisfying because it is able at a blow to account for a large class of initially puzzling or apparently contradictory facts.[25]

Kin selection has until just recently been solely an explanatory theory, the one that seems best able to account for the observations. But an exciting development of the last few years is the development of means to subject it to experimental testing. Kin selection, at least as regards Australian allodapine bees (*Exoneura nigrescens*) and the bacterium *Pseudomonas aeruginosa*, has now been studied in the laboratory.[26] In my mind's eye, I can almost picture Charles Darwin trembling with anticipation were he to have imagined that, one day, not only would we comprehend the mechanisms of heredity (he knew nothing of genes), but that experiments would start to probe the very mechanisms of evolution. So far the first results from these groundbreaking studies broadly confirm the correctness of kin selection.

2.3. Gaia and Evolutionary Stable Strategies

It was into this maelstrom of controversy over levels of selection that Lovelock introduced his Gaia hypothesis. At the very height of the selection wars, in the early 1970s, Lovelock unwittingly chose perhaps the very worst time for introducing a theory proposing selection in favor of the whole biota. While the new wave of evolutionary biologists was busy trying to combat group- and species-selectionist ideas, Gaia went even further:

> Is it possible that the Great Barrier Reef, off the northeast coast of Australia, is the partly finished project for an evaporation lagoon? [to remove salt from the ocean in order to counteract the continued input via rivers, which would otherwise push ocean salinity to ever higher levels]. (Page 98 of Lovelock 1979)
>
> It is worth asking ourselves whether the movements of migratory birds and fish serve the larger Gaian purpose of phosphorus recycling. The strenuous and seemingly perverse efforts of salmon and eels to penetrate inland to places distant from the sea would then be seen to have their proper function [to supply freshwater ecosystems and the land surface with scarce phosphorus]. (Page 105, ibid.)

To most evolutionary biologists such suggestions were and still are preposterous. While one can only stand in awe of the breadth of Lovelock's vision, everything we know about the mechanics of natural selection leads us to the judgment that salmon do not travel inland to spawn for reasons of good citizenship, in order to help the planetary biota as a whole.[27] This migratory behavior must owe more to reasons of individual (or kin) genetic fitness. There must be a good reason for the behavior, even if we do not as yet fully understand it.

The improbability of organisms acting altruistically purely in order to efficiently enact their role as cogs in a great element recycling system becomes apparent when we look at things from the perspective of Evolutionary Stable Strategies (ESS).[28] The essence of this theory is that all organisms adopt those behavioral strategies that maximize the propagation of their genes. This may seem rather obvious and unsurprising and to follow rather straightforwardly from natural selection. However, the beauty of this rather simple idea comes when it is used to compare the likely success of alternative strategies. Consider, for example, the migratory behavior of salmon. The theory of ESSs requires that living at sea but spawning upstream must be an unbeatable or "uninvadable" strategy, because that is what salmon do.[29] For any kind of genetic mutation leading to a different strategy (and the whole concept of ESS assumes that there has been sufficient evolutionary time for very many such mutations to have occurred), the mutants must be less successful than those salmon already pursuing the ESS.[30] Otherwise we would have expected the mutants to have increased in number and for their behavior to have eventually become the ESS. So, for instance, if some salmon were to undergo a mutation that changed their behaviors so that they never migrated out to sea, but instead spent their entire adult lives in the streams or rivers of their birth, then the prediction is that this strategy would on the whole be rather unsuccessful, probably because of a shortage of suitable food in the freshwater habitats (measurements suggest that salmon grow much more rapidly while at sea). Likewise, if a random mutation happened to alter salmon behavior so that they stayed at sea at spawning time, rather than returning up rivers and streams, then the ESS idea predicts that those reproductive attempts would end in failure, possibly because of extremely high rates of salmon egg mortality at sea.[31]

You may already have realized where this line of argument is heading. If, in line with Lovelock's suggestion above, the main reason for salmon migrating inland is to transport phosphorus for the use of other organisms, then this is manifestly not an ESS. If the salmon themselves gain no overriding benefit from their long journeys then mutant salmon that do not make the long journey will start to appear and eventually come to dominate the population over time, because of their dramatic savings in time and energy by not swimming up rivers. Likewise, if corals gain no individual benefit from constructing evaporation lagoons (the opposite is surely the case, that it would be to the detriment of the corals if they made their lagoons so saline that salt started to crystallize over every living surface), then mutants that ducked out of construction duty would be more successful and rapidly become more numerous.

This problem crops up whenever it is proposed that any animal carries out any activity for the benefit of its group (unless inbred), its species, or on behalf of the whole biota. Most such behaviors are susceptible to invasion by freeloaders ("cheats"), mutants who opt out of the hard work for the community but who get to share the benefits (if there are any) all the same.[32] This issue is critical to any consideration of Gaia: how is it possible to get any organisms to do useful work on the behalf of the whole biota when the individual return is vanishingly small, and when the penalties for opting out and becoming a freeloader are likewise small or nonexistent. How could participation in Gaia be an ESS?

2.4. BIOLOGICAL BROADSIDES

The suggestions in the first Gaia book, that organisms would sacrifice their own fitness in order to help regulate the planet, were not well received by evolutionary biologists. First off was W. Ford Doolittle, in 1981:

> [Lovelock's] ideas are inconsistent with everything we now think we know about the evolutionary process. (Doolittle 1981)

Doolittle explained the then-current controversy among evolutionary biologists over what precisely constitutes the unit of selection (gene, individual, or species), but that:

> no serious student of evolution would suggest that natural selection could favour the development in one species of a behaviour pattern which is beneficial to another with which it does not interbreed, if this behaviour were either detrimental or of no selective value to the species itself. (Ibid.)

Even the group selectionists did not go so far as to postulate selection in favor of the whole planetary biota.

Others then continued the tirade, including, in 1982, Richard Dawkins in his book *The Extended Phenotype*:

> I very much doubt that [natural selection leading to Gaia] could be made to work: it would have all the notorious difficulties of "group selection." For instance, if plants are supposed to make oxygen for the good of the biosphere, imagine a mutant plant which saved itself the costs of oxygen manufacture. Obviously, it would outreproduce its more public-spirited colleagues, and genes for public-spiritedness would disappear. It is no use protesting that oxygen manufacture need not have costs: if it did not have costs, the most parsimonious explanation of oxygen production in plants would be the one the scientific world accepts any-

way, that oxygen is a by-product of something the plants do for their own selfish good. (Dawkins 1982)

Lovelock has long since admitted the validity of these criticisms, and the Gaia hypothesis was modified in response, as noted in chapter 1. Subsequent restatements of the hypothesis have omitted any suggestions of self-sacrificial endeavor in favor of the biota as a whole, or of any conscious purpose on the part of the organisms contributing to planetary regulation.

However, the change of tack left a gaping hole: if Gaia did not come about due to enlightened good citizenship then how could it arise at all? The Daisyworld model was developed to fill this hole.

2.5. The Daisyworld Model: A Response to the Evolutionary Biologists

The purpose of the Daisyworld model[33] was to show that regulation at the planetary scale is possible without sacrifice of individual genetic fitness and can be implemented by organisms completely unaware of the larger-scale consequences of their actions. Daisyworld is a hypothetical world whose surface is covered by a mixture of black and white daisies (fig. 2.1). The energy balance (the balance of incoming versus outgoing heat energy) on the Daisyworld planet ignores the effects of greenhouse gases and other factors. Daisyworld is a hypothetical planet without an atmosphere and without clouds. Planetary temperature is controlled only by the warming strength of the sun, the rate at which the planet radiates heat back to space, and by the planetary albedo. The term *albedo* refers to the *shininess* or reflectivity of a surface. It is defined as the ratio of the sunlight that is reflected back off a surface to the sunlight that impinges on it. A planet covered in silver foil would therefore have an albedo approaching 1.0, whereas a planet painted with matte black paint would have an albedo close to 0.0.

The interesting feature of Daisyworld is that its albedo is a life-affected property, because the relative proportions of black and white daisies and bare ground control it.[34] Black daisies absorb most incoming light, whereas white daisies reflect more of it. The rates of spread or dieback of the two types of daisy (and therefore the proportion of the planet surface covered by them) vary as a function of temperature. Both daisies share the same optimum temperature and the same temperature limits outside of which they die away. The dynamic regulation of planetary temperature arises out of the competition between black and white daisies for living space. This competition is preset so that white daisies

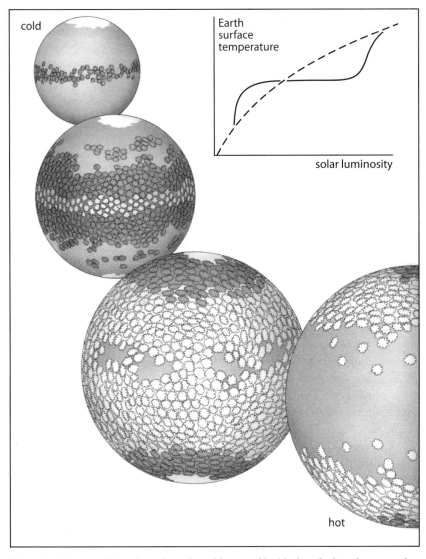

Figure 2.1. Daisyworld is a hypothetical world covered by black and white daisies, each with a different reflectiveness (albedo) to incoming sunlight. The temperature of the planet (T) is influenced by the ratio of the area coverage of the two daisy types. As shown in the inset, this leads to planetary temperature control. As solar luminosity (brightness of the sun) increases over time, as has happened in reality with our own sun, the planet becomes increasingly covered by white daisies and these reflect back more of the sun's warmth, and hence keep the planet cool. Adapted from an original image in (Lovelock 1991) by William Donohoe, http://www.billdonohoe.com.

outcompete black daisies when the temperature rises too high. This occurs because their higher albedo is assumed to reduce the *local* temperature significantly, so that it is not quite so hot where white daisies live. The higher albedo also reduces the global temperature by an infinitesimal amount for each white daisy. Conversely, black daisies absorb more heat and thereby increase their local temperature. They are more successful than white daisies when the temperature is uncomfortably low. A side effect of this is to cause Daisyworld to become more heavily blanketed with black daisies during cold climates, and this warms the planet due to reflecting very little sunlight back to space.

From these simple rules a system of planetary temperature regulation emerges as a by-product (fig. 2.1), even though both the black and the white daisies are each consistently pursuing their own interests. When the Sun is made in the model to get warmer over time, as has indeed happened over Earth history, it can be seen that the daisies unconsciously stabilize planetary temperature over a wide range of solar luminosities (top right of fig. 2.1).

As the title of the original Daisyworld paper suggested, this scheme was only intended as a *parable*, and to this extent it was highly instructive. It countered the criticisms of the evolutionary biologists by showing that a working regulation of the planetary environment is theoretically possible without supposing that natural selection molds organisms to sacrifice their own interests in favor of the global biota. The original Daisyworld has sparked an armada of subsequent extensions: some included multiple species of daisies of different hues, others added rabbits to graze the daisies, and then foxes to graze the rabbits, and so on.[35] The original Daisyworld is an extremely satisfying intellectual invention. It is extraordinary that such a simple mechanism can achieve so much. Perhaps this explains why there has been such an abundance of further work on Daisyworld. The abundance is otherwise surprising, however, given that Daisyworld was posited only as a "toy world" rather than as a dedicated attempt to model the world as it is. Even Lovelock admits as much, although he sees it as his proudest invention nevertheless[36]: "I never intended Daisyworld to be more than a caricature, and accept that it may turn out a poor likeness when we finally understand the world," a view echoed by Tyler Volk:

> In Gaia theory the traditional model has been Daisyworld. This served its purpose long ago to introduce the idea of life creating stability in the face of external change to the environment. But in my opinion Daisyworld is no longer acceptable. It is too far removed from the real world. (Volk 2002)

Some of the later work with more elaborate Daisyworld variants is interesting from a theoretical point of view, but on the other hand is of only limited interest in telling us about how temperature on Earth is really controlled. The central advance Daisyworld represented was to show that regulation can arise automatically as a by-product of organisms pursuing their own ends, and this is certainly applicable to the real world (an example will be discussed shortly). Temperature regulation on Earth, however, appears to have more to do with greenhouse gases than albedo,[37] and so Daisyworld does little to advance our understanding of Earth's actual system of temperature controls.

There are several important simplifications in Daisyworld that prevent it being more than a conceptual model.[38] As just mentioned, Daisyworld ignores the central role of greenhouse gases in the global temperature balance. And this is important, because the same sort of reasoning does not apply to carbon dioxide, for instance, as to albedo.

The key assumption in Daisyworld, from our perspective of how natural selection is or is not compatible with Gaia, is that the daisies have similar effects on the global environment as they do on the local environment. At the same time as a white or black daisy affects the global temperature in a small way, it also affects its own local temperature in a greater way. In helping itself, a daisy automatically also improves the global environment. This assumption is critical because it gets over the problem of individuals gaining no benefit from improving the global environment. In Daisyworld the two go hand in hand. Whenever a daisy improves the global environment it also improves its personal environment. When this key assumption is removed, temperature regulation no longer emerges from the Daisyworld model.[39]

If we look at some of the most important environmental variables in the real world, on the other hand, this key assumption is far from true. The point that Richard Dawkins made for regulation of atmospheric oxygen is also true for atmospheric CO_2, a more important determinant of planetary temperature than albedo. When a microbe in seafloor sediments or in the soil eats a piece of organic debris (part of a polysaccharide cell wall for instance, or a leaf fragment) and as a result turns some solid organic carbon into CO_2, it does not warm its local environment. Likewise, when a bush on land "breathes in" aqueous or gaseous CO_2, it does not get chilled as a result. In this as well as in most other cases there is no parallel between the local and the global effects. This is partly because of how rapidly the atmosphere is mixed. If a large amount of carbon dioxide were to be injected instantly at a single spot on the Earth's surface, it would get spread evenly throughout the whole of the lower atmosphere within just a year or two. Instead of accumulating locally, any CO_2 effects of a single organism get rapidly dissipated into the whole global atmosphere. The same is true, although the mixing time is slower (about a thousand years), for the global ocean. The low viscosities of both air and water allow them to circulate freely and disperse local biotic effects rapidly.

An individual organism cannot benefit from the local effects of its CO_2 alteration because these disperse. It also cannot benefit from the global effects of its CO_2 alteration because of the very small size of most organisms compared to the size of the planet. Even if we take one of the largest organisms, a mature tree, the effects are small. A large tree might assimilate something like one thousand metric tons of carbon into its biomass over its lifetime, but this is still minuscule compared to approximately six hundred billion metric tons of carbon in the preindustrial atmosphere as carbon dioxide. No single organism can significantly alter the global concentration. In reality, therefore, there is a disconnect between the planet-altering actions of individual organisms and the effect they have on their own living conditions.

To summarize, Daisyworld has been successful in showing that, under special circumstances, planet-scale regulation could plausibly arise out of organisms each selfishly pursuing its own best interest. Its weakness is its lack of generality: it is not a realistic representation of how most of the world's important climatic variables interface with the living organisms that affect them.[40]

2.6. WHEN DOES NATURAL SELECTION PRODUCE REGULATION? SOME SCENARIOS

The theme of this chapter so far has been mainly historical, to explain why evolutionary biologists have been so skeptical about Gaia, and how advocates of Gaia have responded. In this next section I will now relate Gaia to our current understanding of natural selection and of how natural selection is observed to generate thermostated environments. I will step through several examples of environmental regulation produced by natural selection, explain how they are compatible with the dominant paradigm of gene-/kin-based selection, and why it appears unlikely that a phenomenon like Gaia could have been produced in the same way.

2.6.1. Physiological Homeostasis within Individual Organisms

Organisms typically regulate their internal chemistry. Single-celled microbes, for instance, use ion pumps to maintain their internal salt content, pH, and other fundamental properties at values that are the most conducive to efficient working of their cellular machinery. This can entail drastic differences between the inside of a cell and the external environment. For instance, single-celled phytoplankton living in seawater (which has a calcium concentration of about

10,800 μmol per liter) typically maintain a much lower internal calcium concentration of < 1 μmol per liter because of calcium toxicity. The coccolithophore phytoplankton, which synthesize shells of calcium carbonate, therefore have a quandary: how can they use large amounts of calcium without becoming overwhelmed by the toxicity? It is not completely clear how they overcome this difficulty, but part of the answer may involve transport of individual ions or of sealed microscopic "packets" of ambient seawater across the cell to the site of coccolith production.[41] Levels of sodium and chlorine (the main constituents of saltiness in the sea) are also much lower inside cells than without, and cells construct specific "osmosolute" (salt-mimicking) molecules to allow them to have low internal salinity without at the same time incurring the strong osmotic pressure usually produced by a salinity gradient across a permeable membrane.

Internal homeostatic[42] regulation is marvelously intricate in single cells but is easier to fully appreciate when we look at larger animals such as ourselves (as examples of multicellular organisms). Along with other mammals we unconsciously execute wonders of exquisite control over our internal body state. If we become too cold then this induces a suite of compensating reactions including shivering, increased blood flow, withdrawal of blood flow to the critical body core that must be kept warm, and so forth. If we start to become too hot then we sweat, we become listless, and so on. Our livers and kidneys are organs whose job is specifically that of maintaining our blood at or near an optimal composition despite variability in what we eat and drink. If we become dehydrated then the kidneys restrict water loss and our urine becomes more concentrated. If we eat less sugar than normal then the liver releases more glucose back into the blood (if you are diabetic, your blood sugar level is imperfectly regulated). Using a combination of organs, hormonal signaling mechanisms, and behavioral responses, our bodies are wonderfully well organized to keep themselves stable at a particular chemistry and temperature. In the face of wide fluctuations in external conditions, the conditions inside a human body are kept restricted to a much narrower range. At temperate latitudes ambient temperatures can vary by 30 or 40°C between winter and summer, but human body core temperatures usually vary by less than 2°C, remaining year-round within the range 36–38°C. This is why even small changes in body temperature are diagnostic of illness: fevers cause body temperatures to rise to somewhere between 38° and 40°C. Core temperatures above 40°C or below 35°C are indicative of heat shock or hypothermia and if prolonged are usually fatal.

Much is made in the Gaia literature of comparing these amazing powers of individual physiological regulation with the supposed regulatory propensities of Gaia:

> The Gaia system shares with all living organisms the capacity for homeostasis—
> the regulation of the physical and chemical environment at a level near that fa-
> vourable for life. (Lovelock 1991)

This is fine as an analogy, but any suggestion of a deeper resemblance is very
misleading, certainly insofar as the genesis of the two systems of regulation is
concerned. Each cell in an individual organism has near-identical DNA, whereas
different individuals have different DNA (clones and identical twins excepted).
It is therefore easy to see why natural selection refines the physiological regulation of individuals. It does so because poorly regulated individuals are not
successful carriers of genes, and therefore those genes disappear from the pool.
Only organisms with genes that code for efficient internal regulation (and
therefore bodies that keep working well in a range of external conditions) survive and contribute genes to future generations. However, from the viewpoint
of natural selection, Gaia is not the same as an individual organism. In order for
the natural selection process to produce something like physiological regulation at the scale of Gaia then, in the words of Richard Dawkins again (but this
point has been made by many):

> There would have to have been a set of rival Gaias, presumably on different plan-
> ets. Biospheres which did not develop efficient homeostatic regulation of their
> planetary atmospheres tended to go extinct. The Universe would have to be full
> of dead planets whose homeostatic regulation systems had failed, with, dotted
> around, a handful of successful, well-regulated planets of which Earth is one.
> Even this improbable scenario is not sufficient to lead to the evolution of plane-
> tary adaptations of the kind Lovelock proposes. In addition we would have to
> postulate some kind of reproduction, whereby successful planets spawned copies
> of their life forms on new planets. (Dawkins 1982)

So, although physiological regulation of individual bodies is admittedly wonderful (fig. 2.2a), natural selection clearly could not have worked in exactly the
same way to produce physiological regulation at the planetary scale.

2.6.2. Extended Phenotypes of Individual Organisms

The genotype of an organism consists of the full set of genes or chromosomes
of that individual, or in other words the complete portfolio of genetic instructions required to make an individual. A phenotype is what gets produced from
the genotype (the precise nature of the resulting phenotype depends mainly on
the genotype but also on the environment). If cars were biological then the
genotype would be the blueprint and the phenotype would be the car. The phenotype consists of both the physical organism (the cell or plant or body) but

also the behaviors the individual expresses. Richard Dawkins pointed out that evolution, in the process of blindly converging toward whatever is most successful, also in some instances engineers an *extended phenotype*. This term refers to birds' nests, rabbits' burrows, spiders' webs, or indeed to any physical structures that sit outside organisms but yet are clearly produced by a genotype in the sense that certain changes to the genotype lead predictably to certain changes in the extended phenotype.

Natural selection has led not only to the evolution of organisms of fantastic complexity and internal regulation, but has also produced individual organisms that, under the influence of their genes, go out and modify the *local environment* toward their own convenience.[43] Another obvious and famous example is that of beaver ponds (fig. 2.2b).

In a story most likely familiar to you, beavers (*Castor canadensis*) engineer their local environments to an extreme degree. They dam small streams using logs, mud, stones, and grass. These dams are frequently a few meters or tens of meters long. The streams back up behind the dams to form ponds, creating a habitat especially suitable for the beavers themselves, which are well adapted to an aquatic lifestyle. The ponds, together with shallow channels the beavers excavate between their ponds and the surrounding forest, possess several benefits for beavers: firstly, beavers can move faster in water than on land, thanks to their webbed back feet and paddle-shaped tail. The channels allow them to swim rather than walk out to trees they want to cut down. Secondly, the cold water in the ponds slows down decomposition, and so the ponds act as a larder within which they can store food (branches and bushes) and keep it fresh for winter. Thirdly, transporting logs and branches is easier when they can float them along their waterways and across the pond. And, finally, within their ponds the beavers are safe from coyotes and other predators.

The benefits to a beaver are obvious. The benefits for other plants and animals are more varied. The shallow ponds are favored habitats for ducks, herons, moose, leeches, otters, and frogs, among others, but the dams prevent salmon migrations. Beavers wreak considerable havoc on trees, not only by felling them in order to gain access to the nutritious (to them) inner bark, but also by waterlogging large areas in an intervention often fatal to the affected trees. Beaver ponds are frequently spiked with tree stumps, or "dead trees standing." Squirrels and other forest dwellers, it is presumed, also suffer from the arrival of beavers in an area. On balance, however, it seems probable that the actions of beavers enhance biodiversity and perhaps even animal biomass within their local, self-made habitats. Over longer timescales, beaver dams probably also help to hinder life-sustaining soils and nutrients from being washed off the land surface and into the ocean.

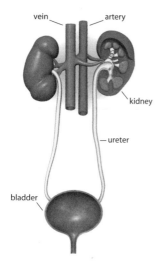

a

Figure 2.2. Agents of homeostatic regulation in the natural world. (*a*) The human kidney is part of the physiological system driving chemical homeostasis in our bodies. (*b*) Beavers build dams (this one is in Alaska) in order to create ponds behind them (maintain a higher water level), although the chemistry and water temperature of these ponds is not then stabilized. (*c*) Termite mounds, such as this one in Australia, resemble mammalian bodies in the sense that the temperature and chemistry within them is controlled. (*d*) The water inside a coral atoll lagoon, on the other hand, is not strongly regulated. (*a*) 3drenderings / Shutterstock.com; (*b*) PhotosByNancy / Shutterstock.com; (*c*) © Can Stock Photo Inc. / Bastian Linder; (*d*) William L. Stefanov, NASA-JSC.

b

c

d

Beavers are perhaps second only to humans in their per-animal impact on the natural world. In the past their impact may have been even more considerable, in North America at least. Perhaps as many as two hundred million beavers once shaped the North American landscape from Mexico to Canada. In the Miocene beavers were twice as big as today, two meters long. We know this not just from fossil skeletons but also from their corkscrew-shaped burrows, which were indirectly preserved because they filled with debris that then fossilized. One can only speculate at the size of the dams and ponds that were constructed by these prehistoric giant beavers.

The beaver dam and pond is an example of a manipulated and regulated environment, one where the changes made are for the benefit of beavers specifically, although other pond dwellers benefit incidentally. If similar dynamics could apply globally then Gaia would perhaps start to become possible. However, there are two major reasons for doubting that this could be so: firstly, the selfishness of genes means that beaver dams are designed for the benefit of beavers only (ground-burrowing rodents presumably do not appreciate the flooded terrain). Beavers probably care little (beaver genes certainly do not care at all) whether their dams are to the overall benefit or detriment of the local flora and fauna as a whole.

Organisms that modify their local environments by laying waste to it are equally likely to succeed, and in fact often do succeed, in nature. Colonies of army or driver ants strip the local forest floor bare of all living animals, then, once there is nothing left, move on to devastate an adjacent plot. Swarms of locusts inflict an even larger toll. Large trees minimize the abundance of forest floor vegetation and therefore wildlife by intercepting the great majority of light and preventing its penetration to the forest floor. Black walnut trees and other plants release allelopathic chemicals in order to poison the local soil against competitors. Lovelock has proposed that natural selection would not favor such despoilers:

> Those organisms which made their environment more comfortable for life left a better world for their progeny, and those which worsened their environment spoiled the survival chances of theirs. (Lovelock 2004)

But the real world seems to think otherwise.

The overriding problem is that natural selection operates according to the simple rule of favoring that which works best in the here and now, with no forethought of future implications or for the wider ecosystem or global impact.

The second main reason for doubt that Gaia could be engineered like a beaver dam is that such selection can only work when actions that benefit the global biosphere also improve the organism's own local environment. As already discussed in the context of Daisyworld, this does not seem possible for many important environmental variables, such as atmospheric O_2 and CO_2 (and therefore global temperature) and various seawater properties, mainly because seawater and air are so fluid and mobile. An organism does not tend to occupy the same parcel of water or air for any great length of time (each fluid rapidly mixes and disperses), and so there are strictly limited personal benefits that can be obtained from optimizing the currently occupied volume. A similar argument can be made about timescales:

> The rewards for good (Gaian) behaviour are as remote as the penalties for bad behaviour. It is difficult to accept that behaviours whose effects on atmospheric or oceanic composition or global temperature will not be felt for thousands of years can be selected for, especially when the first beneficiaries of those effects may be organisms which are not themselves responsible for them. (Doolittle 1981)

The problem is that "good citizen" organisms do not reap the benefits of their actions.

This characteristic of organisms, that they modify their own environment and thereby change the target to which natural selection is trying to evolve them, is also the subject of a recent book titled *Niche Construction*, by John Odling-Smee and others.[44] This book makes the same point to an audience of evolutionary biologists that Schneider and Londer made in the 1980s to an audience of geologists: that there is a coevolution between life and environment rather than a unidirectional arrow of causation molding life to an environment that is unaffected by life's influence. Whereas Schneider and Londer argued that traditional accounts of the geological history of the Earth were incomplete without considering the impact of life, Odling-Smee and colleagues argue that conventional accounts of organic evolution are incomplete, again because they ignore the effects of organisms on their own environments. The evolutionary trajectories of species are likely to have been influenced by the changes they themselves have wrought on their environments. The view that Odling-Smee and colleagues attack is that environments are unaffected by the presence of life. It seems rather clear, even on a cursory consideration of the evidence, that they are correct.

To take an appropriate example from their book, most soils are heavily influenced by biological processes. Soils are formed in large part out of leaf litter and other organic debris, they are penetrated by interlacing roots of trees and other plants that act to bind them together and stabilize them against erosion, they are aerated by various animal burrows and by earthworms, and their organic matter is eventually consumed by fungal and microbial decomposition as well as by earthworms. Soil is the environment in which many animals spend their whole lives. This is true not just for earthworms but also for moles, naked mole rats, and many others. It also forms a crucial part of the environment of most plants. And it is clear that soil is largely a biological construct, in the sense that different soil dwellers have played a major part in determining its structure and characteristics. Other demonstrations of the ability of life to affect the planetary environment will be described in chapter 6.

In this sense niche construction is uncontroversial; however, as discussed above, the inability of individual organisms to affect many important environmental variables within a single lifetime makes it unlikely that evolution can lead to optimization of important global variables such as the oxygen and carbon dioxide contents of the atmosphere. They are so far beyond the capacity of any individual organism to noticeably affect them that there can be no individual fitness benefit in trying to do so. Therefore it is hard to see how evolution could act in a directed fashion on any of these global variables of great importance. For this reason we should not take too far the ideas of niche construction or the extended phenotype.[45] They are of only limited relevance to the possibility of producing Gaia.

2.6.3. Eusocial Living Quarters: Termite Mounds and Bees' Nests

We now come to a third example of environmental regulation observed to take place in nature. One of the great contributions of kin selection (selfish genes) is that it helps us to understand the incredible structures produced by colonies of eusocial ("truly social," see note 23) insects, and the intricate cooperative behaviors they undertake together. Before I describe the regulation of environmental conditions in such colonies, I will look at patterns of relatedness in the colonies, for reasons that will become clear later.

In most eusocial insect species, colony members are very closely related. Most such colonies are, literally, each one big family. This is different from the lower average degrees of kinship between members of packs of wolves and prides of lions, for instance. These latter groups consist of mixtures of siblings, cousins, half-siblings, and even more distantly related individuals. In lion prides the bloodlines become even more mixed when, as mentioned earlier, a foreign male comes in and takes over a pride. In mammals and birds, lifetime monogamy (pair bonding for life, with no infidelities) has proved, after much investigation, to be rather rare, restricted to only a minority of species. This contrasts with the typical situation in eusocial insect colonies. Although by no means universally the case,[46] often in such colonies all reproduction (all egg laying) is carried out by just one mother (referred to as the queen), who mates with just one father during her lifetime. This doesn't necessarily involve the father living for as long as the queen; in some species the queen stores and preserves the ejaculate from one mating so as to last her a lifetime. This enables the single male to continue to be the father of all new colony members, even after he is long dead (similar to the possibilities arising when sperm from human donors is cryo-preserved). For a long time it was thought that the strange "haplodiploid" pattern[47] of gene transfer in insects of the order Hymenoptera (most but not all eusocial insects are hymenopteran), was the main stimulus for the evolution of eusociality in these species.[48] Recently, however, it has been suggested that a more important factor is monogamy,[49] because common parentage of a whole colony ensures close genetic similarity among colony members.

Eusocial insect colonies are fascinating, but also complex. The main point of interest here is that these colonies are close-knit family groups, even for colonies consisting of thousands or millions of individuals. This situation of shared colony parentage (all colony members being offspring of just one, or at most a few mating pairs) holds for the large majority of eusocial species among the bees, the wasps, the ants, and the termites. It is even more marked among the aphids, which at certain times of year reproduce clonally (parthenogenetically), giving rise to colonies of clonal replicates of the founding mother, in each of which all individuals are genetically identical. It also holds for the eusocial naked mole rats, as described above in section 2.2.

This close genetic similarity considerably raises the chances of success in passing on genes by proxy, by caring for and raising siblings, nieces, and nephews, instead of passing on genes directly through the individual's own offspring. For this reason, passing genes by proxy becomes much more profitable, in a probabilistic sense, to the extent that the evolution of permanently sterile helper castes (workers, soldiers, and such) becomes conceivable.

The close genetic similarity also makes it easier to understand why colony members cooperate to build immense (compared to their body size) communal homes. Not all bees and wasps live communally in very large colonies, and some of the communal genera such as the bumblebee (*Bombus*) live in small family nests numbering at most a few hundred individuals. But other genera build immense communal nests such as those constructed by honeybees (*Apis*). These large honeycomb-filled nests can be home to tens of thousands of individuals and, in the case of the giant honeybee (*Apis dorsata*), can be up to three meters in diameter. A natural bees' nest is a wonderful cooperative feat of architecture. It is a complex, collaboratively made structure in which the eggs, larvae, and pupae are protected and fed.

Even more impressive in scale and engineering are termite mounds (fig. 2.2c). Termite colonies can contain millions of individuals, depending on the species, and raise mounds up to eight or nine meters high (equivalent to several times higher than our tallest skyscrapers, when scaled relative to the height of a termite). These mounds are conspicuous features of the landscape in many countries. Termites feed on cellulose, obtained from wood or from grass and leaves, which they digest with the help of microorganisms in their guts, or with the help of "fungus gardens" they cultivate within the mounds. The termites then eat the fungi growing on the provided cellulose. Termite mounds are thus homes to both intra- and interspecific cooperation.

From the point of view of this book, the most fascinating aspect of these insect nests is that their internal environments are strongly regulated toward providing a stable and favorable microclimate for raising the brood, safe from the extremes of weather outside. In the large termite colonies, which form mounds stretching meters up into the sky, the brood is actually raised in subsurface chambers in the earth beneath the mounds. The temperatures there are rather stable, mostly because the construction of the mounds (the thickness of soil between the brood chamber and the outside world) gives them a large heat capacity. These environmentally regulated colonies could potentially point the way to how Gaia could arise out of natural selection, if the same dynamics could work at the global scale.

So-called magnetic termites in northern Australia build their mounds so that they stick up into the air in a shape more closely resembling a blade than a cylinder or a cone (fig. 2.3). They are most decidedly not radially symmetrical about a vertical line through their center. What is more, the orientations of the blades are not random. Nearly all of the mounds "point" to within 10° of due north.[50] The north–south direction of the axis makes a big difference to how they are heated: because they present a flat surface to sunlight coming from the east or west (in the early morning or late evening) they are strongly heated at these times. At noon when the Sun's heat is strongest, by contrast, they present a very thin profile to the Sun and so avoid being overheated. Tests in which mounds were dug up and repositioned along an east–west axis, at 90° to their original orientation, showed a much greater temperature range (up to 42°C at midday) compared to fairly constant temperatures during the day (33–35°C between 10 a.m. and 5:30 p.m.) in the north–south oriented mounds.[51] The end result is that nest temperatures of the latter are kept relatively constant and are protected from the worst extremes of air temperature fluctuations.

In another example (fig. 2.4), from a study of termite mounds in a savannah in Côte d'Ivoire, a narrow range of between 29 and 30.5°C was measured within a mound, whereas external air temperatures varied between 22 and 31°C. As well as being directional solar panels, the spectacular aboveground architectures also act as chimneys and ventilation structures, ensuring that the large amounts of CO_2 respired by the termites, and, particularly, their fungus gardens, do not become overwhelmingly toxic.

Figure 2.3. Field of magnetic termite mounds, Australia. They share a common orientation—the axes of their blades point toward the midday sun so as to minimize their exposure to direct sunlight during the hottest part of the day. Andy Clarke / Shutterstock.com.

Because they are out in the open and of smaller size, temperature in honey-bee nests is inherently less stable and has to be more actively controlled. Honey-bees do this by the social behaviors of (1) when it is too warm, standing at the entrances and fanning air into the nest and also evaporating dilute nectar brought to the nest specially for the purpose, and (2) when it gets too cold, hud-dling en masse around the brood and shivering their flight muscles to warm it. Through these activities bees are able to keep the interiors of their nests within a few degrees of 34°C in the face of much larger fluctuations in the external air temperature. Bee nests and termite mounds can be labeled Gaias in miniature; a large community of organisms for the most part peacefully and smoothly cooperating to keep the nest tidy, aired, and warm. If this can work for eusocial insect colonies, then why not for the planet? Lovelock has suggested that the extrapolation can indeed be made:

> A beehive, for instance, is made of insects and a wooden enclosure; this ensemble is able to regulate its internal temperature homeostatically almost as well as a mammal. A termite colony is another example. Ecosystems are superorganisms—from the anaerobic microbial communities to the great tropical forests. . . . Gaia is the largest and most complex superorganism that we know. (Lovelock 1991)

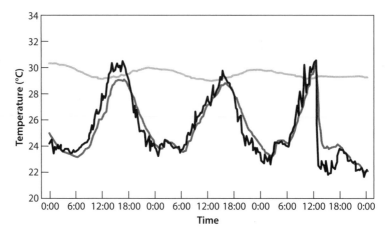

Figure 2.4. Exquisite temperature regulation inside a living termite mound. Temperature fluctuations are larger outside than inside the mound, and temperatures within are kept close to a value favorable for growth and development of the termites and their fungus gardens, at about 30°C. The black line is from measurements of air temperature 1 cm above the mound surface, the dark gray line is from within air channels inside the mound, and the light gray line is from inside the nest itself. With kind permission from Springer Science+Business Media: *Insects Sociaux*, 47: Figure 1 of J. Korb and K. E. Linsenmair (2000).

You may already have worked out the obstacle to this line of argument: the obvious reason why planets cannot work identically to termite mounds is that the planetary biota consists of a completely heterogeneous collection of species and individuals who are on the whole only very distantly related. We can now see that natural selection produces "good citizen" termites and bees because it pays selfish gene-carrying individuals to be altruistic on behalf of their close relatives[52] (carriers of many of their own genes). Natural selection does not favor altruism on behalf of genetic strangers.[53] It does not even always favor altruism to full siblings, as described earlier (siblicide). This is the key difficulty. Because the global biota is not a closely related family group, there is no reason to expect something like eusocial cooperation to operate at the scale of the whole Earth.

2.6.4. Forests, Coral Reefs, and Other Ecosystems

An individual coral is a colony of polyps. Moreover, each coral polyp is genetically identical to all other polyps on the same coral, and so the whole coral can be thought of, from a genetic point of view, as equivalent to one organism. From the point of view of kin selection, if we expect eusocial siblings to cooperate then we should expect even closer cooperation from a group of genetically identical clones. And indeed this seems to be the case: the polyps even share a common gut so that food is shared equally throughout the single coral. When we consider a whole coral reef, however, the situation is completely different. A reef consists of very many individual corals of various species and types, as apparent from the various shapes. For example, branching, tabular, and fan corals all intermingle within a coral reef community. There is genetic homogeneity within each individual coral, but genetic heterogeneity across the reef as a whole. Given that, it is perhaps no surprise that there is, apparently, no homeostatic regulation of the environment of the whole coral reef. In contrast to termite mounds and bees' nests where temperatures are regulated to within a few degrees, water sloshes into, across, and out of a reef without being manipulated so as to keep conditions constant. The temperature of seawater is not controlled within a reef, as evident from the occurrence of bleaching events[54] when the temperature rises too high. It could be argued that it is intrinsically easier to control the temperature in a beehive because of its shape, but we should then remember that the shapes of both reefs and beehives are products of biological activity. Although a coral reef ecosystem is a beautifully complex entity, we have no evidence that it produces anything like physiological regulation of its environment.

The same seems mostly to be true of forests. If the wind direction changes so that a cold wind blows, then the interior of the forest gets cold too, in synchrony with the temperature changes outside it. Although we might see a slight amelioration of extreme temperatures under the canopy, because of the frictional resistance of trees to airflow, we do not see anything like the tight temperature regulation of the mammalian body or of a termite mound. Natural selection has produced temperature regulation in these latter cases by (1) a physical sealing-off of the interior (a covering of skin/fur for mammals, of clay for termite mounds), coupled with (2) physiological or behavioral measures to adjust temperatures when they deviate from the optimum. Coral reefs do not form tight boundaries that restrict water entry and exit (although some fringing reefs around atolls (fig. 2.2d) might appear to form a near-continuous perimeter, the perimeter is not at all watertight) and forests do not effectively close off the forest boundary to wind penetration.[55]

It appears that a coral reef, although also home to an enchanting and vibrant web of noncoral life, can be understood most correctly as simply an area of seafloor where corals are better adapted to thrive than are seaweeds, rather than as a cooperatively maintained living space. Likewise, a forest is best understood as a place where physical and chemical conditions are favorable for the growth of trees, but the inside of a forest is not regulated like the inside of a mammalian body.

2.6.5. Inadvertent Side Effects from Whole Species or Groups of Species

Beaver dams are products of natural selection because beavers carry out considerable (and complex) work in order to construct them, and beavers that construct unstable or porous sieve-like dams will lose out in the struggle for survival. It matters to beavers whether or not their dams work. Mutant beaver genes that lead to their owners building dams parallel to rather than across the current presumably do not get passed on to future generations, because, in the absence of a pool of water, those beavers will starve and/or fall prey to predators. Beaver dams are textbook examples of extended phenotypes.

There is, however, another way in which organisms can produce very large structures that sit outside their bodies. They can do it inadvertently, unconsciously, and without purpose, over many generations, in fact over such a long time that no benefits accrue to the individual organisms creating the structures. In the same way that stone steps in castles have become worn down in the middle from the accumulated tread of hundreds of years of feet, so too have islands such as Nauru (in the Western Pacific) been built up over time by millennia's worth of seabird defecation. The phosphorus-rich guano deposits, until recently busily mined by the megaton for use in fertilizers, were simply the cumulative result of millions of seabirds fouling their nests. Like Nauru, the thick and humus-rich soils of areas like the Russian steppes are the result of many thousands of years of waste accumulation. Plant debris contains a fraction that is resistant to degradation, and this builds up, year upon year.[56] This is a similar idea to that embodied in Daisyworld, and is the way in which many important biologically mediated impacts work on Earth. The essence here is of environmental impact through happenstance, rather than regulation honed by natural selection.

Coral atolls are superb examples of this tremendous power of biology, over time, to change the environment. Plate tectonics causes some parts of the Earth's crust to be gradually and progressively raised up, and others to sink. This holds true for the seafloor as well as for the land surface, and some large areas of seafloor are slowly submerging, taking any seamounts and islands down with them. The rate is usually very slow, of the order of ten to one hundred meters per million years; nevertheless, without the actions of corals in these places, the islands would slowly sink beneath the waves. But the corals intervene and prevent some islands from drowning by depositing coral on top, eventually leading to the formation of coral atolls; in this way they prolong the habitable lifespan of their environment (fig. 2.5).

Coral polyps are animals themselves, but surface corals harbor photosynthetic symbionts and hence require sunlight in order to live.[57] The important side effect of corals that helps maintain their environment is that they leave behind, after their death, a calcium carbonate skeleton, which breaks up to form coral rubble. Whereas most biological products have a matching set of decomposers, this is not true of calcium carbonate, which has no nutritional value for organisms. Whereas it is often the case in nature that one organism's waste is another organism's food, nothing wants to eat coral. Not everything in nature is recycled.[58] The coral rubble left over from dead corals[59] therefore builds up inexorably over time, leading to the formation of great white slag-heaps of a stupendous size. Like icebergs whose true size is concealed because mostly submerged, similarly we cannot easily appreciate the true magnitude of the calcium carbonate refuse heaps that are coral atolls. If such mountains were exposed on land from base to peak, they would be rightfully appreciated for the marvels they are, and would be celebrated as natural wonders of the world. Instead, because of the relative opacity of water compared to air, they are mostly out of sight and therefore out of mind. This production of coral rubble helps a submerging coral atoll (should it happen to be on a part of the ocean crust that is slowly sinking) to continually to "keep its head above water." In fact, the reef keeps its head continually just under the water surface, because of course corals that breach the surface are inhibited from growing. Rubble accretion on a reef therefore ceases as soon as the atoll reaches the surface. Because of this negative feedback,[60] the rate of coral accretion never far outstrips the rate of subsidence (plus sea-level rise) over long intervals, because otherwise the atoll would end up tens of meters above the water surface where corals cannot survive.[61]

There is, however, a maximum rate at which individual corals can grow, die, and leave behind rubble. Therefore, the opposite scenario is possible, that the rate of subsidence (perhaps abetted during a particular time interval by rising sea levels) can outstrip the rate of addition of rubble to the top of the atoll. In this case, once the top of the atoll has descended more than about fifty meters below the surface of the water, where darkness prevents it remaining a viable habitat for corals, then it is probably doomed forever to a submarine existence. There is abundant proof that this inadvertent mechanism for sustaining coral atolls through immense durations of Earth history is not foolproof: in the north Pacific in particular (for some reason they are more prevalent there), the submarine landscape is dotted with scattered failed atolls. There are many underwater mountains, known as guyots, that, even though their peaks are now hundreds of meters below sea level, are covered in a thick layer of ancient coral debris.[62]

I find something stirring and awe-inspiring about this prospect of coral reefs on remote South Sea islands being perched on top of the debris of ancient coral "civilizations." Generations of now-dead corals unwittingly conspired in a vast

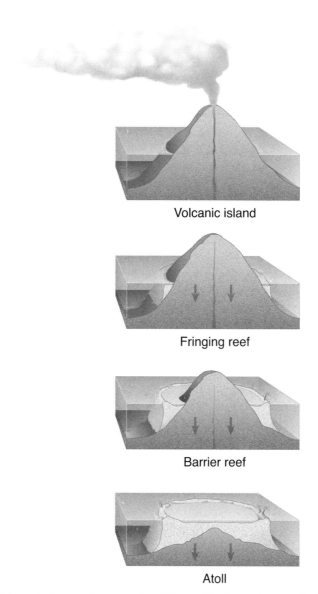

Volcanic island

Fringing reef

Barrier reef

Atoll

Figure 2.5. Schematic showing how a coral atoll forms. Charles Darwin was the first to hypothesize that coral atolls form by the growth of coral reefs on top of subsiding volcanic islands. Sara Boore and USGS.

civil engineering project that now keeps their descendants propped up in the sunlight. Without those hundreds of meters of added calcium carbonate the mountain would now peak out somewhere down in the permanent blackness. Just like archaeological digs on sites of long human occupation, which find layers of medieval rubble on top of Anglo-Saxon rubble on top of Roman rubble on top of Stone Age relics, each civilization having built on top of the debris of the preceding, so the cores down through coral atolls reveal that each generation of corals added its load to the mountain. Here is the real-life embodiment of the sort of inadvertent regulatory process captured in Daisyworld. Each coral colony or coralline alga pursues its own selfish ends, but out of this emerges a mechanism that prolongs the viability of their shared habitat.

After pausing to reflect in wonder at this feat of habitat maintenance through the geological ages, we come back once again to the question of whether or not this sort of process can be extrapolated to the scale of Gaia. Can Gaia be achieved as a wasteworld of by-products,[63] to use Tyler Volk's phrase? The problem with this concept is not entirely one of scale; there is no intrinsic difficulty in such types of feedback operating at a global scale. The problem is one of direction. If the biologically driven feedbacks occur incidentally (rather than being specifically selected for because they have beneficial effects on the global environment) then needless to say they can be either positive or negative, either sustaining or destructive. As will be discussed in later chapters, biological forces of awesome power can alter (and have altered) the planet in unfortunate as well as in fortunate ways.

2.7. Conclusions

It seems, in summary, that natural selection leads to local regulation of environments only in certain circumstances: (1) either inside or in the nearby environs of single organisms, or (2) inside the living quarters of eusocial colonies, in which individual members are each very closely related (possess very similar genes). A general theme that emerges is that tight homeostatic regulation is associated with these "islands"[64] of genetic uniformity or similarity. Tight cooperation and regulation of a communal environment only occurs between organisms that are closely related, although it is by no means guaranteed even then.

This review of environmental regulation in nature therefore offers no obvious clues as to how regulation at a planetary scale could have been produced by or favored by natural selection. Although all life is descended from a common ancestor, there is no way that the biosphere as a whole can now (billions of years later, after innumerable branchings) be considered as anything else but genetically diverse, and we have no evidence that natural selection favors cooperation and "mutual home-building" among genetically diverse entities.[65] The preceding review offers no encouragement that cooperative environmental regulation should evolve at the global scale to produce Gaia.

However, this lack of a proven mechanism for Gaia does not rule it out; after all, Darwin proposed his theory of *The Origin of Species* many decades before anyone (Gregor Mendel excepted) knew anything to speak of about genes and the mechanisms of inheritance. Perhaps some such underpinning will eventually be developed for Gaia, but, as Lovelock agrees, it is certainly not there yet:

> [Critics] are right to insist that a large and still unanswered question remains: if the Earth is indeed self-regulating by biological feedback, how has this come about through natural selection? I like to compare our inability yet to give an answer with Darwin's inability to satisfy those critics who saw the amazing perfection of the eye as something that also could never have arisen by chance natural selection. (Page 278 of Lovelock 2001)

This point was echoed by Bill Hamilton, who became interested in the question of Gaia in later years:

> Just as the observations of Copernicus needed a Newton to explain them, we need another Newton to explain how darwinian evolution leads to a habitable planet. (Quoted in Lovelock 2003b)

I suspect that this is not a realistic hope, for the reasons just set out. However, if anyone is able to achieve it, then they will richly deserve their laurels and accolades.

This chapter has had a tight focus on whether Gaia arises predictably out of natural selection. In the next few chapters we will change tack and examine the plausibility of the Gaia hypothesis in other ways. Even though there will no longer be a specific focus on possible connections between Gaia and natural selection, future chapters will shed some light on this question from different directions. Chapter 6 will examine the transformative effect that some biological processes have had in shaping the Earth environment. In chapter 7 we will consider whether major evolutionary developments have tended to be mostly favorable or injurious to the biota as a whole.

Chapter 3

LIFE AT THE EDGE

LESSONS FROM EXTREMOPHILES

HAVING ESTABLISHED THAT, as far as we can tell, natural selection does not lead automatically to Gaia, this chapter starts using information about the present day Earth system to probe the conjecture that Gaia exists and is at the controls. The chapter focuses on assertion No. 1 (see chapter 1): the suggestion that the suitability of the Earth for life is strong evidence in support of the Gaia hypothesis. This first assertion is only about how suitable the Earth *currently* is for life. The puzzle of how it has *remained* (or *been kept*) suitable for life is saved until later.

In early Lovelock publications, Lovelock went even further than saying the biosphere is suitable, to saying that it is optimal for life:

> The most important property of Gaia is the tendency to optimize conditions for all terrestrial life. (Lovelock 1979)
>
> The greater part of our own environment on Earth is always perfect and comfortable for life. (Lovelock 1988)

In later publications, however, this was modified to suggesting only that the environment is made comfortable for life.

> The first edition of this book used the terms optimum and optimize too freely; Gaia does not optimize the environment for life. I should have said that it keeps the environment constant and close to a state comfortable for life. (Lovelock 1979, in the preface to a revised 1987 edition)
>
> Remarkably, however, conditions in the earth's environment do generally meet the narrow constraints for cell survival. This fact is the strongest evidence for Gaia, the system of life and its environment. (Page 98 of Lovelock 1991)
>
> Present conditions at the surface of the Earth are within the relatively narrow boundaries that eukaryotic, multicellular organisms can tolerate. (Lenton 1998)

These constraints on cell survival, it was suggested, include the requirements that temperatures must be between 0° and 50°C, pH (acidity) of aquatic habitats between 3 and 9, and salinity less than 60.[1]

This chapter scrutinizes whether the good fit between environments[2] and the organisms that inhabit them really is, in fact, strong evidence for Gaia. We will look at whether it helps distinguish between Gaia and the competitor hypotheses. Lovelock called it the strongest evidence for Gaia, but was Lovelock looking at things from the wrong perspective, suggesting that carts push horses rather than horses pull carts? The evidence described below suggests that Lovelock did get cause and effect the wrong way around. From an evolutionary perspective it is not at all remarkable that the conditions in each habitat on Earth are particularly conducive for the life living there. It is well established that evolution forces organisms to closely fit their environments, however awful those environments may be. This can give rise to the misleading impression that those environments have been made to fit the organisms. But in fact the snug fit between organisms and habitats is more a testament to the overwhelming, transforming power of evolution to mold organisms than to the power of organisms to make their environments more comfortable. In the words of Dick Holland: "We live on an Earth that is the best of all possible worlds only for those who are well adapted to its current state."[3]

But before looking at the reasons why organisms and environments tend to be well suited to each other, I will first review the most recent research on extremophiles. What can it tell us about the nature of the most extreme environments that life can withstand? Just how narrow are the constraints for survival mentioned above?

3.1. Habitability Constraints for "Simple" and "Complex" Life

Thanks to the efforts of numerous biologists who, over the years, have trekked out to many strange and exotic locations and analyzed the life there, we now have a much improved idea of the habitability tolerances of different forms of life. These are summarized in tables 3.1 and 3.2. A general rule is that microbes are tougher.[4] On the whole, more primitive microbes are able to withstand (and even flourish in[5]) more extreme environments than are more complex microbes or multicellular life. The more primitive microbes comprise both bacteria and archaea,[6] and are collectively known as prokaryotes. Prokaryotes have simpler cells lacking a true nucleus. Eukaryotes are organisms possessing more complex cells including a nucleus; the eukaryotes include nearly all multicellular organisms and all animals and plants. For (almost) every environmental variable the prokaryotes have wider habitability ranges than do eukaryotes. They can survive in many extreme environments that are deadly to higher life forms such as plants and animals.

TABLE 3.1.
Environmental tolerances of "record-holder" organisms (prokaryotes) in each class of extremophiles

Variable	Descriptive term	Genus/Species of "record-holder"	Minimum	Optimum	Maximum	Natural habitat of organism
High temperature	hyperthermophile	Strain 121*	85°C	105°C	121°C	hydrothermal vents
Low temperature	cryophile	Polaromonas vacuolata	0°C	4°C	12°C	Antarctic marine environments
		Various bacteria[†]	−7°C			Antarctica
High pH	alkaliphile	Natronobacterium gregoryi	8.5	10	12	soils permeated with solfataric gases
Low pH	acidophile	Picrophilus oshimae	−0.06	0.07	4	acid mine drainage
		Ferroplasma acidarmanus[‡]	0	1.2	2.5	
High pressure	barophile or piezophile	MT41	500 atm	700 atm	> 1,000 atm	Mariana Trench (deep sea)
High salinity	halophile	Halobacterium salinarum	150	250	320	hypersaline pools on the Sinai peninsula
		Haloquadratum walsbyi (Walsby's square archaeon)[§]	180			

Note: These are the most extreme conditions in which organisms are known to be able to live and reproduce (not just that they can tolerate temporarily). This table was produced by combining new information with information from Rothschild (2009) and Madigan (2000). It is fairly certain that the record-holder organisms and limits will be revised over time as even more extreme organisms are discovered.

* (Kashefi and Lovley 2003)
[†] (Straka and Stokes 1960)
[‡] (Edwards et al. 2000)
[§] (Bolhuis, Poele, and Rodriguez-Valera 2004)

TABLE 3.2.
Environmental tolerances of "record-holder" eukaryotic organisms

Variable	Descriptive term	Genus/Species of "record holder"	Greatest extreme	Natural habitat of organism
High temperature	hyperthermophile	alga (Cyanidium caldarium)	57°C	hot and acid soils and waters
		Sahara desert ant (Cataglyphis bicolor)	55°C	desert
		Pompeii tube worms (Alvinella pompejana)*	80°C	hydrothermal vents
Low temperature	cryophile	Himalayan midge	-18°C	Himalayan glacier
		Nitzschia frigida[†] (diatom)	-8°C	brine channels in sea ice[†]
High pH	alkaliphile	rotifers, copepods, water boatmen, one fish species (alkaline tilapia, Oreochromis alcalica), flamingos	pH = 9 to 11	Rift Valley soda lakes[‡]
Low pH	acidophile	alga (Cyanidium caldarium)	pH = 0	hot and acid soils and waters
		rotifers, heliozoa, ciliates	pH = 2	Rio Tinto[§]
		vinegar eelworm (Turbatrix aceti)	pH = 3.5	barrels of vinegar, fermenting tree sap
		tambaqui fish (Colossoma Macroponum), Osorezan dace	pH = 3.5	naturally acidic Amazonian black-water rivers,[‖] Lake Osorezan[¶]
High pressure	barophile or piezophile	Sea cucumbers, worms, shrimps	> 1,000 atm	Mariana Trench (deep sea)
		Grenadier fish	700 atm	Japan Trench

High salinity	halophile	Arabian Killifish (*Aphanius dispar*)	175	springs around the Dead Sea, salt ponds**
		Green algae (e.g., *Dunaliella salina*), diatoms	250–330	Dead Sea
		Brine shrimp (genus *Artemia*)	250	Great Salt Lake
		Australian brine shrimp (*Parartemia zietziana*)	353	Australian salt lakes[††]

Note: These are the most extreme conditions in which eukaryotic organisms are known to be able to complete their life cycles and reproduce. Brines (super-salty waters) in sea ice can remain liquid far below 0°C; the dissolved salt endows antifreeze properties. Again, this table should be considered as provisional in the sense that it is only derived from the current knowledge of extremophiles. In many cases it is likely that even more extreme organisms will be discovered some time in the future, for instance as we find out more about chemosynthetic microbes deep within the rocks of the Earth, or as Lake Vostok, lying beneath 4 kilometers of Antarctic ice sheet, is explored for the first time.

[*] It may be that only parts of the worms can tolerate such high temperatures, or that such high temperatures can only be tolerated intermittently.
[†] (Aletsee and Jahnke 1992)
[‡] (Randall et al. 1989)
[§] (Zettler et al. 2002)
[||] (Wilson et al. 1999)
[¶] (Hirata et al. 2003)
[**] (Plaut 2000)
[††] (Bayly and Williams 1973)

3.1.1. Temperature

There is no shortage of naturally occurring cold environments on Earth, as will be discussed in the next chapter. Life is generally scarce in the cold polar regions such as Antarctica. A fundamental characteristic of birds and mammals is their ability to regulate their internal temperatures by burning food to generate heat. Some multicellular eukaryotes, Emperor penguins being a notable example, can therefore survive through even the extreme cold (−50°C) of the Antarctic winter. Reptiles, on the other hand, having only limited control over their body temperatures, are generally at the mercy of the ambient temperature and are not found anywhere on Antarctica. In this case, thanks to their ability to provide their own internal central heating and thereby maintain body core temperature higher than that outside, certain eukaryotes rival prokaryotes in terms of tolerance.

Our knowledge of the hottest environments that life can stand comes in part from geothermal springs, where superheated water from inside the Earth rises to the surface. These springs are typically inhabited only by prokaryotic microbes. Plants and animals are found even in the hottest of deserts, if sufficient water is available, in conditions (temperatures > 40°C) that would cause any "normal" organism to shrivel and die. Mid-ocean ridge hydrothermal vents provide an interesting, alternative, source of information. On land a saucepan of water will come to the boil at about 100°C, depending on altitude; in the deep sea, however, the tremendous pressure allows water to remain liquid as it exits vent chimneys at temperatures up to 400°C. This water cools rather rapidly as it mixes with the much colder surrounding seawater, so much so that temperatures inside some tube worms, living directly on vent chimneys, have been measured at 40°C. While this is very hot compared to the rest of the deep sea (usually within a few degrees of 0°C), it is no hotter than many deserts on land. On occasions, however, the tube worms are thought to be exposed to temperatures greater than 100°C. High temperatures create challenges for life because proteins such as DNA and chlorophyll start to "denature" (unravel) above about 70°C (50°C for some enzymes). Thermophiles have been able to partially circumvent this problem by various strategies including "supercoiling" of the DNA so that it is more tightly packed and less able to unravel, the use of molecular chaperones (proteins that protect other proteins from unraveling), and other strategies. Nevertheless, it seems likely that if temperatures on Earth had risen far above 50°C during the second half of Earth history, the large majority of eukaryotic life would have been presented with a considerable challenge.

3.1.2. Salinity

The term *salinity* relates to the concentration of chemical ions (dissolved salts) in water. Freshwater environments (lakes, rivers) are typically low in salt, but are still home to flourishing ecosystems. Low salinities apparently do not present survival problems.

Seawater in the open ocean has, on average, a salinity of about 35, which corresponds to about 35 grams of salts dissolved in each liter of water. Some species, such as the eels and salmon of the last chapter, migrate between rivers (freshwater) and the sea. Life may well have first evolved in the sea and obviously has no great problem in coping with typical seawater salinities. Even the most extreme degrees of saltiness seem not to present insuperable obstacles to all life. The Dead Sea is the most concentrated natural salt lake in the world, but, despite its name, even it is not quite dead. It has a salinity of about 300, depending on depth and proximity to the low salinity inflow from the Jordan River. The increase in density caused by the intense load of dissolved salt is the reason that bathers find it so easy to float in the Dead Sea, and so difficult to dive down through it. It contains two main groups of life: algae (one species only) and bacteria (prokaryotes). Other salty environments such as the Great Salt Lake in Utah, USA (salinity up to 280), support several specialized salt-tolerant organisms, including multicellular life in the form of brine shrimps (*Artemia salina*). Simple ecosystems are also found in other lakes with less extreme salinity but high concentrations of particular elements. In Mono Lake in California, USA, for instance, a simple food chain thrives in salinity of about 90 and very high concentrations of arsenic (about 200 μmol kg^{-1}).

3.1.3. pH (Acidity)

Our knowledge of the (lack of) viability of highly acidic or alkaline environments comes again in large part from the ecologies of unusual lakes. Soda lakes are natural examples of highly alkaline habitats, with pHs as high as 10.5. A pH of 10.5 is more alkaline than baking soda (pH ~9) and milk of magnesia (pH ~10), and only slightly less alkaline than household ammonia or bleach (pH ~11 or 12). Specialized microscopic animals (copepods) and fish (*Tilapia*) are found in the highly alkaline Lake Natron and other soda lakes of the East African Rift Valley, where pH values can be as high as 9 to 10.5. Large flocks (numbering up to a million or more) of pink flamingos famously live and breed in some of these soda lakes such as Lake Nakuru (pH 10.5), thriving on the photosynthetic cyanobacterium *Spirulina*, which they comb from the water using filters in their beaks. In contrast, the average pH of the present-day surface ocean is about 8.2.

The habitable pH range of microbes (–0.06 to 12, table 3.1) corresponds to a habitable hydrogen ion range from 1.0×10^{-12} to greater than 1.0 mol liter^{-1}; that is to say, spanning more than twelve orders of magnitude. Although eukaryotes on the whole seem to be less tolerant of extreme pHs, nevertheless specialized algae are observed to flourish in the highly acidic (pH = 2) Rio Tinto,[7] which flows through a large, heavily-mined pyrite belt in southwestern Spain. Rio Tinto river water is more acidic than wine (pH ~4) and vinegar (pH ~3) and on a par with lemon juice (pH ~2). Elsewhere, a legacy of past opencast mining operations has left behind lakes with unusual chemistries. Some of these artificial mining lakes are highly acidic, with pHs between 2 and 4. Nevertheless, despite the extreme conditions and despite the lack of time for evolution to have tailored organisms to fit these recently produced lakes, eukaryotic organisms have still been able to colonize them. One example is Mining Lake 111 in the Lusatian mining district of eastern Germany. This lake, formed in 1958, has a pH of 2.6 and high concentrations of sulfate, iron, and aluminum. The sediment surface of this lake is covered by a dense layer of (eukaryotic) diatoms.[8] Another extremely acidic mining lake (pH = 2.7) has a restricted food web consisting of bacteria, two photosynthetic flagellate species, and two species of rotifers that eat the flagellates. More complex animals from higher up in the food chain are absent, presumably unable to tolerate the acidic conditions. Naturally acidic lakes in volcanic craters, such as Lake Katanuma in Japan (pH = 1.8) and Lake Goang in Indonesia (pH = 2.5) also host simplified food webs. These lakes contain algae, insect larvae and even crabs in Lake Goang (although no fish), within waters that are sufficiently acidic to corrode metal fittings on boats.

3.1.4. Pressure

Perhaps surprisingly, high pressures seem to have little effect on habitability. The pressure at sea level is 1 atmosphere, the weight of overlying air in the atmosphere above. Because of the much greater weight of a given volume of water than of air, pressure increases much more rapidly with depth in water. The pressure at 10 meters water depth is 2 atmospheres, double that at the surface, and at 100 meters depth is 11 atmospheres. The pressure increases by 1 atmosphere with every 10 meters of water. Large parts of the ocean floor are between 3 and 6 kilometers deep. A few parts of the ocean, the trenches, are even deeper, up to a maximum of 11 kilometers deep at the bottom of the Mariana trench. As a result, the pressure there is more than 1,000 times as great as at the sea surface. This tremendous pressure, which creates challenging engineering problems for designers of submarines, does not, however, prevent the presence of life. The

Figure 3.1. A whale carcass on the deep seafloor to the west of California. It sits below more than a kilometer of water, but, despite the tremendous pressures, is colonized by a diverse range of life, including eel-like hagfish. With permission of Craig Randall Smith (1999).

main limiter to life on the seafloor seems to be food availability rather than high pressure. Because all photosynthetic life depends on light, of which there is none at the deep seafloor, all species there are either scavengers or decomposers. They all depend for their subsistence on a rain of detritus from the sunlit surface many kilometers overhead.

The only exceptions to this rule are to be found at the mid-ocean ridge vents and a few other locations where geological processes furnish sufficient energy (in chemical compounds that they release) to support specialist organisms able to harness the chemical energy. At mid-ocean ridge volcanic vents, chemically exotic streams of water flow out from the seafloor; tube worms host "chemosynthetic" microbes that are able to generate energy to live on by chemically altering the water flowing out of the vents or seeps (converting one chemical compound to another, for instance via oxidation).

But aside from these scattered oases of locally supported life, everywhere else the seafloor organisms are dependent on the scraps floating down from the "dining table" at the sea surface. Occasionally a rather large scrap floats down. As shown so astonishingly in photos (fig. 3.1) and more recently also in TV

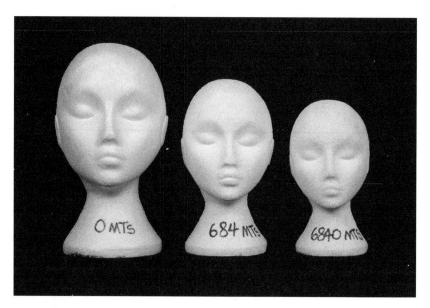

Figure 3.2. The effects of pressure on Styrofoam heads. The left-most head shows the original dimensions, the middle head has been subjected to the pressure of 684 meters of water depth, and the right-hand head to the pressure of 6,840 meters of water depth.

footage, a whale corpse provides a feast of unusual proportions, and the attraction of the feast helps to show the diversity of organisms that can tolerate the great pressure. Even at depths between 1 and 2 kilometers (pressures 100- to 200-fold greater than at the surface) whale corpses attract hagfish, rat-tail fish, sleeper sharks, clams, mussels, crabs, and much more[9] (fig. 3.1). When we add to this the well-known abilities of some of the highest animals (giant squid, sperm whales) to dive to great depths, it is apparent that even very high pressures have been overcome by evolution and do not seriously impinge on habitability. Even though Styrofoam heads subjected to the great pressures of the deep ocean are miniaturized (fig. 3.2), whales resurface unshrunk.

3.1.5. Oxygen

The early Earth, in common with all other planets that we know of, was largely devoid of free oxygen. Its atmosphere is suspected to have been dominated by dinitrogen, as today, but also to have been rich in carbon dioxide and methane instead of oxygen. The earliest life was therefore, necessarily, able to survive without oxygen, or in other words was comprised exclusively of anaerobes. Fairly early on in Earth history, the Great Oxygenation Event occurred (see chapter 7), and the sea and the atmosphere became more oxygenated due to the "oxygen pollution" by photosynthesis. Today the atmosphere contains an enormous 21% of oxygen (approximately one-fifth of all air is oxygen), and the sea is also rich in oxygen nearly everywhere. Despite the prevalence of oxygenated environments and aerobic (oxygen-using) organisms on Earth today, there are nevertheless also many anaerobic environments, populated by anaerobes. These environments often occur where there are barriers to the inflow of air or oxygenated seawater. They are also usually associated with rotting organic matter, because the most efficient decomposing organisms use oxygen and, where there is sufficient matter to decompose, this eventually exhausts the local supplies of oxygen. So compost heaps, the lower reaches of highly stratified productive lakes, and the muddy oozes of estuaries are just some of the places where anaerobes proliferate. A complete lack of oxygen is therefore not at all prohibitive for life as a whole. However, although anaerobic prokaryotes are common, anaerobic eukaryotes are not. The absence of oxygen excludes higher organisms.[10]

3.2. ADAPTED ENVIRONMENTS OR ADAPTED ORGANISMS?

The main point of this chapter is to examine whether the good fit between organisms and environments presents evidence in favor of the Gaia hypothesis. There are three possibilities: either (1) environments fit organisms because the collection of life on Earth (the biota) has manipulated its environments to be especially commodious (Gaia), or (2) evolution has manipulated the biota to be especially well adapted to the environments it inhabits, or (3) a combination of (1) and (2). Obviously, if all or most of the fit is due to evolution, then the good fit is testament more to the powers of evolution than to the existence of Gaia.

The examples of biological coevolution and predator-prey arms races reviewed in chapter 1 are evidence of the overwhelming power of natural selection in shaping the physical characteristics and behavior of organisms. There is no doubt that evolution has molded organisms to fit their niches, including the chemical and physical aspects of those niches such as pH, pressure, and salinity. But at the same time as it makes organisms fit for the prevailing conditions, evolution also makes organisms unfit for conditions that are not encountered.

3.3. HABITAT COMMITMENT

Prokaryotes are able to withstand a range of pH values between −0.06 and 12 (table 3.1). However, the prokaryotes that can tolerate the extreme acidity are, of course, not the same as those that can tolerate an extremely alkaline environment. There are specialists at either end of the range, the acidophiles and alkaliphiles, respectively. Let us assume, for argument's sake, that the optimum pH value for life as a whole is halfway between these two extremes, that is, pH = 6. Or we could assume that it is equal to the pH of seawater, about 8. Either way, the relevance to this chapter arises because neither true acidophiles nor true alkaliphiles can survive at a pH of 6 or 8. The extremophiles do not grow best at the optimum pH for life as a whole. In fact they are so specialized to their extreme pHs that they grow best at pHs that, while not quite at the outer bound of habitable pHs, are pretty near it.

This is also true for temperature and other variables. As discussed later in this chapter, cryophilic animals inhabiting the seafloor around Antarctica, which are used to and can tolerate extremes of cold fatal to all other life, are themselves killed off by a rise to even tepid temperatures. The most extreme prokaryotic thermophiles, on the other hand, grow most rapidly at about 105°C (above the boiling point of water!). Although they can still multiply at temperatures up to 121°C, temperatures below a scalding 85°C are so cold, to them, as to be deadly. For these organisms that are specifically adapted to live in odd environments, odd has become better. After having lived in an extreme environment for generation after generation, the resident microbes become closely adapted to that environment, making the most of a bad job. But they become so closely adapted to it that they lose the ability to grow well in "more normal" environments, environments that one would expect to be more comfortable for them.

The important point here is that organisms come to fit their environments extremely closely, even if, by any objective evaluation, their environment is terrible. Before continuing, it is important to clarify the meaning of the word *objective* in the last sentence. To be objective we obviously need to try and discard any tendencies toward bias that come from our belonging to the particular species *Homo sapiens*. To an extent the "goodness" of an environment is all in the eye of the beholder: so the cold of Antarctica feels terrible to a human being but not to an emperor penguin, a salty ocean feels terrible to a freshwater crocodile but not to a saltwater crocodile, normal river water feels terrible to an algal cell from the Rio Tinto, and so on.

Although difficult, it is possible to overcome these subjective biases by confining ourselves to objective measures such as the species diversity and productivity of a given ecosystem. So, for instance, a salt lake may objectively be considered a terrible environment for life if only one or a few species thrive there, or the Antarctic continent may objectively be considered terrible because of the very low biomass and biological productivity. These lines of argument are much more reasonable, but even then we should be wary of possible pitfalls. For instance, an individual sulfur pond could have low biodiversity because it is one of a kind far from any other similar ponds from which it could be colonized. Alternatively its diversity could be reduced because its small size precludes higher trophic levels because of the relative scarcity of food.

To get back to the main argument, what the research into specialist extremophiles tells us is that any environment (as long as it is not so poor as to be completely uninhabitable) will appear to be a utopia, if we judge it solely according to how suitable it is for the organisms that inhabit it. The process of evolution by natural selection not only fits organisms to the environments in which they live, but also, it appears, makes them unfit for other environments. When an extreme environment is consistently extreme, for instance always very hot or always at very high pressure, then there is no selection pressure for the extremophiles living within it to also be able to tolerate less extreme habitats. If they never encounter less extreme conditions, they do not need to be able to survive them. And indeed we can see from the example extremophiles above that evolution does cause them to lose their ability to tolerate less extreme conditions.

Let us call the above phenomenon "habitat commitment," whereby a species in a fairly static environment tends to become tightly fitted to that environment at the expense of becoming fit for no other. How does habitat commitment occur? We can get a satisfying insight into the process from work on limpets, clams, and fish living in Antarctic waters.[11] Antarctic waters are special in at least two ways: firstly because they are so cold, and secondly because temperature varies little throughout the year, rising only slightly above freezing even in summer (for instance, seawater temperatures at 15 meters depth near the Rothera Antarctic base were found to vary only between $-1.9°C$ and $+1.4°C$ in the years 1997–2000). This unusual constancy of temperature has led to inhabitation by organisms that are unusually sensitive to only minor shifts in temperature. Scientists investigating the temperature tolerances of several Antarctic seafloor-dwelling species have found that many of these species have a "survival envelope" only 6–7°C wide, compared to 15–25°C wide for comparable species from other parts of the world. Scientists from the British Antarctic Survey extracted some Antarctic seafloor animals from their natural habitat and then tested their temperature tolerances in the laboratory. They found that Antarctic limpets (*Nacella concinna*) were all incapacitated by temperatures above about 6°C (half were incapacitated above 2°C), burrowing clams (*Laternula elliptica*) by temperatures above 5°C (2°C), and scallops (*Adamussium colbecki*) by temperatures exceeding 2°C.

What was the underlying reason for this intolerance of even slightly warmer waters? Why do these mollusks start to "overheat" at water temperatures so cold that sailors falling overboard would die of hypothermia within minutes? In this case detailed physiological work has suggested an answer. It involves mitochondria, the energy factories within each cell. Each mitochondrion usually works rather slowly at very cold temperatures (part of a more general phenomenon we will return to in the next chapter). However, when temperatures increase, the mitochondria start upping their work rate. Each mitochondrion uses up oxygen while producing energy, and when all the mitochondria in one of these Antarctic seafloor animals are working rapidly, the oxygen demand starts to outstrip the possible supply into the animal. Oxygen demand appears to increase more quickly with temperature than does oxygen supply. Anaerobic (oxygen-scarce) conditions start to develop in the animals, but these cannot be sustained indefinitely and the animals therefore die, through a process analogous to asphyxiation, if "high" temperatures are sustained. Although oxygen demand also follows temperature in organisms from elsewhere, those living on the Antarctic seafloor, because neither they nor their recent ancestors were ever exposed to high temperatures, have not been equipped by evolution to cope with them.

Of course this is all well and good if their environments remain stable. But these extreme cases of habitat commitment may spell trouble for them in a future warmer world. Because of their narrow survival envelopes with respect to temperature, Antarctic marine species are thought to be at particular risk from global warming.

If habitat commitment is a universal property of evolution, then we should expect to be able to discern its action in the evolutionary history of larger animals, in addition to those of the "lowly" microbes and clams. And so we can. A striking example is the famous history of the dodo (*Raphus cucullatus*) on the remote island of Mauritius, far out in the Indian Ocean to the east of the much larger island of Madagascar.[12] Dodos were flightless, ungainly, and, according to paintings, rather ugly birds that were restricted solely to Mauritius. The relevance to our story becomes apparent through recent DNA evidence that has illuminated their ancestry. This shows that they were members of the pigeon family, and more specifically most closely related to (almost certainly descended from) the Nicobar pigeon (*Caloenus nicobarica*). There are obvious obstacles in collecting DNA evidence from an extinct species, but luckily some remains of a dodo are kept in the Ashmolean Museum in Oxford, and it proved possible to extract some DNA from the bones of this specimen. What is interesting is that the Nicobar pigeon is a flying pigeon with a nomadic lifestyle well suited to the colonization of remote oceanic islands. These separate pieces of information allow us to construct a probable history of the creation and then destruction of the dodo species.

Some time in the far past a severe storm must have swept roaming Nicobar pigeons (at least two, or else a female bearing eggs) to the remote island of Mauritius. It is not surprising that the first colonists were flying birds, given the drastically lower chances of flightless birds reaching such a far-flung island. Once arrived at Mauritius, habitat commitment came into play, and the lack of any large predators removed much of the advantage to the power of flying. Mutant Nicobar pigeons possessing only stunted wings and/or withered breast muscles must have gained in other ways (deploying their limited resources to other ends, such as a larger and stronger beak or more eggs) and have become more successful in the competition for survival. Not only that, but as the Nicobar pigeons evolved into dodos they also became tamer, given that there was little or nothing on Mauritius they needed to run away from. The reflexive instinct to flee from people or other large animals was maladaptive (causing them to expend energy for no benefit) on Mauritius and was lost.

Portuguese sailors first arrived on Mauritius in 1507, but real trouble for the dodos seems to have been delayed until around 1600, when Dutch mariners first started using Mauritius as a way station. From the need to replenish ships' supplies or just for enjoyment, the visiting sailors culled dodos in large numbers. Deprived of the power of flight, in both senses of the word, the dodos never stood a chance. It is likely that their already poor prospects were worsened even further by the introduction of pigs and monkeys, which ate their eggs, presumably laid directly on the forest floor because the dodos couldn't fly up to the treetops. Within a human lifetime, probably in the 1660s, the last dodo had been taken. Habitat commitment on an island free of mammals not only converted Nicobar pigeons to dodos but also rendered them helpless when humans discovered the island. This same story has been repeated on many other distant islands originally without mammals. They also possess(ed) flightless birds descended originally from flying ancestors such as pigeons or rails. Many large birds disappeared following human colonization of New Zealand (moas and others), the Hawaiian Islands (for example, moa-nalos), Madagascar (for example, the elephant bird *Aepyornis*), and elsewhere.

Another example, although a humbler and less well known one, represents one of the most complete records of evolution so far found in the fossil record. It is also another record of regressive evolution (habitat commitment), showing again that what evolution gives, evolution can also take away. Mike Bell, a paleontologist, painstakingly examined more than five thousand stickleback fossils spread throughout 10,000 years of lake sediments dating from the time since the peak of the last ice age. This wonderful dataset shows what happened after sticklebacks first colonized a lake in Nevada. The sticklebacks that arrived in the lake were originally, courtesy of evolution, heavily armored for protection against the larger fish that normally eat them. But in the new lake there were no stickleback predators. An analysis of the fossils shows with abundant clarity how the average number of dorsal spines declined over time as the generations passed since their first arrival in the lake.[13]

Other beautiful examples exist. In the previous chapter the coevolutionary arms race between cuckoos and the birds they parasitize was described. In Africa one such arms race is between weaverbirds and cuckoos, and it has led to each clutch of weaverbirds possessing distinctive colors and patterns so that the cuckoo eggs (which cannot be individualized to this degree) stand out. In essence, the weaverbirds label their eggs, and this minimizes their likelihood of being tricked into raising cuckoo eggs. The link to habitat commitment comes with a study[14] of weaverbirds that were introduced to the islands of Mauritius and Hispaniola, where there are no cuckoos, one hundred and two hundred years ago, respectively. They now produce less patterning of their eggs than do the weaverbirds in Africa. Just like the sticklebacks that lost their spines, the island weaverbirds are seen to be losing the distinctiveness of their eggs now that it no longer provides a benefit.

Habitat commitment is also apparent in other evolutionary histories, such as the modification over time of the ancestors of hippopotamuses into whales. Evidence from DNA[15] has surprised taxonomists by showing that dolphins and whales (cetaceans) are descended from ancient hippo-type animals, and in fact are the closest living relatives to hippos. Whales are even closer relatives to hippos than hippos are to cows, sheep, and other ruminants who are their next closest relatives. It appears that the evolutionary lineages of cows and hippos split apart *before* the evolutionary split between cetaceans and hippos. The dramatic physical differences between the two are testament, again, to the power of habitat commitment. Hippos spend most of their time in the water but some of it also on land. They therefore retain legs capable of bearing their weight, together with a whole suite of other adaptations sustaining the ability to walk. Whales and dolphins, on the other hand, have completely forsaken the land and become totally committed to life in water. It is therefore no surprise that the fundamental change in habitat has been accompanied by a fundamental change in the structure of the creature. Whales and dolphins no longer possess legs, although rudimentary vestigial leg bones can still be discerned deep inside the bodies of some species.[16] Large tail flukes have been evolved, whereas hippos possess none, and there are other differences. Presumably, once the hippo-like ancestors of whales started living completely in water, mutants with smaller (and/or more finlike) legs were more successful (because energy and resources no longer spent on useless legs could be employed elsewhere in the body, or toward reproduction).

The point of relevance to this argument is that evolution can be seen not only to accrete useful adaptations but also to discard those that are no longer

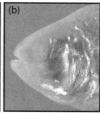

Figure 3.3. Degeneration of eyes in cavefish. Both fish are of the same species (*Astynax mexicanus*), the left-hand one from a surface-dwelling population and the right-hand one from a blind cave-dwelling population that live their whole lives in total darkness. The population within the Mexican caves differs also in other attributes such as lack of skin pigmentation. Reprinted by permission from Macmillan Publishers, *Nature*, 431:844–847 (2004): Yamamoto, Stock, and Jeffrey.

useful. In the process it fits the organisms to the environments they presently inhabit, while degrading their fitness for environments they are no longer exposed to.

The hippos to whales story is far from unique. In fact it is repeated all over the animal kingdom. Whereas most evolutionists (and Rudyard Kipling) have concerned themselves with how new organs are constructed, how organs are lost is an equally valid area of study. Speleologists in particular are interested in this question of "regressive evolution" in order to explain the loss of eyes in animals (for example, fig. 3.3) living permanently deep in cave systems where no light ever penetrates. Darwin, a forerunner here as so often, pondered this question of blind cave dwellers in *The Origin of Species*.[17] He also discussed blind mole rats (loss of sight following a shift from an aboveground to a within-ground lifestyle), lizards losing legs as they morphed into snakes (during the transition to a slithering lifestyle; boa constrictors still possess a rudimentary pelvis and legs), beetles losing wings after colonizing small windy islands (so as not to be blown out to sea), adult baleen whales no longer possessing teeth even though their embryos still do, and so on. In each case the evolution to a radically new habitat has evolved loss or degeneration of previous adaptations as well as the acquisition of new ones. Evolution tears down as well as builds up. It fits organisms to their present environment only, heedless of any consequent degradation of exquisite capacities built up over many millions of years. With some imagination we can even see it in the transition many hundreds of millions of years ago when some mutant lungfish took the first tentative steps onto the muddy edges of the land. All of the mammals (including, unfortunately for them, whales, dolphins, and seals) have long since lost the gills possessed by their ancestors that would otherwise have allowed them to breathe underwater.

3.4. "All these worlds are yours, except Europa . . ."[18]

This work on extremophiles and others combines to tell us that organisms come to "love" their environments, however harsh.[19] This makes it well-nigh impossible to judge the optimality of an environment just from looking at its suitability for the resident organisms. If inhabitable at all, we will always expect the match to be good, thanks to the work of natural selection.

Let us suppose, for example, that life is discovered at some future date in the icy ocean of Jupiter's moon Europa.[20] If such a discovery is made, then the life there will be found to be supremely well suited to survival in the Europan ocean, while at the same time being spectacularly ill adapted to survival in the oxygen-rich oceans of Earth. Presumably, according to the same logic Lovelock used for Earth, such a discovery could be taken as evidence that the Europan environment had been molded to make it favorable for Europan life. It could be proposed that the Europan biota, with different preferences, had manufactured a different kind of planetary environment. However, it is likely that the Europan environment, if it is capable of sustaining life at all, will be an impoverished one. Sunlight cannot penetrate the kilometers of ice thought to cover the surface of Europa, and so photosynthesis cannot be possible. Hydrothermal vents may exist on Europa, with scattered communities living off the mineral-rich waters. But even if this is so then the overall rather limited supply of energy is likely to make Europa at best only a meager habitat for life.

3.5. Conclusions

So how does the work reviewed in this chapter contribute to the question at the heart of this book? Lovelock suggested that the suitability of the planetary environment for life is strong evidence in favor of Gaia. The evidence presented here suggests that only a much weaker claim is possible. The fact that the biosphere is comfortably attuned to the biota inhabiting it is to be expected just through evolution. We have abundant evidence of the incredible power of evolution, and we know its mechanism. Habitat commitment demonstrates that evolution alone will tend to produce a comfortable environment for the organisms inhabiting it, by adjusting the environmental preferences of the organisms.

The author Douglas Adams (of *Hitchhiker's Guide to the Galaxy* fame), when asked to comment on Gaia,[21] is reported to have replied:

> Imagine a puddle, waking up in the morning, and examining its surroundings [a brief pause here, to let the audience grapple with this rather odd image]. The puddle would say, "Well, this depression in the ground here, it's really quite comfortable, isn't it? It's just as wide as I am, it's just as deep as I am, it's the same shape as I am . . . in fact, it conforms exactly to me, in every detail. This depression in the ground, it must have been made just for me!

The point here being, of course, that while the puddle suspects the depression was made to fit it exactly, in fact gravity makes the puddle fit the depression exactly. Cause and effect are the opposite way around. In Adams's view, the same is true for organisms and their environments, relative to the Gaia hypothesis.

Evidently evolution does the same job to organisms as gravity does to puddles. In neither case can we uncritically accept the proposition (that the environment has been made for the organism or the depression for the puddle). But, on the other hand, we should not overstate the strength of this evidence: of course it does not rule out a role for the biota in making the environment comfortable. What we can say with confidence is that there is no need to invoke biological regulation of the environment to explain an Earth that is comfortable for its biota.

A comfortable Earth is explicable without Gaia, but what about continued habitability? The Earth is presently habitable for life, and has also remained so for billions of years.[22] While reading this chapter, perhaps you have accepted that habitat commitment can mold organisms to fit environments but noted that this process of molding organisms to habitats does not in itself explain why the Earth has remained habitable for so long. There are after all limits to habitability, which probably resemble those in tables 3.1 and 3.2. And you would be correct; habitat commitment tells us nothing about why or how the Earth has stayed habitable for life since the Archean. This topic will be dealt with in chapter 9, and is more complicated and thought-provoking than you might expect.

One aim of this book is to try and discriminate between the three hypotheses about how planet and life interact (chapter 1). The fact of the habitability of the Earth, and of the good fit between its organisms and environments, is fully compatible with Gaia. However, it is also at the same time equally compatible with the coevolutionary and geological hypotheses. If life inhabits an environment out of its control, at the mercy of geological and cosmic dynamics (the geological hypothesis), then, through habitat commitment, we would still expect the environment to appear to "fit like a glove," as long as the geological/

cosmic forces don't take the planet on a trajectory toward a state so extreme that evolution can no longer adapt life to cope with it (again, this is the subject of chapter 9). Likewise, if life and the environment are continually molding each other (coevolution), but in a haphazard rather than a directed way, we should nevertheless still expect the appearance of an environment made especially for life, because of habitat commitment.

The overall conclusion of this chapter is that all three hypotheses are equally well supported by this evidence. The fact that Earth's environment satisfies the survival constraints[23] for life on Earth (see the Lovelock quote at the beginning of this chapter), and furthermore is well-suited to the life inhabiting it, does not favor Gaia over the other two hypotheses because the biota has been tailored by evolution to fit the current state of the Earth.

Chapter 4

TEMPERATURE PACES LIFE

THIS CHAPTER CONTINUES the theme of the last. The mission remains the same, to probe more deeply the Gaian assertion (the first assertion listed in chapter 1) that the Earth is exceptionally well suited for life. This claim will now be subjected to more detailed scrutiny in two companion chapters, this one and the next, focusing on one environmental variable in particular—temperature.

One of Stephen Schneider's comments to James Lovelock about the Gaia hypothesis was: "Jim, how can you possibly believe in Gaia when there are ice ages? If there was a Gaia it would stop them from happening." Lovelock's riposte was that an ice age may "now be the normal comfortable state of the Earth," with the implication that it is the warmer interglacials that represent repeated failures of regulation.[1] So, is Schneider right or is Lovelock right? Schneider suggests that ice ages are harmful for life, whereas Lovelock implies they are favorable. Over the course of the next two chapters I will show that the data give a clear answer: ice ages are unfavorable for life as a whole. During ice ages the Earth is far too cold.

But the discussion is not just about ice ages. It is also about the related question of whether Earth's present-day (interglacial) temperature is optimal for its biota. To answer both of these questions we need to look in some detail at how temperature affects life.

In the previous chapter it was shown that the "mold-fitting" tendencies of evolution mean that the Earth is necessarily exceptionally well suited to the needs of the organisms currently inhabiting it. It would seem on the face of it that this habitat commitment places an insurmountable obstacle in the way of the current aim. How can we evaluate objectively how favorable the Earth is for life, when habitat commitment shapes life to be uniquely fitted to Earth as it is, not Earth as it may be? But all is not lost. In this and the next chapter some clever and elegant means for overcoming this obstacle are described.

In this chapter, the first of the two companion chapters, the question of optimal temperatures is addressed primarily by looking at the evidence from physiological and ecological studies—evidence from studying living organisms. I summarize what is known about the effects of temperature on growth, metabolic rates, primary production, biomass, and biodiversity. By building toward a mechanistic understanding of temperature effects, we can overcome the habitat commitment obstacle. The chapter that follows will conclude this discussion of how comfortable Earth is for life. In that chapter I will take an Earth history perspective. By comparing the life that evolution produced during earlier climates that were either particularly cold or particularly warm, we will overcome the complication that today's life is inevitably well matched to today's planet.

4.1. Temperature and Biological Rates

Statements of the Gaia hypothesis have been vague about what exactly is optimized. It is unclear whether Gaia is proposed to optimize primary production or biodiversity or abundance.[2] We will therefore review information on how each of these relate to temperature. This chapter looks at some case studies, datasets, and fundamental mechanisms, all of which illuminate the effects of temperature on physiology, metabolic rates, growth rates, ecological processes, and biodiversity.

4.1.1. Temperature and Growth

As set out in the previous chapter, those organisms that grow best in a particular environment, however objectively poor it is, come to dominate in it. Even if an environment is so extreme as to be nearly uninhabitable, then its resident organisms will still "prefer" it to more moderate conditions, if they never normally encounter them. So both hyperthermophiles and psychrophiles find tepid water lethal, one because it's too cold, the other because it's too hot.

If each environment is to some extent a utopia for the organisms it hosts, how then can we determine whether any environment is "better" than another? We could assume that the center of the habitable range (the midpoint between lethal limits) is optimum, and therefore that the environment closest to the midpoint is the most hospitable. Unfortunately, however, the midpoint is not universally the most hospitable point, particularly for temperature. An alternative way is to collect information on maximum growth rates of a whole variety of different organisms, each adapted to grow best in a different sector of the range of habitability, and then plot out maximum growth rates as a function of the variable of interest (temperature in this case). Just such an approach was taken in this next piece of work.

4.1.2. The Eppley Curve

Results from this approach are shown below (fig. 4.1a) for phytoplankton growth rates and temperature, taken from a classic paper Dick Eppley published in the 1970s. The beauty of Eppley's work is that he looked at a whole variety of different phytoplankton species, not just one. He compared both cold-dwelling and warm-dwelling phytoplankton in his plot. Whereas many individual species tolerate only a rather narrow band of temperatures and grow quickest somewhere near the middle of that band, a completely different picture emerges when all of the individual species' responses are summed together in a composite plot. The habitable band is much larger for phytoplankton as a whole than for individual species, and, most importantly, we see that the fastest phytoplankton growth rates occur at the highest temperatures. The optimum is skewed way over to the high end of the temperature range. It is not near the middle.

If we assume that the phytoplankton community composition will come to be dominated by those species that grow fastest at each temperature, then there is clearly an exponential dependence of community growth rate on temperature. If the species composition is allowed to change as temperature changes, the thick line in figure 4.1b can be used to predict community growth rates at different temperatures.

Eppley's curve is special because it has allowed us to overcome the obstacle presented by the organisms found in situ being always adapted to the prevailing conditions. By summing up over many species living in many different environments, his curve gives an unfettered view of how temperature affects community growth rates. This plot is invaluable as a starting point in this quest to understand how life fares in both cold and warm climates.[3]

4.1.3. All Hail Arrhenius?

The exponential dependency in Eppley's plot is not surprising from a chemical point of view. All chemical reactions except radioactivity are temperature sensitive. They go slower when it's cold, and faster when it's hot. As in the Eppley curve, the rate increases exponentially with temperature. This was first realized by the Dutch and Swedish chemists Jacobus Henricus van't Hoff and Svante Arrhenius in the 1880s.[4] Arrhenius proposed that reaction rates (k) are proportional to an exponential function of the temperature (T, expressed in kelvins), giving rise to what is now referred to as the Arrhenius equation:

$$k = A.e^{\left(\frac{-E_a}{R.T}\right)} \qquad \text{(equation 4.1)}$$

whose appearance is shown in figure 4.2 (A, E_a, and R are various constants).

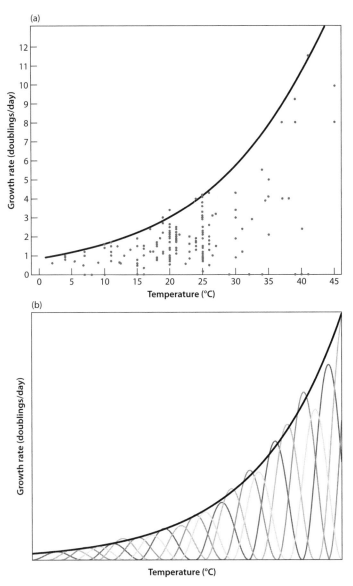

Figure 4.1. Composite growth curve for phytoplankton at different temperatures. (a) Measured growth rates plotted against the temperatures at which they were each grown in the laboratory, for many different species originally collected from a variety of both cold and warm seas. (b) Schematic explaining the pattern of the data in panel a. Each species can usually tolerate only a limited range of temperatures and is best adapted to a specific optimum temperature. Growth rate declines above or below the optimum temperature for that species. It is only when multiple individual species curves are drawn on the same plot, and a line drawn through the peaks (assuming that in nature the fastest-growing species at each temperature will be the one that prevails), that an exponential response to temperature is obtained. (a) From R. W. Eppley, *Fishery Bulletin*, 70:1063–1085 (1972).

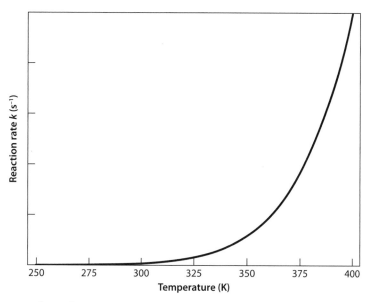

Figure 4.2. The Arrhenius equation in graphical form. The speed at which a reaction proceeds is plotted against the temperature in kelvin (the equivalent temperature in degrees Celsius is obtained by subtracting 273.15).

Most forms of life, including without doubt all phytoplankton (in fact all plants, both on land and in the sea), are ectotherms. Ectotherms do not control their own internal temperatures, and so their internal body (cell) temperatures are forced to vary in synchrony with the ambient temperature. This contrasts with endotherms (birds and mammals), which control their internal temperatures by generation of body heat.[5] Biochemical reactions are no different in this regard from other chemical reactions, and so the metabolisms of ectotherms tend to speed up or slow down according to the local temperature. A rise in temperature of 10°C, for instance, often causes about a twofold ramping up of metabolic rate. This fundamental of life explains why ectothermic reptiles across the world, such as iguanas, snakes, and turtles, become sluggish in the cold, and why phytoplankton growth rates increase exponentially with temperature. The effect of temperature on biology can be seen rather nicely in figure 4.3, which shows a comparison between how quickly different activities are performed by various Antarctic organisms (living in very cold, ~0°C, water temperatures) and the rates at which related species living in warmer waters perform the same activities.

This plot is especially interesting because, while it supports the general rule of biological rates being temperature dependent, at the same time it shows that

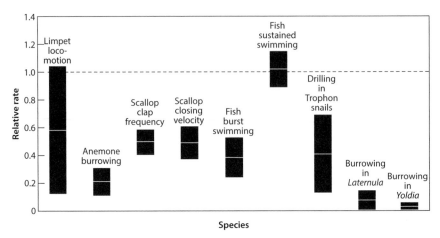

Figure 4.3. Comparison of activity rates of various Antarctic marine organisms with those of equivalent temperate species. Each bar shows the range of observed relative rates (rate at which activity is carried out by an Antarctic species divided by the rate at which the same activity is carried out by a sister species living in warmer waters). Several different behaviors are seen to be carried out at slower speeds by organisms living in colder water. Sustained swimming speeds in fish are an exception. With kind permission from Springer Science+Business Media: *Polar Biology*, 27:357–367. Fig. 8, Peck et al. (2004).

the Arrhenius relationship is not universal. Sustained swimming speeds in fish are an exception: fish living in cold water are able to achieve similar sustained speeds to fish in warm water. How can this be, given that the fish concerned are not endothermic? How can the dictates of the Arrhenius equation be avoided?

The question of how these fish combat Arrhenius's law will be returned to later, but the important point for now is that factors other than temperature can also influence rates of chemical reactions. Temperature does not have an absolute monopoly over reaction rates. For instance, the concentrations of chemicals can also be important. Consider the rate at which a spoonful of sugar will dissolve in a mug of coffee. It will indeed tend to dissolve more quickly in a hotter cup of coffee, all else being equal. However, if many spoonfuls of sugar have already been dissolved in the hotter cup of coffee, it is going to be harder to dissolve in yet one more.[6] Again we see that it is not just the temperature but also, in this case, the composition of the medium that controls the reaction rates.

The potential power of the Arrhenius law—but also the possibility of other factors affecting rates—leaves us with an intriguing question: to what extent does temperature have an overriding influence on biological rates in the real world?

4.1.4. A "Universal Metabolic Rate Model"?

Several scientists have approached this question from another angle. Geoffrey West, Jim Brown, Brian Enquist, and other colleagues focused originally on metabolic rates in many different animals and plants, and how this varies with body size of the organism. They provided an explanation[7] for why, as seen in data collected over very many years, the speed at which bodies "tick over" (their resting metabolic rates) is correlated in a particular and unexpected way with their size. Specifically, they explained why resting metabolic rates are typically proportional to the body mass raised to the power of 0.75, a relationship first noticed by Max Kleiber and now known as Kleiber's rule. There would be no need of a special explanation if total body metabolism was seen to vary linearly with body size, such that a doubling in body size leads to a doubling in metabolism. That would coincide precisely with what we would intuitively expect anyway. The observed 0.75 relationship is much more puzzling.[8] It was not at all clear why metabolism should scale with the three-quarter power of body mass, but West and colleagues proposed an explanation. The reason, they suggest, that metabolic rates do not usually scale directly with body size (equivalent to a power of 1.0), is down to the difficulty of supplying oxygen (and/or other substances) rapidly to all parts of larger bodies.[9]

While a fascinating scientific question in its own right, their explanation of this peculiar result ignored the additional role of temperature. This omission was rectified in later work[10] by the same authors working with James Gillooly, when they expanded the analysis to include temperature. This later work has proposed a more general model to characterize the combined effects of both temperature (T) and body mass (M) on metabolic rates (B):

$$B \approx M^{3/4} . e^{\left(\frac{-E_i}{k.T}\right)} \qquad\qquad \textit{(equation 4.2)}$$

where E_i and k are constants. You have probably noticed that the temperature term takes the same form as in the Arrhenius equation of the previous section.

Data from a wide range of organisms of vastly differing body sizes (a range of more than fifteen orders of magnitude!), from both endotherms (which control their body temperatures by generating heat internally) and ectotherms (which do not), from both animals and plants, support the dual roles of body size and temperature in controlling the metabolic rate of organisms at rest (fig. 4.4). Note that the reason why metabolic rates of birds and mammals (endotherms) are also seen to be correlated with temperature (fig. 4.4g) is that the plot shows metabolic rates against *body* temperatures (including those of hibernating animals—see the caption), not ambient temperatures.

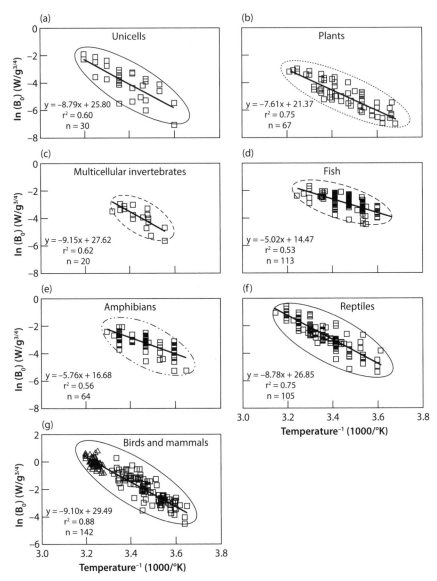

Figure 4.4. Temperature controls mass-normalized metabolic rate. Temperature (*x* axis, hotter to the left) is seen to exert a strong influence on the metabolic rate (*y* axis) of numerous types of organism, when the latter is corrected for body size. The observed relationship is consistent with the expectations from equation 4.2, and is seen to hold true for a wide variety of different types of organisms. Note that birds and mammals are normally endothermic and hence maintain their normal body temperatures (triangles in panel g) within a narrow range; during hibernation or torpor (squares in panel g), however, body temperatures can drop to much lower values. From Gillooly et al. *Science*, 293: 2248–2251 (2001); reprinted with permission from AAAS.

The body temperatures of endotherms, when they are not in hibernation or torpor, are more or less independent of and unaffected by ambient temperatures. Figure 4.4g therefore says nothing about how fast they normally metabolize in hotter or colder climates. For ectotherms, on the other hand, figure 4.4 shows that living in a hotter climate imposes a requirement to eat more in order to satisfy the increased "costs of living." This runs counter to our own experience as humans, of course. For us, because our internal body temperatures are maintained at a similar high value regardless of the climate we live in, the energy costs of thermoregulation are less when internal and external temperatures are more closely matched. Therefore, if anything we need to eat less in summer than in winter, less in a hot climate than in a cold climate. For ectotherms, the complete opposite is true.

It logically follows that for some organisms there are actually advantages to living in the cold. For sedentary organisms, such as many passive filter feeders living attached to rocks on the seafloor, it is easier to avoid starvation in the cold, when internal energy reserves are run down much more slowly, than in warm conditions when metabolism ticks over a lot faster and it is necessary to eat more just to stay alive.

The data in figure 4.4 are only for resting metabolic rates, but later work has proposed that a similar relationship applies among temperature, body mass, and the rate at which embryos develop in eggs,[11] the rate at which ecosystems respire CO_2, and the rates at which many other biological and ecological processes proceed. These advances are all interesting. They also inspire curiosity as to what else may be similarly controlled. How general might the relationship be? Does it apply also to rates of photosynthesis and growth, and to active (rather than resting) metabolic rates?

If it is completely general, then it raises a host of interesting questions. To consider just one example: are reptilian brains and therefore thought processes equally affected; that is, do reptiles think more quickly on warm days than on cold days? We tend to think of cold rainy days crawling by, but perhaps the subjective experience is completely the opposite for a lizard, with cold days whizzing by, whereas warm days drag on and on. Perhaps the whole subjective experience of time differs according to temperature for ectothermic animals which, unlike us, don't thermoregulate their brains.

Such flights of fancy aside, the reality in terms of temperature effects is more complicated than just one simple law. The mass-temperature-metabolism relationship outlined above turns out not to be completely general.[12] For most organisms its applicability also breaks down at temperatures close to or below the freezing point of water, given that life requires liquids to remain liquid. Vital action tends to come to a halt near 0°C, not near 0K.[13]

While increased warmth allows ectotherms to be more active, heat is not in all regards a good thing. As just discussed, ectotherms starve more quickly when warm. Also, as ambient temperatures rise above about 40°C, increasing proportions of complex biological molecules such as DNA start to unravel (unless protected by molecular chaperones). The unraveling of DNA makes organisms inefficient. Eventually, if repair is unable to keep pace with unraveling, it will kill them. Therefore, while the overall pattern is of exponentially increasing activity with temperature, this only holds true below some upper limit.

Another very important caveat is that growth rates are often dictated to a greater extent by a shortage of resources (food, nutrients, light or water) than by temperature and body size. If the "building materials" for reproduction are simply not available then the growth rate will be zero, regardless of other factors such as body size and temperature. If water is scarce in a desert then growth cannot be achieved until water is obtained. This may help explain why temperature (together with body size) has proved especially successful in predicting resting metabolic rates and speed of egg hatching, neither of which depends strongly on external resources, but has proved less successful so far in predicting other biological rates.

And there is yet one more caveat to bear in mind, one that modifies rather than contradicts the general pattern that cold slows down biological processes. Evolution is not a passive bystander. If the ability to act swiftly will improve fitness, then special adaptations may evolve toward this end. For instance, although each individual mitochondrial unit in muscle tissue must work more slowly in the cold, some polar fish have evolved a greater density of mitochondria in their muscles,[14] which accounts for their higher-than-expected swimming speeds shown earlier in figure 4.3. Building more units has compensated for the reduced efficiency of each mitochondrial unit. As will be discussed again later on in this chapter, evolution has developed various adaptations in response to the problems of low temperature, but these adaptations incur a cost, such as the increased energy and resources needed to build more mitochondrial units. And this is hardly surprising. If there were no associated costs, more densely packed mitochondria would have evolved rapidly in all species.

4.1.5. Complications at the Larger Scale

Continuing this theme of how temperature affects biological rates, we have just looked at how temperature affects the rates at which individuals perform different activities, but what about rates across whole ecosystems? A recent example of temperature and body mass both being important, but not providing the whole picture, is seen in datasets of ecosystem respiration that have been obtained in recent years. Ecosystem respiration is the combined "whole community" rate of food consumption, as measured by the release of carbon dioxide from the ecosystem as a whole. In daytime it is hard to assess, because photosynthesis and respiration compete against each other, with the former consuming carbon dioxide and the latter releasing it. At night, on the other hand, only respiration operates, and it can therefore be measured in isolation without worrying about the confounding effect of photosynthesis on the net flux. These measurements have been carried out using so-called flux towers, which measure the vertical distribution of CO_2 concentration over time, both within and above the local vegetation. The rate at which the local ecosystem is "breathing in" or "breathing out" carbon dioxide can be computed from knowledge of the vertical and temporal gradients in CO_2, together with wind speed and other information. Some of the data coming out from these flux towers, at many sites and on many different nights of varying warmth, is shown in figure 4.5.

On the one hand this data supports the great importance of temperature, with the general trend at each individual site being toward increased respiration release of CO_2 on warm nights relative to cold nights. When each individual site is considered in isolation, the trend with respect to temperature is what we would expect. Every single site shows a trend toward higher respiration in warmer conditions, reduced respiration in the cold. The complication comes when different sites, situated in different climates, are intercompared. Viewed from this perspective, the importance of temperature is not so great, because some warmer sites emit less CO_2 than expected, compared to the colder sites.[15] The differences in average CO_2 release between sites are found not to be a simple function of temperature, and are also not successfully predicted when total biomass and average organism size are also factored in. Some other, currently mysterious, factor must be influencing the respiration of the whole ecosystem. Nevertheless, these community-scale data also support the supposition that rates of activity are generally higher at higher temperatures.

4.1.6. The Effect of Temperature on Primary Production

This section continues to examine the effect of temperature on ecosystem-scale rates, looking in particular at the effect of temperature on primary production (rate of carbon uptake for photosynthesis), both on land and at sea.

We first look at a dataset of real-world measurements of phytoplankton phytosynthesis. This study gives a second example of how control over rates is determined by much more than just body size and temperature. The amount of photosynthesis carried out in each liter of seawater depends on two factors: (a) the density of phytoplankton (how many cells are photosynthesizing in each liter of seawater), and (b) the rate at which each cell is photosynthesizing individually. The rate per liter is then the product of (a) and (b).

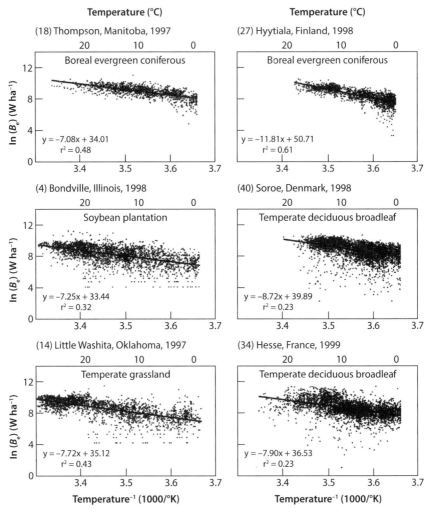

Figure 4.5. Effect of temperature on ecosystem respiration. The numbers along the top axis of each plot indicate ambient temperature (°C), with hotter conditions to the left of each plot; those on the vertical axes are the natural logarithm of the respiration rate of the ecosystem (watts per hectare), with higher rates toward the top of each plot. The latitudes of the study sites vary from 35°N (Oklahoma) to 62°N (Finland). Reprinted by permission from Macmillan Publishers: *Nature*, 423: 629–642, Enquist et al. (2003).

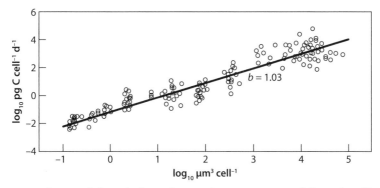

Figure 4.6. The rate of phytoplankton photosynthesis is not successfully predicted by the 3/4–power rule. Parallel measurements of cellular photosynthetic rate (pg C d^{-1} cell^{-1}) and average size (μm^3) of cells, across various marine ecosystems, show a relationship in which cellular photosynthetic rate does not vary according to size to the power of 0.75, but varies instead at a rate closer to size to the power of 1.0. Phytoplankton photosynthesis does not, therefore, follow the "universal" scaling rule. © (2007) by the Association for the Sciences of Limnology and Oceanography, Inc.: Marañón et al. (2007).

This particular study, by Emilio Marañón and colleagues, investigated the second of these factors. More specifically, they investigated how well cell size predicts cell photosynthetic rate using Kleiber's rule (0.75 power). It was found that it doesn't work very well. When the measured rate of photosynthesis by cells of a given size was divided by their abundance, it was found again (fig. 4.6) that the predictive equation above did not agree with the data. While this study throws doubt on the generality of equation 4.2, the additional effect of temperature was not investigated in this study.

A positive correlation between temperature and photosynthesis could nevertheless be brought about if warmer waters usually contained more phytoplankton (factor (a) above). However, the exact opposite is true. Phytoplankton are, on the whole, more abundant in colder waters, because cold waters usually contain more nutrients. The plentiful nutrients in colder waters contrast with the prevailing scarcity of nutrients elsewhere in the surface ocean. If there are no nutrients, there is no possibility of any phytoplankton growth, in the same way that a brick house can't be built without any bricks. The geographical differences in nutrient scarcity outweigh geographical differences in temperature as a controlling factor over phytoplankton abundance.

Given that the intensity of aquatic photosynthesis is not governed primarily by direct effects of temperature, it is therefore not much of a surprise that the spatial pattern of primary production in the oceans does not exhibit a maximum near the equator and minima toward the poles. There are strong similarities between the distribution of biomass in the oceans (see fig. 4.8) and that of primary production. Although there are in some places rather rapid rates of primary production near the equator, there are also equally rapid or even more rapid rates at many higher latitudes, such as in the temperate North Atlantic. These waters are much colder. Some of the lowest primary production values are typically found in the very warm but generally nutrient-impoverished low-latitude mid-ocean gyres between about 10 and 35 degrees of latitude. In contrast, primary production rates are much higher near the equator even though seawater temperatures are not all that different. We know that the higher rates close to the equator are mainly due to upwelling of nutrients there that allows the phytoplankton to bloom.

These last pieces of knowledge show clearly that the "universal" scaling law is not always applicable, and that other information (nutrient availability in this case) is needed in order to correctly predict primary production rates at sea.

Temperature is of secondary importance in controlling primary production at sea but is of greater importance to primary production on land. Measurements of crop yields in different locations often show a positive relationship with temperature, revealing again the underlying potential for the Arrhenius's relationship to control biological rates. However, the dynamics of plants under agriculture—heavily fertilized with nutrients, frequently irrigated with water, and growing unshaded in open sunlight—are not necessarily representative of what limits natural unattended systems. Forests, tropical rain forests in particular, are among the most productive natural ecosystems on Earth. Only wetlands produce a greater weight of new biomass per unit area per year.[16] As can be seen from figure 4.7, some of the most productive ecosystems are found in those parts of the world that are both warm and moist. The conclusion from these studies, and also from the planetary-scale distribution of vegetation, is that the more luxuriant and productive ecosystems occur predominantly in warm and wet regions.[17]

In contrast to the situation at sea, nutrients seem to play only a minor role on land, a suggestion reinforced by the observation that the Amazon rain forest sits on top of very old, heavily weathered rocks. It is likely that these old rocks are by now only a very weak nutrient source, and the Amazon is thought to sustain its nutrient requirements largely from recycling, with any losses having to be replenished by dust blown in from afar[18] or sediments washed down from the Andes. Despite what has to be a very weak nutrient source, rates of primary production in the Amazon are tremendous. But even though nutrient scarcity does not tend to be so critical on land, a separate resource scarcity is much more frequently of prime importance. On land it is water availability that in many places overrides body size and temperature as the major control.

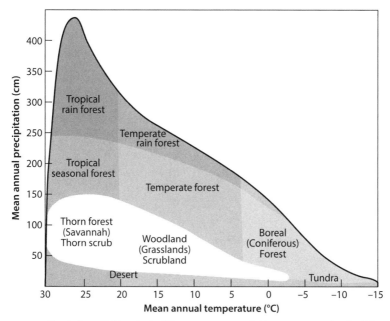

Figure 4.7. The relationship between climate and vegetation type. With the exception of wetlands (not shown on this diagram), rain forests are the most productive vegetation type; they also contain the highest biomass per unit area. They occur in the warmest parts of the world where rainfall is also high. In areas where the local climate is colder and/or drier, vegetation is sparser. From *The Great Ice Age: Climate Change and Life*. R.C.L. Wilson, S.A. Drury and J.L. Chapman, ©2000, Routledge and the Open University. Reproduced by permission of Taylor and Francis Books UK: Wilson, Drury, Chapman (2000).

4.1.7. Summary of Temperature Effects on Biological Rates

To summarize this section: while it is clear that all rates are potentially subject to the ubiquitous dictates of the Arrhenius law, temperature only sets the maximum potential rate. Other factors or limitations are often important in practice in setting the actual (realized) rates. Nevertheless, many physiological processes will proceed more slowly on a colder planet, more rapidly on a hotter planet. Mammals and birds aside, a slower tempo of life is likely to be the rule on a cooler planet.

4.2. TEMPERATURE AND BIOMASS

So much for rates, what about biological abundances? In the next chapter I will describe how careful paleobotanical work around the world has documented a much lower global land biomass during the peak of the last ice age than was present in preindustrial times. Herbivores, with relatively less food to live on, were presumably also less abundant in ice age times, and so in turn carnivores. In this section I look briefly at the distribution of biomass on the world today and see what it tells us about effects of temperature.

On land, average biomass per unit area is highest in moist tropical forests, lower in temperate climates (although still quite high where there are forests), lower again in high-latitude tundra environments, and lowest of all (nil, because of absence of any vegetation at all) over bare rock and ice.[19] The abundance of phytoplankton in the surface waters of the globe can be seen in figure 4.8.

Phytoplankton typically live for only a few days, and for this reason we do not as a rule tend to see large patches of living but not actively proliferating phytoplankton, because such patches would soon be consumed by zooplankton and disappear. The smaller phytoplankton in particular can be grazed at incredible rates, with, on occasion, more than 50% of the population being eaten every day. Populations can only be sustained under such conditions if growth rates are equally high or even higher, which they often are.[20] Because of the ephemeral nature of phytoplankton blooms, and because blooms that are no longer rapidly growing tend to get swiftly consumed, biomass and primary production are strongly related in the oceans. As is the case for primary production, biomass tends to be low in low-latitude oceans in general, and in the mid-ocean gyres in particular, because of nutrient scarcity.

The distribution of plants on land is also not a simple function of temperature. Vast swathes of land experience warm or hot climates but harbor relatively little vegetation: the interior of Australia, the Sahara and Kalahari deserts in Africa, deserts across Arabia and the Middle East, the Gobi desert in Asia. In all of these cases the lack of vegetation is due to a lack of water. Comparison of global maps of terrestrial vegetation with those of precipitation demonstrate rather clearly that where there is a combination of warmth and moisture lush vegetation is found; where either is lacking so too is vegetation. The most luxuriant vegetation of all (rain forests) is found where it is both hot and there is plentiful water, although well-stocked forests also cover many higher-latitudes areas including Scandinavia, Siberia, and Canada.[21]

4.3. Temperature and Biodiversity

The day has passed delightfully. Delight itself, however, is a weak term to express the feelings of a naturalist who, for the first time, has wandered by himself in a

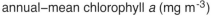

annual–mean chlorophyll *a* (mg m⁻³)

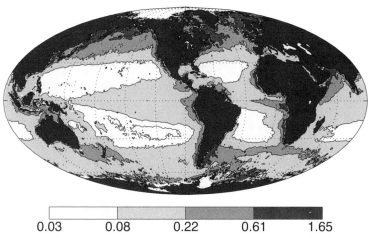

| 0.03 | 0.08 | 0.22 | 0.61 | 1.65 |

Figure 4.8. Distribution of photosynthetic biomass across the global ocean. This composite image (compiled using a year of satellite images, 1997–98) shows the variation across the ocean surface in the amount of plant matter, as calculated from the concentration of pigment detected in the water (mg chlorophyll-*a* m⁻³). From *Ocean Dynamics and the Carbon Cycle: Principles and Mechanisms*, Figure 2.18(a), © 2011, Richard G. Williams and Michael J. Follows. Reprinted with permission of Cambridge University Press.

> Brazilian forest. The elegance of the grasses, the novelty of the parasitical plants, the beauty of the flowers, the glossy green of the foliage, but above all the general luxuriance of the vegetation, filled me with admiration. A most paradoxical mixture of sound and silence pervades the shady parts of the wood. The noise from the insects is so loud, that it may be heard even in a vessel anchored several hundred yards from the shore; yet within the recesses of the forest a universal silence appears to reign. To a person fond of natural history, such a day as this brings with it a deeper pleasure than he can ever hope to experience again.
>
> —Charles Darwin, *The Voyage of the Beagle*

Darwin's initial observations on the variety and exuberance of life in a Brazilian rain forest, as he arrived for the first time from England's more sober climate, can now be seen as part of a wider picture. According to modern textbooks: "Perhaps the most widely recognised pattern in species richness is the increase that occurs from the poles to the tropics"[22] (see also fig. 4.9), and, "Most striking of all is the decline of diversity from equator to poles."[23] The same pattern

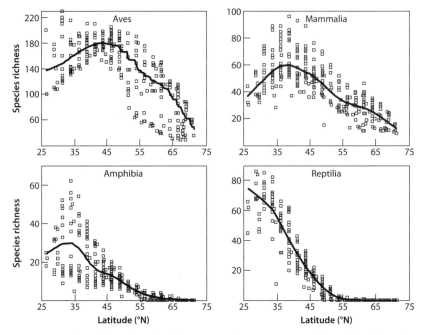

Figure 4.9. Effect of latitude on biodiversity. Variation in species richness (number of species per 2.5° × 2.5° quadrat) as a function of latitude in USA and Canada. All graphs show a decline in number of species toward high latitudes. The species richness of reptiles and trees (ectotherms) increases consistently toward the equator, whereas that of birds and mammals (endotherms) shows a mid-latitude peak. Reptiles are absent farther north. From David J. Currie (1991). *The American Naturalist*, p. 34, figure 2. Published by the University of Chicago Press for the American Society of Naturalists. © 1991. All rights reserved.

also impressed itself on another of the great scientific travelers, Alexander von Humboldt:

> Those who are capable of surveying nature with a comprehensive glance, and abstract their attention from local phenomena, cannot fail to observe that organic development and abundance of vitality gradually increase from the poles towards the equator, in proportion to the increase of animating heat.
>
> —translation from *Ideen zu einer Physiognomik der Gewächse*, 1806

The pole-to-equator pattern in biodiversity is abundantly clear. The underlying cause of this relationship with latitude is, however, harder to get at. It has variously been proposed as being due to: (i) differences in seasonality (climate being more stable year-round near the tropics, which, it is claimed, is more conducive to biodiversity); (ii) differences in long-term stability (low latitudes less affected by ice ages); (iii) greater intensity of predation at low latitudes (assuming predators crop the most abundant species especially heavily and thereby promote scarcer ones); (iv) differences in sunlight energy input; (v) higher rates of primary production near to the equator (on land); and (vi) differences in surface area at different latitudes (caused by the greater girth of the Earth at the equator than toward the poles; the tropics also get an additional boost because the northern and southern hemisphere tropics abut: a greater contiguous area of an ecosystem is assumed to increase the rate at which new species come into being but to reduce the probability of each individual species going extinct).[24]

Only some of these hypotheses are able to account simultaneously for a second great observation in biodiversity: that it also decreases with increasing altitude (for example, fig. 4.10). The total number of different species per unit area declines as mountains are ascended, even though sunlight availability, for instance, is unaltered.

One environmental factor that covaries with both latitude and altitude is, of course, temperature. It is apparent (next chapter) that temperatures close to 0°C impose considerable stresses and that subzero temperatures are associated with marginal environments. Could it be that an increase in temperature away from life-threatening values opens up an increasing number of niches and thereby increases biodiversity? More rapid growth rates in warmer conditions, leading to shorter lifespans[25] and therefore more generations per century, may also promote biodiversity through accelerated evolution.[26] These are speculations, however. No one has yet come up with a satisfactory general theory to explain all of the observed the patterns in biodiversity, but temperature may well play a major role.

As with other proposed factors, however, temperature, while important, is not the whole story. There is a well-substantiated pattern on the seafloor toward higher animal diversity in cold muds of the deep sea than in warmer muds of shallow seas. Diversity does not, in this case, simply mirror temperature. Diversity of seabed-inhabiting organisms does appear to decrease poleward (and hence correlate with temperature) when the survey is carried out all at shallow depths,[27] but not when shallow and deep diversity are compared.

4.4. Conclusions

That now completes this overview of how physiological rates, biomass, and biodiversity are each related to temperature. In each case the overall tendency is a positive correlation with temperature—higher values at higher temperatures,

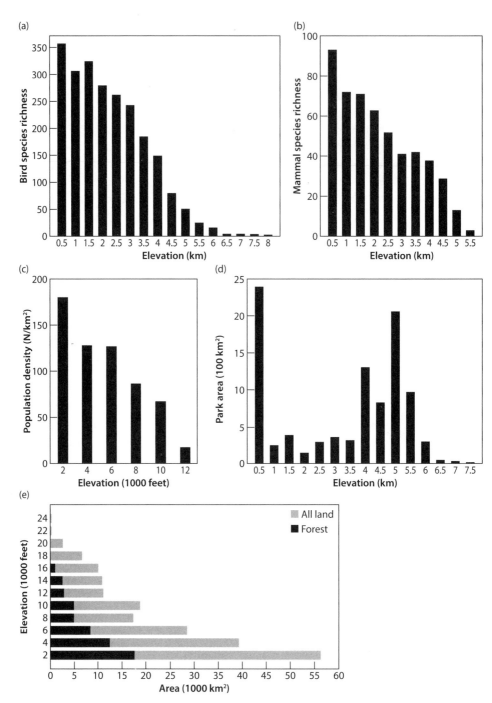

Figure 4.10. Effect of altitude on biodiversity of birds and mammals in Nepal. Nepal contains land from near sea level (70 meters) up to an altitude of 8,850 meters. Forest cover and species richness (number of different species per altitude band) decrease with increasing altitude, in particular above 5,000 meters. Note that some elevations are in meters, others in feet. Reprinted by permission from John Wiley and Sons: Hunter and Yonzon *Conservation Biology* (1993). Published by Blackwell Publishing for the Society for Conservation Biology.

lower values at lower temperatures. There are however plenty of exceptions, such as biomass and primary production in the sea and in deserts, and biodiversity in shallow versus deep seafloor muds.

The next chapter, the second of these two companion chapters, will now provide further insights from rather different perspectives into the current question of interest: is the Earth too cold to be optimal for life?

Chapter 5

ICEHOUSE EARTH

THIS CHAPTER, A companion to the last, continues the investigation into whether life fares better on a colder or a warmer planet. This in turn is part of an examination of the Gaian assertion that the Earth is a comfortable habitat for life. Whereas the last chapter looked mainly at effects of temperature on present-day organisms and the rates at which they carry out activities, now the approach is to compare the life that evolution produced during past cold and warm climates of long duration. This allows us to overcome the hindrance that today's life is inevitably well matched to today's planet, due to habitat commitment. From this and from other lines of inquiry, we will see that the Earth is too cold to be optimal. As end products of evolution ourselves, the Earth may feel just fine to us, but, I will try to show, on a more objective assessment, the Earth is too cold.

This chapter ranges across a number of topics: a description of the near absence of life on Antarctica is followed by an investigation of life during ice ages (dealing directly with the debate between Schneider and Lovelock as to whether or not they are more favorable) and during the hot Cretaceous. These and other subsections throw light from different directions on the central question of how warm or cold the Earth *should* be, if optimal for life. Finally, a second perspective on Earth's suitability for life is given by considering nitrogen availability on planet Earth.

5.1. ANTARCTICA

Another way of circumventing the obstacle of habitat commitment is to compare the flora and fauna naturally occurring in warm and cold parts of the world. Starting with cold parts, biological activity and abundance is more or less completely prevented by the presence of ice sheets. Nowhere is this seen more clearly than in Antarctica. It is the fifth largest continent on the planet (larger than Australia or Europe) and yet largely devoid of life. It is possible to travel hundreds of miles across Antarctica and never once see a single blade of grass, leaf, bush, tree, or animal. There are some exceptions that prove the rule, such as the Emperor penguins of chapter 3 that overwinter on Antarctica with their eggs. Other exceptions are ice algae that discolor the surface of the ice in a few places, and slow-growing lichens in some of the dry valleys of Antarctica. There are also a few species of plants and insects scratching a meager living on the Antarctic Peninsula, which sticks up north into slightly more clement conditions. But not a single land mammal or reptile is able to make a living off the land of Antarctica. The seals, whales, penguins, and other mammals and birds that grace the deep south in such large numbers are living off the plankton and other marine life; if they visit the land at all, it is only as a place to rest or nest. The emperor penguins, for example, raise their chicks far from predators but live off the sea. The list of permanent residents of Antarctica includes only fungi, lichens, and a few tiny insects and simple plants.

Even hardy animals such as reindeer, which thrive in the frigid conditions of Lapland and Alaska without ever visiting the sea for sustenance, are not found on Antarctica. It is only in slightly less cold climates, such as on the island of South Georgia in the Southern Ocean, 1,000 kilometers closer to the equator than is the northern tip of the Antarctic Peninsula, that such animals can survive year-round (they are not native, but were introduced there in the early 1900s by Norwegian whalers). Antarctica certainly has no indigenous large herbivores today, even though it did once, long ago in the geological past when the climate was warmer. And, given the absence of anything on land to eat, this is hardly surprising. The overwhelming poverty of life on Antarctica has very little to do with its geographic isolation from other continents. Isolation can explain a distinctive and less diverse flora and fauna, such as is found on Australia, but cannot explain the near absence of life on Antarctica. This is due to the extreme cold.

The reason that many species are unable to live in these colder climates relates to a scarcity of nutrients and liquid water on ice sheets, and also to the effect that cells rupture as ice crystals grow within them. The continued growth of an ice crystal within a cell will eventually burst the cell's walls. With the exception of occasional ice algae, photosynthetic plants do not grow on ice, and there is therefore an absence of a food source to fuel the rest of the food chain. For reptiles and other ectotherms there is the additional problem that all enzyme activity slows down at the very low temperatures.

It is no wonder then that the eye gives a true picture of the desolation of Antarctica. Biological surveys and investigations confirm what is already obvious to the unaided eye, that Antarctica is an icy windswept desert, more or less barren of life. It is simply not a congenial habitat for life. At the same time as we savor the austere beauty of its classical icescapes, we can also appreciate that it is utterly hostile as a place to live.

Antarctica is by far the largest single area of permanently ice-covered land on Earth, comprising nearly 9% of the total global land area. The world's other major ice sheet covers most of Greenland. Both Antarctica and Greenland are almost completely carpeted in ice, and in both cases the ice is often incredibly thick; in many places it is several kilometers from the top of the ice sheet to its base on solid rock. All in all, 11% of the total land area of Earth (~16 million km²) is presently unavailable for life because of permanent ice.

5.2. Life in a Cold Climate: Ice Ages

We may wonder why, if the Earth environment is supposed to be the end result of a process keeping it comfortable for life, more than 10% of the land surface is presently off limits for life, because of the extreme cold. This inconsistency only becomes the more glaring when we step back and look at recent Earth history in wider perspective. For the last 12,000 years, known to geologists as the Holocene period, the planetary climate has been in an interglacial phase, or in other words a warm interval between ice ages. But this is not the usual climate state. Interglacials occur only about once every 100,000 years or so and last for 10,000–50,000 years each. The remainders of each cycle, occupying in total more than three-quarters of the last half million years, have been ice ages. We know these durations rather accurately from the ice core records, which are shown and described more fully in chapter 8. The last ice age lasted for about 100,000 years, from circa 110,000 years ago to circa 12,000 years ago. Taking this wider view, the ice age state is clearly the norm and the interglacial state the anomaly over Earth's recent history. Our current climate is rather benign in comparison with the ice age; interglacials are only temporary reprieves from the cold that characterizes the long-term average.

5.2.1. Effect of Ice Ages on Available Living Space

During the colder periods of ice ages global temperatures averaged ~10°C (~5°C colder than now), and about 45 million square kilometers (~25% of the land surface area of the Earth at that time—table 5.1) were covered in thick glacial ice (fig. 5.1). In addition to the Antarctic and Greenland ice sheets we have today, the Laurentide, Innuitian, and Cordilleran ice sheets covered northern North America, and the Scandinavian and Barents ice sheets covered large areas of northern Eurasia. Even allowing for the fact that lower sea levels led to exposure of the continental shelves (providing more than twenty-five million square kilometers of potential extra living space; compare the two parts of figure 5.1),

TABLE 5.1.

Comparison of the Earth surface during the present interglacial and at the height of the last ice age

			Earth Surface Area (millions of km²)			
	Deep Sea (water depth > 200 m)	Shelf Sea* (water depth < 200 m)	Land			
			total	ice sheets	ice-free	% covered by ice
Last Glacial Maximum (LGM)	330	7	173	45	**128**	26%
Modern	336	**28**	146	16	**130**	11%
Change as ice age ends	+6	**+21**	−27	−29	**+2**	

Note: Statistics for the Last Glacial Maximum (LGM), the coldest period of the last ice age, which preceded the present interglacial, are for 21,000 years before present, and for the modern are calculated for the preindustrial world. Calculated using the ICE-5G (VM2) reconstruction (Peltier 2004).

* Supporting the number calculated here, other analyses suggest that 22 million km² (Rippeth et al. 2008) or 24.5 million km² (Montenegro et al. 2006) or 60% (Lerman and Mackenzie 2005) of present-day shelf seas were land during the Last Glacial Maximum. However, the calculation of shelf-sea area is limited in accuracy by the 1° × 1° resolution of the Peltier dataset. Many shelves are less wide than a whole degree of latitude or longitude. Many 1° × 1° grid cells will in reality therefore include both land surface and shallow sea, or both shallow and deep sea, and the topography within the grid cell will not be fully accounted for by the single number in the dataset, which can only show average elevation.

on balance bioavailable (ice-free) land surface declined. More than half of the new land that "rose" out of the sea during the ice ages was situated at high latitudes (> 50°) and was consequently covered in ice or tundra. These sites today are extremely productive seas (compare fig. 5.1a and b).

Some of the most productive parts of the ocean are the continental shelves, where nutrients supplied from rivers and active recycling of nutrients in shallow sediments allow the seas above to be unusually fertile. Something like 30% of global ocean primary production occurs on the shelves, even though they currently account for less than 10% of global ocean area.[1] Many of the world's major fisheries are in shelf seas. During the ice age the Bering Sea shelf, to take an example, rather than hosting large populations of whales, crabs, and walruses, as it does today, was instead an icy and desolate land bridge between Asia and America. Similarly the North Sea, the Baltic Sea, Hudson Bay, the Patagonian Shelf, and the vast shelves around the Arctic Ocean were all transformed from extensive highly productive seas to ice sheets or other frigid land. Lower

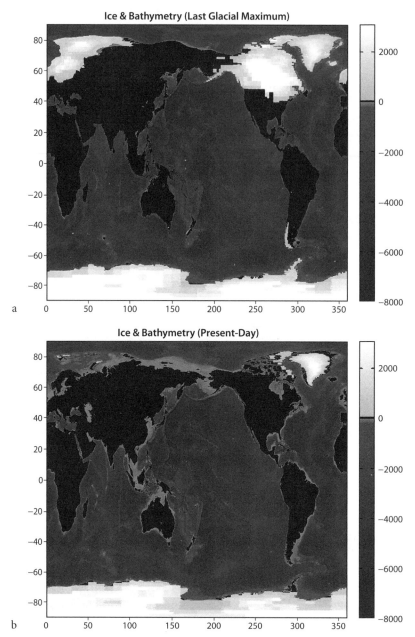

Figure 5.1. Earth surface reconstructions from (*a*) the height of the last ice age (21,000 years ago), when large areas of the northern continents were submerged under kilometers of ice, including the present-day locations of Oslo, Chicago, Copenhagen, Stockholm, Hamburg, Moscow, Boston, and Montreal; and (*b*) the Holocene inter-glacial (the preindustrial era before human-induced changes). Black = land, white = land-based ice sheet, gray = ocean, with shade varying according to depth (shelf seas are light gray). Depths in meters. Not shown: permanent sea ice, seasonal sea ice, seasonal snow and ice on land, permafrost, tundra. Calculated using the ICE-5G (VM2) reconstruction. Peltier (2004).

sea level meant that the oceans were robbed in order to give to the land,[2] but in many cases nothing much could grow on the new land. Other extensive shallow seas that lie at lower latitudes, such as the Yellow Sea, the Arafura Sea, the Gulf of Thailand, and the Java Sea, were all most likely transformed from the productive seas of today to rain forests or other highly productive terrestrial ecosystems at the height of the last ice age. A calculation of the net effect of all these changes is given in table 5.1 below.

The most obvious drawback of ice ages is the burying of large areas of land under ice, but the unfortunate impacts are by no means limited to those caused by glaciers and ice sheets. Cold temperatures, even if not quite freezing, still limit organic growth and turnover rates. In the absence of an increase in resource availability, average metabolic rates must have been reduced during ice ages, for reasons set out in the last chapter. Frosts kill or injure many plants, and colder average temperatures would have led to shorter growing seasons in temperate latitudes. Belts of permafrost (frozen solid ground) and tundra (frigid treeless plains) expanded out from the poles.[3] Much of western Europe (the portion not covered by ice sheets) was a treeless tundra.

5.2.2. Effect of Ice Ages on Life on Land

These many detrimental impacts led to a much lower global amount of vegetation during the last ice age (fig. 5.2). Or, to put it another way, as the last ice age drew to a close, the amount of vegetation burgeoned in response to the increasing warmth as it was released from the oppressive grip of the ice and cold. The geographical pattern of vegetation across the globe at the height of the last ice age has been mapped in considerable detail, based on a variety of evidence including the distribution of fossilized remains of different plants and soils preserved from that time. The picture that is revealed of the icy world is one in which even larger areas than today were covered by sparse tundra vegetation, deserts, and grasslands, and one in which thick forests (including tropical rain forests) were reduced in extent. When one compares this map to an estimate of the natural vegetation and soils that existed on Earth after the last ice age but before widespread agriculture and industrialization, it is estimated[4] that the total amount of carbon held in plants and soils together more than doubled following the termination of the last ice age, from 960 up to 2,300 Gt C.

Part of the explanation for the reduced glacial vegetation was the greater extent of ice and near-freezing conditions, but it was not only temperature that was responsible for the reduced vegetation. There was ~80 parts per million less carbon dioxide in the atmosphere then (concentrations of 200 parts per million or less), and the carbon dioxide scarcity inhibited land plant growth (obviously,

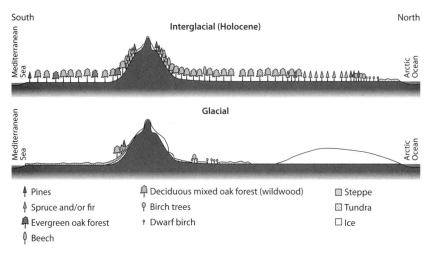

Figure 5.2. Schematic illustrating European vegetation cover changes between inter-glacials (top) and ice ages (bottom). From *The Great Ice Age: Climate Change and Life*. R.C.L. Wilson, S.A. Drury and J.L. Chapman, ©2000, Routledge and the Open University. Reproduced by permission of Taylor and Francis Books UK: Wilson, Drury, Chapman (2000).

in order to support their photosynthesis, plants need to "breathe in" CO_2 and therefore suffer from an equivalent to asphyxiation if CO_2 levels fall too low). It was also, the evidence suggests, a drier world (hence the expanded area of deserts and arid scrublands, including the Sahara), that imposed yet another limitation on plants.

On land the fossil evidence leaves little doubt: Schneider was correct and Lovelock was not—the abundance of life was severely curtailed during ice ages. Ice ages were unfortunate events for the land biota as a whole.

5.2.3. Effect of Ice Ages on Life in the Sea

Next we examine the state of affairs for life in ice age oceans. What does the evidence say about the productivity of the oceans during the coldest parts of the last ice age? As far as the open ocean (the deep ocean, away from any coasts) is concerned, it would seem that Lovelock may be correct. Although we do not have any very straightforward sources of evidence that can be trusted to tell us reliably (as ice core bubbles do about atmospheric composition, for instance) about past rates of marine primary production, a compilation of evidence from several less reliable proxies, examined at many places around the globe, suggests that rates of open ocean primary production per unit area were, in most places, slightly higher at the end of the last ice age than they are now.[5]

Even though open-ocean primary production appears in most places to have been slightly more intensive at the last glacial maximum, it is still not quite clear whether summed ocean primary production was overall higher or lower than now. For a start, the total area of open ocean was slightly less than it is now due to sea-level fall (table 5.1). The decrease in area available for phytoplankton growth was more important than suggested by table 5.1, because sea ice also extended much farther away from the poles during the last ice age.[6] Ice is only partially transparent to sunlight, and thick sea ice blocks virtually all light and so prevents phytoplankton growth underneath it. Both winter and summer sea-ice cover must have been more extensive during the ice age. The oceans were also deprived of most of the highly productive shelves during ice ages, because of lower sea level. The total area of shallow seas (0–200 meters water depth) was only about a quarter of what it is now (table 5.1). At present it is not possible to know precisely how the shrinkage in ocean area and the loss of shelves weighed in the balance against increased productivity in the remaining area, but it seems likely that the overall total was not greatly different from that today.

Returning to Lovelock and Schneider's disagreement, the ocean looks to have been more or less equally as productive as today, but because the land surface was so much less productive, overall Schneider was right. He was correct in suggesting that ice ages represent much less favorable conditions for life. Our current understanding leads to rejection of Lovelock's proposal that ice ages are better for life.

5.3. Life in a Warm Climate: The Cretaceous

Although our understanding of how life on this planet fared during ice ages is far from complete, it seems undeniable that these long-lived cold periods were relatively inhospitable for life. The next question to spring to mind is: what about when the climate was warmer than today? Unfortunately, the wonderfully precise records trapped in ice are of no use this time. This is because the last few million years have been times when climate has been either similar to now (interglacials), or, for most of the time, much colder than now (ice ages). During the last few million years there appear to have been no times of excessive warmth. The warmest interval we know of was the Eemian interglacial (~130–115 thousand years ago) when temperatures on Earth averaged about 1°C warmer than the interglacial we now inhabit. Of greater interest, because they were warmer and lasted for longer, are intervals far back in time, when climates were far hotter than now. But unfortunately no ice survives from this long ago and so there are no ice core records from these times.

An intensively studied period is the Cretaceous,[7] a long warm interval of geological time between 143 and 65 million years ago (Mya). How did life fare on Earth during this hot interval? Was the Earth at this time a barren, parched desert? Did life survive only in oases in a landscape otherwise burned to a crisp? Apparently not—the actual picture fossil data reveal is rather different.

5.3.1. The Cretaceous on Land

We do not have a full picture of life and environments from this far back in time, and so any reconstruction has large gaps and uncertainties. It would also be a mistake to consider that such an enormous interval of time as represented by the Cretaceous, which lasted for about 80 million years, had a constant climate throughout. Nevertheless, there is no doubt that it was a time of generally warm climates, and the Mid- and Late Cretaceous especially were much warmer than today. The information we currently possess suggests that although interiors of continents were sometimes arid at lower latitudes, there was also plenty of luxuriant vegetation and teeming life, spread over a greater latitudinal range than today.

Most remarkable of all, an accumulating wealth of paleobotanical and paleontological data makes it clear that relatively rich communities of organisms were present at high latitudes. The Antarctic Peninsula (which was not a peninsula at that time, because Antarctica and South America were connected) was home to deciduous forests, as were northern Alaska, northern Canada, and northeast Russia[8] at about 70°N (well within the Arctic Circle and near to the edge of the Arctic Ocean) (figs. 5.3a, b, c, d). In fact, an interesting topic of research at present is how deciduous trees were able to survive the pitch-dark winters near the poles; they managed to survive despite months of darkness. Even more surprising is that fossil remains of dinosaurs,[9] crocodiles, and freshwater turtles[10] have been found in these locations (fig. 5.4). These findings are all the more remarkable because reptiles are cold blooded (and dinosaurs may well have been). Either these ancient species had evolved the ability, like mammals, to regulate their internal temperatures (although if so then these traits did not get passed on, because all reptiles today are cold blooded), or else the comfortable warmth allowed them to migrate to within several hundred kilometers of the poles. Fossilized fruit and leaves from the breadfruit tree, *Artocarpus dicksoni*, were discovered as long ago as 1883 in western Greenland. Today the relatives of this species are found only in the tropics. The fossilized breadfruit were found in sedimentary rocks of Mid-Cretaceous age, but it is only much more recently that a large spectrum of plant and animal fossils have confirmed that the breadfruit fossils fit into a coherent picture. There is no doubt that some of the presently rather barren Antarctic and Arctic wastes were, at times during the Cretaceous, home to flourishing ecosystems that are no longer seen there today.

Although there is a popular conception that a much warmer Earth would also be a drier Earth (because deserts occur in hot places), it is in fact more likely that the Cretaceous was also on the whole a wetter Earth. The rate of evaporation of water is strongly controlled by temperature, and so evaporation rates from the Cretaceous ocean must have been higher than today. What goes up must eventually come down, and so rainfall rates must also have been higher then than today. On the other hand, rates of evaporation from the land surface are also tied to temperature. Cretaceous soils must, on average, have received more rain and also evaporated more water (dried out more quickly). Their moisture will have been dependent in some way on the balance between the two, as well as by how quickly water drained into rivers. A further complication is that warm air can hold more moisture than cold air before it condenses out, implying increased water transport from ocean to land during warmer epochs. Taking all these factors together, it remains uncertain whether Cretaceous soils were on the whole more moist or more arid than soils today.

Most importantly of all, and to some extent finessing this question about moisture of soils, we have tangible botanical evidence of the vegetation that was present during the Cretaceous, in the form of the fossil record of spores, seeds, and leaves (occasionally even tree trunks or other pieces of wood) preserved in rocks. These fossils are more abundant in some places than others, reflecting both the thickness of the local vegetation at that time and also the propensity of the local environment at that time to bury fossils. In swampy environments, for example, a relatively large fraction of the organic material falling to the ground is preserved as fossils. In nonwaterlogged environments, most plants remain out in the open air after death, which is conducive to their decay. This evidence, abundant in some places but sketchy in others, leads to the conclusion that on average through the particularly hot Late Cretaceous (100–65 Mya), vegetation was luxuriant at high latitudes but more variable at lower latitudes. While near-polar and temperate latitudes seem to have been fairly densely vegetated, as was an equatorial humid band, at midlatitudes (for example, between 10 and 40°) it appears to have been typically rather arid, particularly in the interiors of continents away from the coasts. The reconstructions of vegetation closer to the equator are not as reliable as the reconstructions nearer to the poles, however, because of a relative lack of fossil plant data.

Cretaceous vegetation maps are compiled from a variety of evidence. Fossilized dinosaur bones from a locality imply reasonably abundant vegetation. Fossilized leaves possessing specialized adaptations for water retention (or the

a

b

c

d

Figure 5.3. High-latitude fossils dating from the Cretaceous: (*a*) tree stump, (*b*) leaves from a conifer and a broad-leaved plant, (*c*) cycad leaf, (*d*) tree stump. (*a*)–(*c*) were found in Alaska, (*d*) was found on the Antarctic Peninsula (James Ross Island). Despite continental drift, Antarctica remained close to where it is today throughout the Cretaceous. (*a*)–(*c*) © Robert A. Spicer; (*d*) www.photo.antarctica.ac.uk.

presence of pollen from the same plants) suggest on the other hand a relatively arid environment. The argument for thin continental vegetation at midlatitudes can also come from some other evidence: there was a lack of coal formation in these areas during the Cretaceous. However, there is also a general lack of peat and coal formation today in the Amazon, despite the luxuriant rain forest. Instead, plant matter seems to decay rapidly in the hot and humid conditions of the Amazon, perhaps in part because there is seasonal flooding alternating with seasonal dry periods. Most environments are not waterlogged year-round. Whatever the exact cause, soils and coals do not tend to accumulate in the Amazon. There is a curious juxtaposition of luxuriant vegetation rooted in only a rather thin layer of soil. This observation about the Amazon helps us better understand the Cretaceous pattern of vegetation; it illustrates that a lack of coals is not by itself reliable proof of thin or absent vegetation.[11]

Reconstructing ancient landscapes is a fascinating detective puzzle in which multiple pieces of evidence have to be collected, evaluated, weighed up against one another, compared for consistency and finally synthesized into an overall assessment. The picture this story reveals for the Late Cretaceous is only partially supportive of the argument I have pursued so far in this chapter and the last. While ice ages were clearly less favorable for vegetation, and for life in general, the much warmer climates of the Late Cretaceous were not obviously

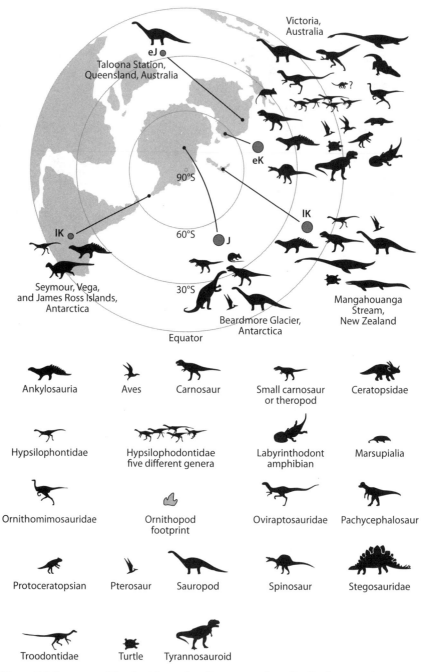

Figure 5.4. Summary of fossil evidence that dinosaurs and reptiles lived near to the poles during the Cretaceous and Jurassic periods. Numerous fossils from these times have been found at paleo-latitudes of up to 70° or 80°, well within the Arctic and Antarctic Circles. Modified from Rich, Vickers-Rich, and Gangloff (2002). Artist: Draga Gelt.

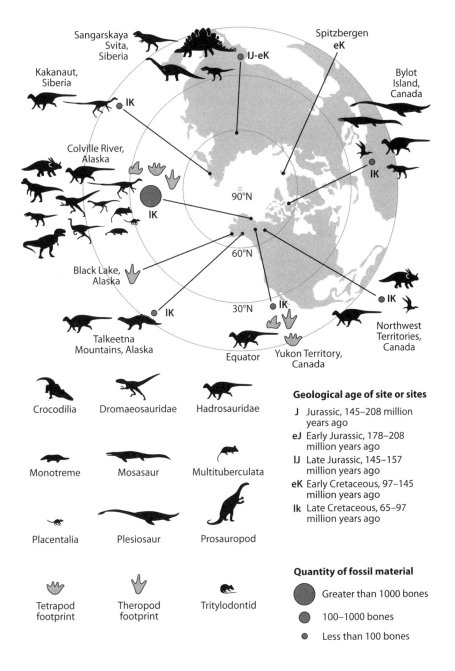

Sangarskaya Svita, Siberia

Kakanaut, Siberia

lK

Spitzbergen
eK

Bylot Island, Canada

lJ-eK

Colville River, Alaska

lK

90°N

lK

60°N

Black Lake, Alaska

lK

30°N

lK

lK

Talkeetna Mountains, Alaska

Northwest Territories, Canada

Equator

Yukon Territory, Canada

Crocodilia

Dromaeosauridae

Hadrosauridae

Monotreme

Mosasaur

Multituberculata

Placentalia

Plesiosaur

Prosauropod

Tetrapod footprint

Theropod footprint

Tritylodontid

Geological age of site or sites

J Jurassic, 145–208 million years ago

eJ Early Jurassic, 178–208 million years ago

lJ Late Jurassic, 145–157 million years ago

eK Early Cretaceous, 97–145 million years ago

lk Late Cretaceous, 65–97 million years ago

Quantity of fossil material

Greater than 1000 bones

100–1000 bones

Less than 100 bones

much more favorable to land life than the climate is today. No continents or large expanses were covered in ice through most of that time, and tundra was also probably largely absent, but scrubby savannah-like vegetation and arid landscapes may have been more prevalent than today.

Higher sea levels during the Cretaceous (probably about 100 meters higher than present[12]) led to the flooding of large areas of what is now dry land. Europe was fragmented into several pieces by shallow seas covering most of France, Germany, and Italy; today large parts of these countries sit atop Cretaceous-age sediments such as chalks and limestones that can only form under water. North America was likewise split in two by a north–south shallow sea running from the Gulf of Mexico up to the Arctic, and much of northern Africa was also submerged.

On balance it seems that terrestrial life flourished in the warmer Cretaceous, albeit on a smaller total land surface because of the higher sea level. The average density of life on land (per unit area) was probably similar to that today, although our knowledge is quite uncertain.

5.3.2. The Cretaceous at Sea

Next we need to consider how life fared in the sea. Much of the flooded land became shallow seas, the combined area of which was much greater than today. Evidence of the productivity of these ancient shelf seas can be seen in the thick layers of chalk rocks laid down in the Late Cretaceous, made up principally of the calcium carbonate shells of coccolithophores, a type of phytoplankton, and foraminifera, a type of zooplankton. The thickness of the strata (today forming chalk cliffs along the south coast of England and the north coast of Denmark, for instance) suggests that local conditions during the Cretaceous were conducive to a thriving marine biota, or at the very least conducive to the chalk-forming organisms. The calculated rate of accumulation (about 15 centimeters of chalk sediment per 1,000 years) is large compared to the average rate of sediment accumulation at the bottom of the deep ocean today (order 2 centimeters per 1,000 years).

There is strong evidence that the Arctic Ocean, in which the marine food web is impoverished or largely absent at the present time because of sea-ice cover, was a productive basin during most of the Late Cretaceous.[13] Cores from the Arctic Ocean seafloor north of Ellesmere Island have retrieved Cretaceous-age marine sediments and found some of them to contain abundant diatoms and dinoflagellates (bloom-forming phytoplankton), occasional fish remains, woody fragments, leaves, spores, and pollen. There is also a complete lack of dropstones (indicative of icebergs) in the sediments. An indirect temperature estimate at this location during the Late Cretaceous, based on the chemical make-up of a specific set of microorganisms, suggests sea-surface temperatures of about 15°C.[14] The overall picture that we obtain is of flourishing sea life in an ice-free Arctic Ocean that was much warmer than today. This evidence suggests that at high latitudes, both seas and land were more productive during the Late Cretaceous than they are today.

In terms of the ocean as a whole, the layers of seafloor sediments that piled up during the Cretaceous are remarkable not only for the rapid accumulations of chalk but also for occasional black bands interspersed between sediments of a more usual color. These organic-rich black shales are evidence of low oxygen conditions in the deeper parts of the sea. A lack of oxygen inhibits complete decomposition of the plankton remains that rain down from the sea surface. In terms of ocean productivity, this evidence of anoxia is rather equivocal; it could be indicative of either higher or lower rates of primary productivity than today.[15]

An ambitious technique[16] has been applied to try and calculate levels of both marine and terrestrial primary production at different times back through the last 300 million years. This technique, based on intricate details of how oxygen isotopes are fractionated by different terrestrial and marine processes, includes many assumptions and steps in the calculations. It cannot be considered as straightforward or direct evidence of past levels of primary production. Nevertheless, while keeping these caveats in mind, it is of interest that this technique calculates that rates of marine primary production during the Mid-Cretaceous (100 million years ago) were about twice as high as today, and during the Early to Middle Eocene (50 million years ago, another greenhouse era) were nearly three times stronger than at present. Terrestrial primary production at 100 and 50 million years ago is estimated by this same technique to have been about 150% and 200% of present rates.

Other evidence, also rather indirect, again suggests that Late Cretaceous seas, including those at lower latitudes, were probably fairly productive. The largest teleost fish seen anywhere in the fossil record (*Xiphactinus*, which grew up to five meters long) comes from the Late Cretaceous. The Late Cretaceous was also remarkable for giant swimming reptilian pliosaurs (up to ten meters or more in length) and mosasaurs (some were longer than 15 meters). Given that metabolism increases with both body size and temperature (section 4.1.4) these large animals living in warm seas must have consumed smaller marine life at a prodigious rate. The presence of these giant marine animals is not a straightforward indicator of the productivity of the seas, but it does suggest that the seas must have been at least reasonably fertile.

While it is not possible at present to form a definitive picture, on balance it seems most likely that the productivity of the global ocean in the Cretaceous was either similar to or greater than it is today. Taking the evidence from both land and sea together, the Cretaceous appears to have been a favorable climate for life.

5.4. Do Sea Dwellers Like It Cold?

This evidence of productive seas on a warmer planet is important to the wider question of the climate sensitivity of marine productivity. Lovelock's main defense to the criticism about ice ages is that ice ages were actually better for life on the planet overall, because the seas were more productive during ice ages. The truth of this assertion was questioned in section 5.2.3, based on available proxy evidence.

Before further addressing this question of the temperature dependence of marine productivity, it is worth remembering that organisms living in the sea share similar physiological propensities to those living on land. Apart from the few swimming warm-blooded mammals (whales, dolphins, seals, and so forth), birds (for example, penguins), and one or two exceptions among the fish,[17] marine life is ectothermic and cold slows them down (see figures 4.3 and 4.1). In the perennially very cold waters around Antarctica crabs, starfish, and other organisms grow very slowly and are relatively long-lived.[18] They also possess special adaptations fitting them better for life in the extreme cold.[19]

Lovelock's argument that a cold planet is better for marine life is based on the assumption of reduced "stratification" in the oceans. Stratification is the separation of a liquid into vertically stacked layers of different density. When the vinegar and the oil in French dressing separate into different layers, then the dressing has become stratified. In the ocean the sun's heat warms up surface waters, reduces their density (hotter water is lighter than colder water), and thereby induces them to float on top of the heavier colder waters below. Stratification is important for marine life because it tends to inhibit stirring of the ocean by wind or other factors. Deeper waters have higher nutrient concentrations, and when they rise up to the surface they bring nutrients with them.

As discussed in the previous chapter, the lack of nutrients is the critical factor holding back phytoplankton growth in most places in the ocean, and hence strong stratification (considered over long timescales and at the global scale) is antagonistic to phytoplankton growth. This is demonstrated nicely by the observation (see fig. 4.8) that some of the highest concentrations of phytoplankton occur west of Peru, west of Southern Africa, and also west of Mauritania (northwest Africa). These areas are all places where "upwelling" occurs, that is to say, where deep, nutrient-rich waters are brought to the ocean surface.

The argument for elevated marine productivity on a colder planet comes from the expectation that when air temperatures are cold, surface waters will also be cold, and will therefore be less buoyant relative to the deep water. Because surface and deep waters will be more similar in temperature (and therefore density), the degree of stratification will be reduced, strong winds will more easily stir the water to a greater depth, and hence there will be an increased rate of supply of nutrients from the deep, allowing phytoplankton to thrive.

This argument is weakened, however, once it is appreciated that the temperature of the deep ocean eventually follows the temperature of the surface ocean, because the former is formed from the latter when it sinks. Deep-water temperatures therefore also track climate, albeit with a lag of a thousand years or so, the time that it takes the deep ocean to be completely replenished by water sinking from the surface.[20] The upshot of this logic is that any effect of climate change on ocean stratification and ocean primary production will only be temporary.

The suggestion that marine life prefers it cold thus seems to be at odds with both theory and data. From a theoretical perspective the climate effect on stratification is probably not long-lived. And according to the available data, marine life apparently coped reasonably well with both ice ages and Cretaceous warmth.

5.5. What Heat Do Homeotherms Choose?

We have so far looked at three ways of overcoming the considerable obstacle to our understanding that is presented by habitat commitment (by organisms adapting over evolutionary time to their environments). The great advance represented by Eppley's plot (see fig. 4.1), firstly, was the idea of plotting the temperature responses of very many species all together on one plot. Secondly, life on Antarctica is known to be greatly impoverished compared to life in the tropics. A third means of sidestepping the obstacle, employed in this chapter, has been to compare life during ice ages with that of the present day and with that during the Cretaceous. We will now consider a fourth and final way. Because homeotherms (such as ourselves) can regulate internal temperatures at values distinct from the external environment, the chosen values may give an idea as to what the automatic process of evolution finds to be optimal.

The large majority of microbes, plants, and animals share a common characteristic: they have no internal central heating. The collective name for organisms unable to control their body temperatures through burning food is "ectotherms." Their body temperatures are therefore at the mercy of fluctuations in ambient conditions. This is why reptiles are much more numerous in tropical climates and become less and less abundant the closer one gets to the poles. It also explains the basking behavior common to many reptiles; by sunbathing they absorb heat from their surroundings and use it to raise their own internal temperatures. In this way they can achieve a limited degree of behavioral thermoregulation. And

Figure 5.5. Climate impact on reproductive strategy: (*a*) in a warmer region of Australia the young of Collett's Snake (*Pseudechis colletti*) hatch out of eggs, whereas (*b*) in a cooler climate the young of the related Common Black Snake (*Pseudechis porphyriacus*) are born live, within thin membranous sacs from which they soon emerge. Republished with permission of Annual Reviews, Inc.: R. Shine, *Annual Review of Ecology, Evolution, and Systematics*, 36: 23–46 (2005). Permission conveyed through Copyright Clearance Center, Inc.

TABLE 5.2.
Core temperatures of endotherms (birds and mammals)

Endotherm Group	Core Temperature (°C)
Mammals:	
Placental mammals	
- humans	37
- other terrestrial	37–40
- whales	35–36
Marsupials	35–37
Monotremes	30–32
Birds	38–42

Note: Body temperatures are for healthy animals not engaged in strenuous activity. Although some animals, such as humans, achieve quite a tight thermoregulation, with internal body temperatures varying little between the Arctic and the Equator, other animals are less fastidious, and internal body temperatures can vary somewhat according to ambient temperatures. Camels, for instance, can tolerate a rise to greater than 40°C during the desert daytime. In order to conserve food during cold periods some endotherms are able to enter special cold-body states for a limited period, dramatically lowering their body temperatures, for instance to as low as 8°C during hummingbird torpor, or down to close to freezing during winter hibernation of the Arctic ground squirrel.

it is vitally important to them whether or not they are warm. Given that they can't heat their own muscles or organs, if they are cold they must of necessity remain sluggish and immobile. And if they are cold then growth, along with all else, is retarded, and this applies just as much to reptilian embryos as it does to mature adults. Whereas most reptiles lay eggs, there are many instances where, out of two closely related reptile species, one lays eggs and the other instead gives birth to live young. What is interesting is that when the habitats of the related species are compared, the species giving birth to live young is generally found to live in a colder area than the sister species that lays eggs[21] (fig. 5.5a, b). This is most probably because embryos left out in eggs grow too slowly in colder climates, whereas those kept inside their mothers' bodies presumably benefit from the somewhat elevated temperatures due to basking and related behaviors and due to (limited) metabolic heat.

Insects are likewise ectotherms and constrained by how warm it is. Although just one or two insects survive on the Antarctic Peninsula (in a dormant state for most of the time), because of the extreme cold their whole lives are conducted in slow motion. A grylloblattid insect may take six years to grow from egg to larva to adult as it ekes out a slow-motion existence on the Antarctic Peninsula.

Birds and mammals have adopted a different strategy, one that has most probably been evolutionarily favored by the gradually cooling climate over the last fifty million years or more (chapter 8). Birds and mammals (endotherms), including ourselves, burn large amounts of food in order to keep their (our) bodies warm. We are profligate energy users. The benefits are obvious: for instance, when we wake up in the morning we don't need to sunbathe before we can get going. But the costs are also high: at least half of all the food we eat is used to heat our bodies.[22] The great implications for energy use were also apparent in figure 4.4, as you may perhaps have noticed. The scale on the y axis for panel g (birds and mammals) is shifted relative to that of all other panels. At any given body temperature, birds and mammals expend a lot more energy than do reptiles, amphibians, or fish of a similar size.

This brings us to an interesting point. If endotherms have control over their internal temperatures, such that evolution has been able to "choose" how hot they should be, independent of their external environment, then how hot does evolution appear to like it?

Has evolution optimized endotherms to be warmer or cooler than the average external conditions? What has evolution got to say about whether bodies work better when chilled or when warmed? Toward what endotherm body temperature has the process of evolution converged?

Table 5.2 shows the typical core temperatures of a range of birds and mammals. Given that the average temperature on the surface of the Earth is about 15°C, clearly bodies work best at temperatures higher than those of the external environment. When mutations in the evolutionary past led to some endotherms regulating at colder body temperatures and some at warmer, the warmer animals must have had greater success and been selected for, despite the extra food intake such a strategy demands.

While it is true that evolution has had more or less free rein to adjust internal temperatures of endotherms, there are metabolic costs to the difference choices. Because of this we do not have a completely unbiased view of exactly how hot evolution likes it. The warmer the body core is, compared to the external environment (or in other words, the greater the temperature difference between the inside and outside of an animal's body), the faster the body loses heat, and therefore the faster it must take in food (fuel) in order to maintain its body temperature at a constant value. Endotherms are capable of marvels of central heating: a prime example is the male Emperor penguin who not only keeps himself alive through the extremes of the Antarctic winter but also simultaneously incubates an egg, at times maintaining a temperature within his brood pouch that is 80°C warmer than the outside. But the metabolic cost of this miracle of thermoregulation is that male Emperor penguins waste away to only half of their late summer body weight during the long cold winter without food.[23] Despite these occasional marvels, the exigencies of thermodynamics make it metabolically much cheaper to keep the body core only a few degrees warmer than outside. Evolution must have automatically performed some sort of trade-off between optimal body warmth and the increased likelihood of starvation associated with great disparities between internal and external temperatures. We can only speculate whether, if it had proceeded on a hotter planet, evolution would have produced endotherms with even warmer insides.

The thermoregulated nests of social insects are also relevant to the preceding discussion. Termites and bees also tend to keep the insides of their nests warmer than the prevailing climate; for instance, bees' nests are thermoregulated at close to 34°C (fig. 2.4, section 2.6.3).

5.6. Nitrogen-Starved While Bathed in Nitrogen

In this chapter and the last I have discussed a variety of evidence leading to the conclusion that Earth is too cold. But this is not the only aspect of planet Earth at which one can look askance, from the perspective of a fault-finding engineer expecting the Earth to be managed for the benefit of its life. There are other aspects where one can say: "If this is supposed to have been optimized, couldn't it have been done better?" Only one further example will be briefly touched upon here—that of nitrogen starvation.

The problem for much of the biota is akin to that of the Ancient Mariner: "Water, water, every where, nor any drop to drink." The analogy is quite appropriate. Life on Earth is bathed in nitrogen, but, just like the Ancient Mariner's water, not of the right sort. The most abundant molecule in the atmosphere is dinitrogen (N_2: making up ~78% of the total, more than three-quarters of every lungful of air we breathe in), and this dinitrogen is also one of the most abundant constituents dissolved in seawater. Only the major dissolved salts such as sodium, chlorine, magnesium, potassium and calcium are more abundant than dinitrogen in seawater. Despite this superabundance of dinitrogen in both air and water, organisms regularly die of nitrogen starvation. It's a surprising arrangement.

The problem with dinitrogen is that it is chemically inert and therefore difficult to use. Dinitrogen molecules consist of two nitrogen atoms tightly bound together. The tightness of the "triple bond" between them (the two atoms are electromagnetically linked by three bonds, rather than the normal one) requires a large input of energy to break. When organisms (and also industrial processes) fix nitrogen, the expenditure of large amounts of energy is required in order to fix dinitrogen into forms more easily used by organisms, such as nitrate, nitrite, and ammonium. The latter bioavailable forms are collectively referred to as "fixed" nitrogen. This energy demand is one reason why so few organisms can fix nitrogen. Another reason is that it appears that only one enzyme, nitrogenase, has ever evolved which is capable of this biochemically complex feat.[24] Nitrogenase has the unhelpful property of being deactivated by exposure to oxygen. Because oxygen is also superabundant in the atmosphere and surface ocean, use of nitrogenase often requires the nitrogen fixers to undertake "physiological gymnastics" in order to keep oxygen and nitrogenase apart.

These handicaps help to explain why most organisms do not engage in fixing nitrogen (in fact most organisms are physiologically unable to perform it). What is harder to explain is why fixed nitrogen is in such short supply in the biosphere. In fact it is a critically scarce resource for many organisms. One set of organisms, the N_2 fixers, controls the rate of supply of fixed nitrogen. In the process of living and dying, they indirectly add to the stocks of fixed nitrogen. It works in the following way: when they die and are decomposed, the nitrogen atoms they initially absorbed from the surrounding environment in the form of N_2, which became incorporated into their biomass, get returned to the surrounding environment again, but this time in the form of fixed nitrogen. Given that life (the N_2 fixers) controls the rate of creation of new fixed nitrogen, why isn't the biosphere organized in such a way that there is a sufficiency of fixed nitrogen for all? The major fluxes in the nitrogen cycle are nearly all biologically driven, including N_2 fixation and the major loss fluxes of fixed nitrogen from the biosphere.[25] If the nitrogen cycle were really controlled by a mechanism that worked on behalf of the biota, we would expect either N_2-fixation rates to be higher or fixed nitrogen destruction rates to be lower[26]; we would expect to see higher fixed nitrogen levels rather than widespread scarcity.

As a consequence of this general shortage of fixed nitrogen, rates of plant growth are most strongly and most immediately limited by nitrogen in many terrestrial ecosystems.[27] And this occurs despite three out of every four molecules breezing past them being pure nitrogen, but of the wrong form. Multitudes of aphids and other herbivores also die every day because the plants they eat are too dilute in fixed nitrogen for them to be able to synthesize the enzymes they need to live.[28] Likewise, when oceanographers on research cruises carry out experiments on bottles of seawater, collected from the ocean surface waters many hundreds of miles away from land, they typically find that the strongest phytoplankton growth response is elicited when they add fixed nitrogen nutrients (in the form of ammonium or nitrate) to the bottles.[29] Their experiments show that in most places proliferation of phytoplankton is being most strongly prevented by a shortage of fixed nitrogen. And this is despite every teaspoonful of seawater containing copious quantities of dinitrogen. To my mind this paradox of nitrogen starvation while being bathed in nitrogen is one of the strongest arguments against the Gaian idea that the biosphere is kept comfortable for the benefit of the life inhabiting it.

There are a couple of arguments that can be raised against this conclusion of "unnecessary" nitrogen starvation, but neither is serious. The first is that a high concentration of dinitrogen in the atmosphere is necessary for preventing Earth's plants from burning up. Without dilution of oxygen by dinitrogen, the atmosphere would be > 90% oxygen, and it is certainly true that fires would then occur much too frequently to allow land plants to survive. However, this argument is easily dismissed once we realize that diverting enough nitrogen to remove nitrogen starvation everywhere would only make a very small dent (much less than 1%) in the atmospheric N content and therefore would not alter fire frequency.[30]

A second argument suggests that nitrogen stocks of the oceans are in fact optimally managed. When deep ocean water upwells to the surface, bringing life-sustaining nutrients with it, the fixed nitrogen and phosphorus nutrients in that water are typically at a ratio of 15 or 16:1, which is also the ratio in which the growing algae need those nutrients. It is indeed a point of wonder that there is such a tight correspondence between need and supply. However, to call this situation optimal is to imply that phytoplankton are better off facing a critical scarcity of two nutrients (the eventual upshot of exact equivalence in supply, as nutrients get used up following upwelling) rather than just one. Under a situation of proliferating phytoplankton competing for resources one nutrient must eventually becoming limiting. But there is no need for a second to also become limiting at the same time. It is a case of "so near but yet so far": the supplies of fixed nitrogen are almost sufficient to exceed demand and eliminate scarcity, but not quite.[31]

The overall picture that emerges from considering this nitrogen paradox is that it would be wholly perverse to design a system this way, if the aim were to try and make conditions comfortable for the biota. A truly Gaian biota would include denitrifying bacteria that, out of consideration for their fellow Earth dwellers, ceased their destruction of fixed nitrogen as soon as stocks of it began to run low. Either that or else the other phytoplankton (those not able to fix N_2) would continually self-limit their uptake of other nutrients in such a way as to leave behind enough for the exclusive use of N_2 fixers, so that the N_2 fixers could keep levels of fixed nitrogen in excess of need. Of course neither of these phenomena are observed in nature. The phenomenon of ubiquitous nitrogen starvation amid plentiful dinitrogen can readily be understood from a perspective of survival of the fittest (organisms competing for resources and using the most convenient forms first), but not from a perspective of optimization of the biosphere by the biota.

5.7. Conclusions

The theme of these last three chapters has been to examine the proposition that the Earth is an optimal crucible for life, or, if not optimal, at least highly suitable. I have probed the suggestion that the planet has been "optimized by and for the biota" and looked to see whether it withstands scrutiny.

In my opinion the evidence does not stack up. I have focused in particular on temperature. The dominant climatic state of the last several million years has been ice ages. Lovelock has argued, puzzlingly, that these are a more favorable state for life on Earth than the present interglacials. This does not, however, sit well with various observations. During ice ages: (i) there was less land free of ice (about a quarter, that is, one part in every four, was smothered by ice); (ii) much of the most productive parts of the ocean, the shelf seas, were turned into dry land by lower sea level; and (iii) the total mass of carbon locked up in vegetation and soils was only about half as much as today.

Not only were ice ages unfortunate episodes for life on Earth, but also, conversely, the Earth during the Cretaceous was possibly even more congenial than it is today,[32] although the evidence is not conclusive on this point. Even hotter was possibly even better, for life in its entirety. During the Cretaceous, green land stretched from pole to pole, with little or no snow covering the poles. The green carpet of continuous forests extended up close to 90° north and south, and Antarctica harbored vibrant ecosystems. Although low-latitude land distant from the sea may have been too dry to support much life, nevertheless the Cretaceous planet was probably, on the whole, slightly more favorable for life.

Of course it is true that a Cretaceous climate would not best suit the biota that resides on the planet today, adapted as it is to the present icehouse Earth. If today's biota were to be instantaneously transplanted en masse to the Cretaceous, a big die-off would follow. However, if evolution were given time to produce a hotter-adapted biota, then life on the hotter Earth would, arguably, be more fertile, more verdant, more vigorous, and more diverse than on the present cold one. This need for time to adapt to climate change is an important reason why the ongoing global warming is likely to be so harmful for life on Earth, even though a warmer planet might, after thousands of years for biology to adapt, become more hospitable.[33]

The excessive coldness of our icehouse Earth, coupled with other features such as unnecessary nitrogen starvation, suggest strongly that the Earth is far from perfect for life. The data argue for rejection of this Gaian claim. The world as we see it does not correspond to any idea of optimality; it is not an uncannily good home for life. It is clearly adequate and suitable, but it is by no means perfect.

Chapter 6

GIVEN ENOUGH TIME . . .

6.1. Introduction

Now I turn to examining Lovelock's second assertion (chapter 1); namely, that the Earth's atmosphere is a biological construct that is distinctly different from any expected abiotic chemical equilibrium. This claim can be broadened to the wider claim that Earth's environment bears the definite and considerable imprint of biological processes and is distinctly different from the environment that would be present if Earth did not possess life. We know that over geological time all of the water in the ocean passes through gills, all of the air in the atmosphere through lungs, and all of the soil through earthworm guts. Life clearly therefore has the capacity to alter the planetary environment, but this is different from showing that it has actually done so. Lovelock claimed that life does modify the environment. Life is not simply a passive passenger living within an environment set by physical and geological processes over which it has no control. The biota have not only lived within the Earth's environment and processed it but also, it is suggested, have shaped it over time.

The opposite view has been put very elegantly by Kevin Laland (although he doesn't agree with it):

> I will begin by drawing a distinction between two kinds of causation that may be instrumental within biology, *linear* and *cyclical*. Linear causation is exemplified by the hammer and nail. Appropriate use of a hammer drives the nail into some substrate such as wood, but nails do not reciprocate by propelling hammers. Here, for clarity, I am ignoring the force exerted by the nail on the hammer. Most people would feel comfortable with the description of the hammer as *causing* the nail to enter the wood. Chickens and eggs, on the other hand, represent cyclical causation. Chickens produce eggs, and eggs hatch into chickens. Neither could exist without the other.
>
> The conventional view of the evolutionary process is closer to the linear than the cyclical notion of causation. Hot climates select for heat dispensing adaptations, such as sweat glands or large ears. But no matter how much an animal sweats it is not regarded as affecting the local temperature to any significant degree. As a

result of natural selection, the properties of environments shape the properties of organisms, but not the other way around (except in cases such as coevolution where other organisms are the environment). (Laland 2004)

Lovelock's claim, challenging the view Laland so well articulated, will be given only brief attention here, because there isn't any controversy on this point. There is no doubt that Lovelock is correct, and few now disagree. When Lovelock first formulated the Gaia hypothesis there were many scientists who would have defended an abiotically determined world; now there are hardly any. Through the efforts of Lovelock and many others, there is now widespread agreement on the important role that life has played in shaping key features of the environment.

6.2. Earth's Atmosphere Is Not Inert

One of the most important of these features is that Earth's atmosphere contains abundant oxygen in the simultaneous presence of many compounds that combine with it. In every book that Lovelock has written about Gaia he has included the story of how, as a consultant to NASA in the 1970s, he argued that there are fundamental differences between the atmospheres of Earth and of its neighboring planets, and that these differences are due to the ongoing activity of life on Earth. The differences can be detected from space[1] and therefore provide a potential means of remotely determining whether there is abundant life on a planet, without the need to visit it.[2] This was not, needless to say, welcome news to the NASA scientists who were at that time working out how to send the Viking robotic spacecraft to touch down on Mars and look for direct evidence of life. Lovelock's approach argued that there was no such need. Regardless of any uneasiness at NASA, the discussion had the enormous benefit of generating a realization that life affects Earth's atmosphere at the largest scale.

If we compare the atmosphere of Earth with that of its two sister planets, Mars and Venus (table 6.1), we can see some clear differences. The most notable is that the atmospheres of Venus and Mars are dominated by carbon dioxide but contain very little oxygen, whereas Earth's atmosphere is more or less the opposite way round.[3]

6.2.1. Abundant Oxygen Even Though Continually Consumed

The reasons for the atmospheric compositions of Mars and Venus are quite easy to understand; they are in chemical equilibrium with the rocks and soil of the planetary surface. The Martian atmosphere and the "regolith" seen in the pic-

TABLE 6.1.
Comparison of planetary atmospheres

Gas	Venus	Mars	Earth (life)	Earth (no life)
CO_2 (%)	96.5	95	0.03	98
O_2 (%)	0.0	0.13	21	0.0
N_2 (%)	3.5	2.7	79	2
Total Pressure	90 bars	0.0064 bars	1 bar	55 bars
Av. Surface Temp.	460°C	–53°C	13°C	warm/hot (> 50°C)

Note: the last column is a speculative reconstruction of the possible composition of the early Earth's atmosphere, prior to the evolution of any life, and prior to the condensation of the ocean. The total pressure and CO_2% were calculated by assuming that all the carbon currently in crustal calcium carbonates and kerogens (organic matter in rocks) was initially in the atmosphere as CO_2. There is a lack of surviving geological evidence to prove or disprove this speculative reconstruction of the earliest Earth.

tures from the rovers Spirit and Opportunity (for example, fig. 6.1) are chemically compatible with each other. After millions of years of contact, neither now engenders any chemical changes in the other. The atmosphere is inert with respect to the land surface. If a portion of Martian air was to be left for 10,000 years in an airtight box, together with a sample of Martian regolith, we would expect both to emerge unchanged.

The same cannot be said for the Earth. If a portion of the Earth's atmosphere and a representative sample of Earth rocks and soil were likewise left in a sealed container together for even a few decades, oxygen in the air would oxidize the rocks and soil (iron in the rocks would be turned to rust, for instance) and in so doing large amounts of oxygen would be drawn out of the air.

The continued exposure of long-buried rocks to air acts to remove oxygen from Earth's atmosphere. And it is not only the exposure of rocks to air that removes oxygen from the atmosphere, but also the chemical reactions that take place as gases emitted by volcanoes emerge to meet an oxygen-rich atmosphere for the first time. Considerable quantities of water vapor, carbon dioxide, and sulfur dioxide (SO_2) are emitted from volcanoes. Sulfur dioxide subsequently combines with oxygen and water to form sulfuric acid, that is, acid rain, most of which eventually ends up in the ocean as dissolved sulfate, SO_4. In the process of converting SO_2 to SO_4, oxygen is removed from the atmosphere. Lesser quantities of other oxygen-reactive gases, such as carbon monoxide (CO, oxidized to CO_2), hydrogen sulfide (H_2S, oxidized to SO_2), and hydrogen gas (H_2, oxidized to H_2O), are also emitted and contribute to running down oxygen reserves.

As Lovelock and others have noted, it is only possible to maintain a high oxygen concentration in the face of these continual depredations if it is also

Figure 6.1. Martian panorama. The Mars exploration rover Spirit took the photos to construct this panorama over six Martian days in late November 2005. Seams between individual frames have been eliminated from the sky portion of the mosaic to better simulate the vista a person standing on Mars would see. NASA/JPL-Caltech/Cornell University.

continually produced. The main process responsible for producing O_2 in these large quantities is photosynthesis. Whereas we, like other animals, breathe in oxygen and exhale carbon dioxide, plants do the opposite. Trees, grasses, algae, and other plants all harvest sunlight to make carbohydrates, and in so doing they take in carbon dioxide and release oxygen as a waste product. Although plants themselves use up some of their own carbohydrates to power their metabolisms, the total amount of photosynthesis (oxygen production) on Earth slightly exceeds the total amount of respiration (oxygen consumption), leading to an ongoing net injection of oxygen into Earth's atmosphere. This source of oxygen is sufficient to match the sinks due to weathering of newly exposed rocks and chemical alterations of newly emitted volcanic gases.[4] In this way the atmosphere is kept rich in oxygen.

6.2.2. Significant Methane Even Though Continually Consumed

In addition to being out of equilibrium with Earth's rocks and volcanoes, Earth's air is not even in equilibrium with itself. If a sample of Earth's air were left unaccompanied in a sealed box for a thousand years, it would emerge slightly altered. Although the Earth's atmosphere does consist primarily (78%) of non-reactive N_2 gas, with the even more inert noble gas argon making up a further 1%, the large majority of the rest (~21%) is oxygen. Oxygen cannot be considered an inert gas. On the contrary, it is able to form chemical bonds and combine with almost all other elements. It combines with methane (CH_4), oxidizing it to form carbon dioxide (CO_2) according to complex reactions that can be represented in simplified form as:

$$CH_4 + 2O_2 \rightarrow CO_2 + 2H_2O \qquad (equation\ 6.1)$$

Although methane is not abundant in the atmosphere, it is present in nonnegligible quantities (currently slightly less than 2 parts per million, fig. 6.2), and it is this that is initially puzzling, given its strong propensity to combine with the ever-present oxygen. As Lovelock pointed out, from expectations of static chemical equilibrium we might expect there to be no methane at all in Earth's atmosphere, because it should all have been mopped up by the abundant oxygen. This apparent paradox becomes even more puzzling when we realize that the current situation is not a one-off associated with anthropogenic impacts, but rather that small but significant quantities of methane have persisted in the natural atmosphere since long before human industries started polluting the planet. Methane gets trapped in bubbles as accumulating snow turns to ice, as does CO_2, and ice cores therefore contain a record of past methane[5] as well as of CO_2; methane concentrations varied naturally in the preindustrial atmosphere (synchronously with the long swings in temperature between ice ages and interglacials) between about 0.3 and 0.7 parts per million.

Given the tendency for methane and oxygen to mutually annihilate, their continued joint presence must be due to ever-present sources. The nonnegligible methane concentration can then be explained by way of a dynamic, as opposed to a static, equilibrium. If it takes a short but finite time before methane is destroyed by oxygen, then massive incessant production will be associated with a small but continual residual concentration made up of the methane that has been produced but not yet destroyed.

It is clear that there is indeed a massive incessant production of methane, from many sources. Small amounts of methane are released annually from volcanoes. However, by far the largest releases of methane are biological. The largest sources of methane are from decaying organic matter, either in the ground or in animal guts. If organic matter is broken down anaerobically (in a place where oxygen is not available, such as in bogs, wetlands, the inner reaches of cow stomachs or termite guts) then the bacteria able to live there use a different metabolism that releases methane as a by-product. Methane generation follows

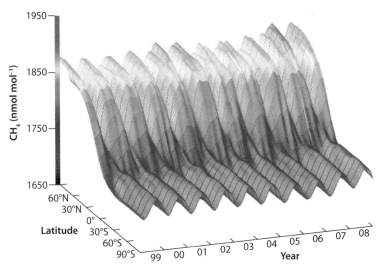

Figure 6.2. Global distribution of atmospheric methane in the marine boundary layer. This plot shows the human-induced rise in atmospheric methane between 1993 and 2002, together with the seasonal and spatial variations and the difference between northern and southern hemispheres (due to the greater amount of land area and hence terrestrial vegetation in the northern hemisphere). Barry Reichenbaugh and NOAA.

death and decay in oxygen-depleted environments. This very large ever-present source of methane is responsible for its small but nonnegligible concentration in the atmosphere, despite its rapid removal by oxygen (the average lifetime of a methane molecule in the atmosphere is about ten years).

The essential point here is that large quantities of methane are produced annually on Earth by different classes of organisms. The otherwise anomalous presence of methane in an oxygen-rich atmosphere is another symptom of life's large-scale impact on the natural environment.

6.2.3. Where Did All of the Oxygen Originally Come From?

Not only is life important in maintaining an oxygen-rich atmosphere against continual depredations, but it was also responsible for its genesis. A more detailed coverage of the "Great Oxidation Event," the period when Earth was transformed from an anoxic to an oxygenated planet, will be given in chapter 7. Here I will just describe the evidence that "life did it."

It is likely that Earth's initial atmosphere was not so dissimilar to that of its sister planets (table 6.1). Assuming this to be the case,[6] where did all of the initially absent O_2 subsequently come from? The answer is pleasingly simple: from photosynthesis. Or, to be more precise, from the burial of organic material formed through photosynthesis.[7] Every time that photosynthesis takes place in an algal cell or in a plant leaf, sugars are formed from sunlight and carbon dioxide, with oxygen released as a by-product. When those sugars are subsequently respired or decomposed, oxygen is consumed and CO_2 released, and the original reaction is completely negated. The net biogeochemical effect is then zero. In other words, there is no summed effect. But if those sugars (or some other form of organic carbon derived originally from CO_2) are buried in rocks without being broken back down, the original reaction is not counteracted. The small decrement to atmospheric CO_2 and increment to atmospheric O_2 persist. Over geological time, as more and more organic carbon became incorporated into the crustal rocks of the Earth, more and more CO_2 was drawn down out of the air and more and more O_2 added to it.

When geologists have attempted to audit the rocks of the Earth's crust, part of the audits include a summing up of the total amounts of organic matter remains (including organic carbon) dispersed throughout the sedimentary rocks of the crust.[8] These include organic carbon in concentrated forms such as fossil fuels (coal, oil, and gas) and shales, although these constitute only a small proportion of the total. By far the largest contribution to the total comes from thinly spread kerogens (more disseminated forms of organic matter). But in any case, the point of interest here is that these audits come up with numbers that roughly agree with the amount of organic carbon that we would expect to have been buried in order to generate the observed amount of oxygen in the atmosphere.[9] The bulk of the oxygen in the atmosphere most probably came from this biological source, given that there is so much organic carbon dispersed in sedimentary rocks, and that this must have left behind oxygen upon its production.

The discussion so far has looked at how, over the vast stretches of geological time, the never-ceasing operation of biological processes has led to radical changes in the atmospheric composition of Earth, and how this in turn is interwoven with changes to the very composition of the rocks that form the Earth's surface underneath our feet. The specific evolutionary events that led to the rise in atmospheric oxygen (the evolution of oxygenic photosynthesis, the first spread of trees across the land surface) will be described in more detail in chapter 7. Here we have concentrated on showing that "life did it," and even more that "life still does it." If all life were to be removed from the Earth then methane in the atmosphere would rapidly decline because it would no longer be produced in large amounts, and the oxygen would more slowly[10] drain away to oxidize new rocks and gases emerging from the interior of the Earth.

Lovelock's second assertion is thus abundantly supported by this understanding, and there is no difficulty in accepting that it is correct. The Earth's atmosphere is in large part a biological construct. The degree to which this assertion supports Gaia, or favors it over competing hypotheses such as coevolution of life and planet, is however a separate question to which we will return at the end of the chapter. Before that we will briefly examine one other example (among several that could have been chosen) of how the activities of life on Earth, when taken in bulk, and applied over enormous reaches of time, have led to regulation of the planetary environment toward a state very different from that which would have been produced if life were absent.

6.3. Ocean's Nitrate Is Biologically Controlled

At the end of the previous chapter I mentioned that there is a striking correspondence between the ratio in which nitrogen and phosphorus are needed to make new plankton, and the ratio in which nitrate and phosphate are present in deep seawater. The biological factors that have produced this correspondence will now be described.

6.3.1. Deep-Sea and Plankton N:P Ratios

Oceanographers have lowered hundreds of thousands of bottles into the ocean, brought back samples of water from various depths, and then measured the nutrient concentrations in those samples. Below the surface layer in which the phytoplankton live (they cannot live any farther down because of a lack of sunlight) a remarkable correspondence is observed between the concentrations of nitrate[11] and of phosphate. Whereas other nutrients tend to rise and fall together but not always in the same proportion (they are not linearly coupled; fig. 6.3b and c), nitrate and phosphate tend to rise and fall together always in the same relative proportions (fig. 6.3a), or in other words they are tightly and linearly correlated. Any change in phosphate is almost always accompanied by a matching change in nitrate of the same sign but about fifteen times larger in magnitude. This is apparent in scatterplots of nitrate against phosphate concentrations because the points form more or less a straight line with a slope of close to 15.

In deep water the two nutrients nitrate and phosphate are typically present in a ratio of close to 15:1.

Intriguingly, the deep-water ratio is also very close to the ratio at which nitrogen and phosphorus are present in phytoplankton. Although there is considerable variability,[12] most measurements of the elemental composition of plankton in the real ocean have produced N:P ratios in the range 12–20, with an average close to 16.

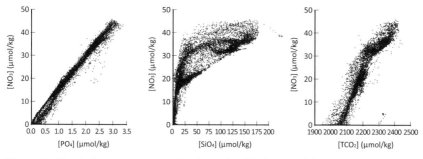

Figure 6.3. Nutrient:nutrient scatterplots from the global ocean. (*a*) nitrate concentration vs. phosphate concentration, (*b*) nitrate vs. silicate concentrations, and (*c*) nitrate vs. dissolved inorganic carbon concentrations. Each point on each scatterplot corresponds to one water-sampling bottle from which simultaneous measurements of two nutrients were made. The plot as a whole is then constructed from very many such pairs of measurements, from different locations and depths in the ocean. Points closer to the origin tend to be from surface waters (typically lower in nutrients), points more distant from the origin from deeper waters (typically higher in nutrients). Reprinted from Tyrrell, in *Encyclopedia of Ocean Sciences*, 677–686 (2001). © 2001, with permission from Elsevier.

6.3.2. A Sublime Coincidence?

As Alfred Redfield first pointed out, this means that deep water, when it upwells to the ocean surface, comes replete with nitrate and phosphate in almost exactly the ratio at which phytoplankton need it:

> One fact of great general interest emerges. . . . It may be noted that as the quantities of nitrate and phosphate diminish, as they do as one approaches the surface layers of the sea, they diminish simultaneously in such proportions that no great excess of one element is left when the supply of the other has been exhausted. . . . It appears to mean that the relative quantities of nitrate and phosphate occurring in the oceans of the world are just those which are required for the composition of the animals and plants which live in the sea. That two compounds of such great importance in the synthesis of living matter are so exactly balanced in the marine environment is a unique fact and one which calls for some explanation, if it is not to be regarded as a mere coincidence. (Redfield 1934)

As noted in the previous chapter, the deep waters are not *exactly* suited for phytoplankton needs, containing slightly less nitrate than required to use all the phosphate. There is nevertheless a close similarity, and Redfield bequeathed the challenge of determining whether this resemblance is purely a coincidence or whether it is maintained by some stabilizing process.[13] A computer model[14]

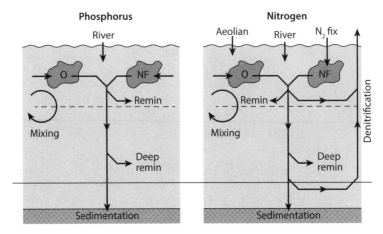

Figure 6.4. Structure of minimal model of nitrogen and phosphorus cycling in the ocean. The model has only two boxes (surface and deep), but includes both N cycling and P cycling, and also, critically, N_2-fixing phytoplankton (NF) competing against nonfixing phytoplankton (O). Some organic matter decays and is remineralized back into the surface ocean, some into the deep ocean after sinking under gravity, and some particles resist decay for long enough to sink to the seafloor and become permanently buried in marine sediments. Denitrification returns nitrate to N_2, and rivers and atmospheric (aeolian) deposition supply fixed nitrogen to the oceans.

suggests stabilization. Within this model, emergent negative feedbacks automatically keep the seawater N:P ratio at or near to the organic N:P. In the same way that thermostats turn on heating or air conditioning to stabilize room temperatures, automatic processes probably stabilize the N:P ratio of the sea.

6.3.3. Reason for the Correspondence

I will not describe this model in detail here, but a schematic is shown in figure 6.4. The model simulates the main features of the cycling of nitrogen and phosphorus in the ocean and incorporates the competition between nitrogen-fixing and other phytoplankton.

The model stabilizes the N:P ratio of the sea. When the model ocean is knocked away from equilibrium by an instantaneous one-off injection or removal of nitrate, it subsequently recovers back to normal (fig. 6.5a). An injection of nitrate is countered by a subsequent imbalance of nitrate inputs and outputs leading to net removal; a removal of nitrate is countered by net creation of new nitrate. Regardless of where it is started from (randomized initial conditions of the two phytoplankton types and of the two nutrients), in all cases it converges back toward the same final state (fig. 6.5b). The model, although highly

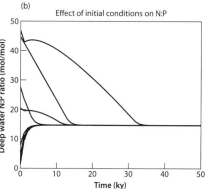

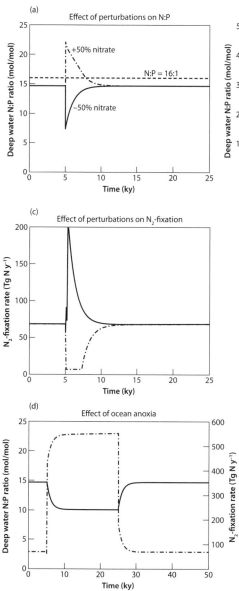

Figure 6.5. Stabilization of deep water nitrate to phosphate (N:P) ratio in the model. (*a*) Recovery from perturbations. The solid line shows how the model deep water N:P ratio recovers from instantaneous removal of half of all nitrate from the deep ocean at 5ky. The dash-dot line shows the response to instantaneous addition of extra nitrate (+50%) at 5ky. The dashed line shows the N:P ratio in phytoplankton organic matter. (*b*) Convergence from a variety of random initial conditions to the same final equilibrium deep N:P ratio. Each line is the trajectory of deep water N:P in a different model run. (*c*) Nitrogen-fixation fluxes over time from the same model runs as shown in (*a*). (*d*) Separate model run showing the response of the model deep water N:P ratio (solid line) to an imposed tenfold increase in the amount of nitrate removal through denitrification, sustained throughout the period starting at 5 ky and finishing at 25 ky. The dash-dot line shows the nitrogen-fixation flux over time, and that it increases to compensate for the higher rate of nitrate loss between 5 and 25 ky.

simplified,[15] offers a plausible explanation of how what seems to be an unlikely coincidence could in fact be the result of an automatic stabilizing process.

Why does the model exhibit this behavior? The reason for the emergent property of N:P stabilization is that nitrogen fixation (a process that leads indirectly to the creation of new nitrate[16]) is stimulated by nitrate scarcity and repressed by nitrate abundance. Because of the costs associated with N_2 usage, it tends to be used only as a last resort, when all other sources of N have been exhausted. In these circumstances, because all potential competitors are handicapped by the lack of an essential ingredient,[17] N_2 fixers easily win out. When this is coupled to the realization that growth of N_2 fixers results in creation of new nitrate, then the operation of the feedback becomes clear (fig. 6.6a, b, c). A reduction in nitrate to below equilibrium levels sets in train a sequence of events that eventually pumps more nitrate into the system and restores nitrate to previous levels.

The N:P ratio of the ocean is most likely regulated over long timescales according to this general principle even though the ecological competition at the heart of the model is a little more complicated in reality.[18]

6.3.4. Nitrate Destruction Counterbalanced by Nitrate Creation

One of the interesting properties of this model is that changes to rates of nitrate destruction are automatically counteracted by changes to the rate of nitrate creation through nitrogen fixation. If denitrifying or anammox bacteria are made to become more active and start liberating more nitrate[19] back to N_2 in the deep sea, then this lowers the nitrate to phosphate ratio of upwelled water and makes conditions more conducive for N_2 fixers (fig. 6.5d). The end result, once equilibrium has been reestablished, is that the extra output of nitrate is matched by an equivalent increase in input of nitrate.

Such a scenario has probably played out on Earth many times. During the Cretaceous there were several oceanic anoxic events (OAEs)—intervals when the deepest ocean waters became depleted in oxygen, leading to widespread mortality of many organisms living in and on the seabed. In contrast, both denitrifying and anammox bacteria thrive in low-oxygen environments and will therefore have instead been much more numerous during the OAEs. Rates of nitrate destruction must have been extremely high. The sedimentary rocks that formed on the seafloor at that time contain telltale biochemical markers indicating that nitrogen fixers were also especially abundant at those times[20] in the surface waters above (fig. 6.7).

What is also apparent from figure 6.5d, however, is that the seawater N:P ratio in the deep ocean is not stabilized to exactly the same value as before, in

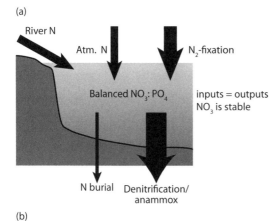

(a)

River N

Atm. N

N$_2$-fixation

Balanced NO$_3$: PO$_4$

inputs = outputs
NO$_3$ is stable

N burial   Denitrification/
anammox

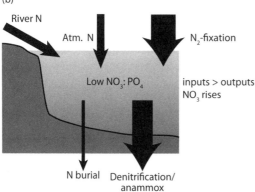

(b)

River N

Atm. N

N$_2$-fixation

Low NO$_3$: PO$_4$

inputs > outputs
NO$_3$ rises

N burial   Denitrification/
anammox

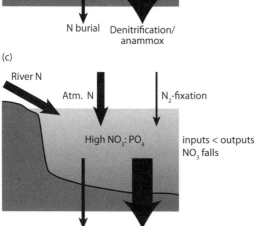

(c)

River N

Atm. N

N$_2$-fixation

High NO$_3$: PO$_4$

inputs < outputs
NO$_3$ falls

N burial   Denitrification/
anammox

Figure 6.6. Control over seawater N:P ratio: (*a*) the total input (sum of all input fluxes) balances the total output at the equilibrium seawater N:P ratio; whereas (*b*) an increase in N$_2$ fixation occurs when there is nitrate scarcity at low N:P ratio, leading to total input exceeding total output to raise the seawater N:P ratio in a negative feedback; and (*c*) a decrease in N$_2$ fixation occurs when there is nitrate excess (high N:P ratio), leading to a situation in which total input is less than total output, thus lowering the seawater N:P ratio, again in a negative feedback.

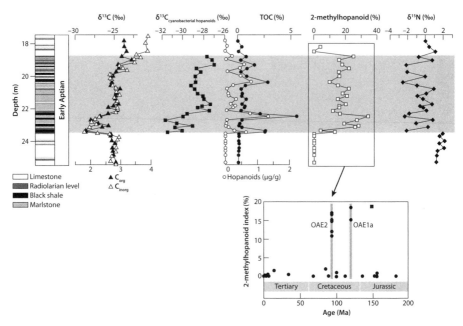

Figure 6.7. Greater numbers of cyanobacteria when the deep ocean was anoxic. The plots show the measured variation over time of the 2-methylhopanoid index in organic-rich rocks. This index indicates abundance of cyanobacteria relative to other phytoplankton. The plot on the lower-right shows the long-term record from 200 million years ago (Mya) to the present. High values are restricted to two oceanic anoxic events (OAE1a, ~120 Mya, and OAE2, ~93 Mya) and one sample from an atypical environment (a lagoonal microbial mat deposit, the datapoint at ~150 Mya). The record from OAE1a is expanded in the upper part of the plot (area shaded gray). Republished with permission of Geological Society of America, from "N$_2$-fixing cyanobacteria supplied nutrient N for Cretaceous oceanic anoxic events," Kuypers et al., *Geology*, 32: 853–856 (2004); permission conveyed through Copyright Clearance Center, Inc.

this case where deep-water nitrate destruction is increased. During OAEs it is likely that there was not as much similarity between deep-water N:P and organic N:P as there is now. Upwelling deep water would have been much more deficient in N rather than being almost perfectly matched to requirements as it is today.[21] On reflection, the imperfect control is not surprising. The critical implementers of the feedback are the nitrogen-fixing phytoplankton, which by necessity of their photosynthetic lifestyle inhabit only surface waters. Therefore, while the perturbation (nitrate removal) takes place in the deep where oxygen is depleted, it is only counteracted once that deep water reaches the surface. Because nitrogen fixers can only respond to the surface water situation, the place where the critical feedback takes place is spatially distant from the deep ocean where the perturbation takes place. It is therefore not surprising that the regulation of deep N:P ratio is imperfect.

6.3.5. Biological Processes Control Nitrate Availability

To return to the theme of this chapter: it is not very important in the context of Lovelock's second assertion exactly how this automatic feedback operates, or even whether it operates perfectly or not. It is immaterial here whether the N cycle is or is not perfectly stabilized. What makes the nitrogen cycle so interesting for our purpose is that everyone agrees that the nitrogen cycle is under biological control. In all of the long debate about whether the marine nitrogen cycle is balanced or not, neither side questions that biology rules. The largest fluxes in the nitrogen cycle (denitrification, anammox, nitrogen fixation, nitrification, ammonification, and so on) are all carried out by different types of bacteria.[22] If there is a controlling feedback then it must be a biological one, probably nitrogen fixation. Any control over the levels of nitrate in the ocean is due to biological processes, not to inorganic ones. And it does appear that the level of nitrate is controlled.

I have focused on the marine N cycle, but it is not only nitrate whose levels are under biological control. Silicon is also biologically stabilized, according to another model similar to that just described.[23] It is the fifth most abundant element dissolved in river water flowing down into the sea, but by contrast is largely absent from seawater due to intense biological removal.[24] Biological removal also explains why carbon and calcium, the two absolutely most abundant dissolved constituents of river water, are so relatively scarce in seawater (fig. 6.8). They are also actively removed from seawater within biological residues that fall to the seafloor and get buried; both are removed in $CaCO_3$, and carbon is also removed in soft organic tissues. The lack of intense biological removal of elements such as sodium and chlorine leaves them to completely dominate seawater salinity, despite their relative scarcity in dissolved salt in river water. In general, the biologically utilized elements are anomalously scarce in seawater (fig. 6.8). What this goes to show is that the plankton do not sit in a medium that has been predetermined by external forces. In fact, the opposite holds true, at least for certain chemical elements. It is the plankton themselves (or, rather, innumerable generations of the ancestors of today's plankton) that have determined the chemical composition of the medium they inhabit.

6.4. Conclusions

The question of this chapter is whether Lovelock was correct to assert that biological forces can shape the planetary environment. The evidence covered in this chapter, from atmosphere and oceans, is sparklingly clear in terms of the

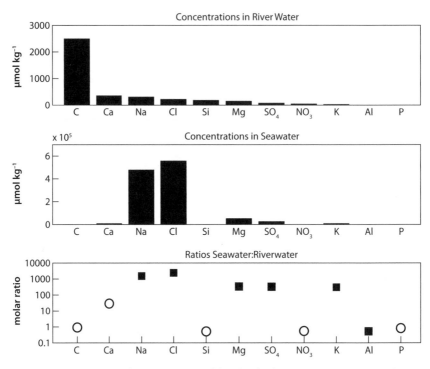

Figure 6.8. Comparison of concentrations of dissolved substances in river water and in seawater. The top panel shows concentrations in river water, the middle panel concentrations in seawater, and the bottom panel the ratio of the two (mol/mol, on a log scale). Open circles denote elements that are widely used by marine organisms, either to make new soft tissues or else to make hard parts (out of calcium carbonate, $CaCO_3$, or opal, SiO_2). Solid squares denote elements that are not heavily utilized by living organisms. All of the main biologically utilized elements are scarce in seawater compared to expectations based on river supply and comparison to lightly utilized elements. Aluminum is also surprisingly scarce, in this case because it tends to stick to falling particles and thereby gets rapidly removed from seawater. Based on data in Sarmiento and Gruber (2006).

implications that can be drawn from it. And this is without considering many other examples that could also have been used to prove the point, such as the effects of vegetation in creating and stabilizing soils, the effects of plant transpiration on the cycling of water, the fossil evidence for the effects of diatoms on silicon concentrations,[25] and so forth.[26] Lovelock was clearly correct to claim that the coexistence of oxygen and methane in Earth's atmosphere is evidence of life, and that life can alter the planet.

Lovelock's assertion can be accepted, but to what degree does this evidence by itself prove Gaia? Once we accept Lovelock's assertion, where does it take us?

How powerful is it in discriminating between the three candidate hypotheses laid out in chapter 1, and is evidence for a biologically modified planet also evidence that a Gaia-like mechanism was behind it?

If we consider just two of these competing hypotheses, Gaia and coevolution, then evidence of a biologically molded planet is not helpful. Both hypotheses assume that the biota can influence their environment. The difference is that whereas Gaia assumes that the influence will be toward improving the environment, coevolution makes no such claims. To distinguish between these hypotheses we need to consider whether biological interventions improve the environment for the biota, a question that was considered in the preceding chapters and that is also returned to in the next chapter. Accepting that the biota *alters* the environment is no help in comparing these two hypotheses; what we need to know is whether it does so beneficially.

The evidence just described is, on the other hand, very helpful when it comes to comparing Gaia to the third hypothesis, the "geological" hypothesis introduced in chapter 1. This third hypothesis posits that life is a passive agent, living on the Earth's surface but with no influence over it. According to this hypothesis, the nature of Earth's environment is determined exclusively by geological and astronomical processes. Clearly this last hypothesis is strongly contradicted by the evidence just discussed and can be rejected without reservation.

Chapter 7

EVOLUTIONARY INNOVATIONS AND
ENVIRONMENTAL CHANGE

HAVING ESTABLISHED THAT life has the power to shape the Earth, we can now move on to look more closely at the exact nature of life's impact on Earth. The nature of this impact is most clearly seen following some major evolutionary developments, when a radical addition has taken place to the collection of life on Earth. What sort of environmental alterations do biological changes produce?

In this chapter, we ignore externally forced changes and focus solely on internally generated changes due to evolutionary inventions of new forms of life. Sometimes change or variability in Earth properties is externally driven. For example, glacial/interglacial cycles are paced by Milankovitch oscillations[1] in the way that the sun heats the Earth. Another type of external forcing is the great disturbance the impact of an asteroid or comet causes, such as the asteroid that brought about the demise of the dinosaurs while bringing the Cretaceous to an end. But such responses to external events are irrelevant to this chapter. We are interested instead in change or variability that results from the evolution of new organisms with novel means of manipulating the environment. Some key milestones in biological evolution, such as the advent of multicellular organisms, are not, as far as we know, associated with any revolutionary environmental change. But others are. We examine the effects of the two evolutionary innovations that have most obviously shaken the world: (i) the evolution of oxygen-yielding photosynthesis, and (ii) the colonization of land by the first forests.

From the perspective of evaluating the Gaia hypothesis, these internally driven changes are of the greatest interest. A well-regulated planet could hardly be blamed for being buffeted about by the vagaries of celestial mechanics and collisions, and can even be congratulated for its multiple recoveries from the terrible devastations of extraterrestrial impacts. However, when the sources of variability are from within the Earth system itself, then it is instructive to consider whether the evolutionary changes led to a "better" or "worse" (more or less hospitable) world.

According to the Gaia hypothesis life has ensured that the planetary environment has stayed comfortable and habitable. We might therefore expect evolutionary advances to improve planetary stability and growing conditions. Gaia proponents have indeed suggested something similar, that only environmentally favorable innovations persist:

> In Gaia theory, organisms change their material environment as well as adapt to it. Selection favours the improvers, and the expansion of favourable traits extends local improvement and can make it global. (Lovelock 2003b)
> At this stage, it seems that natural selection can form an integral part of planetary self-regulation and, where destabilizing effects arise, they may be less likely than stabilizing effects to attain global significance or persist. (Lenton 1998)

However, others disagree:

> Gaian feedbacks can evolve by natural selection, but so can anti-Gaian feedbacks. (Kirschner 2002)
> The answer seems to be that an innate lethality is a side-effect of the main factor leading to life itself—the process that we call evolution. . . . (Ward 2009)

So what is the reality and who is right? The key issue in this chapter is whether major evolutionary developments have transformed the planet in a way that is beneficial overall for the life living on it. Let's scrutinize the evidence, focusing on these two fundamental innovations that have each had large impacts on the planet.

7.1. The Great Oxidation Event—How the Evolution of Oxygen-Yielding Photosynthesis Remade the Planetary Atmosphere

The largest and most profound of the biologically induced changes to the Earth was the construction of an oxygen-containing atmosphere. In the last chapter we saw (table 6.1) that neither of Earth's sister planets in the solar system possesses more than trace concentrations of oxygen in their atmospheres. The early Earth was almost certainly the same. Although the Earth may have gained some oxygen via hydrogen loss to space (H_2O split by sunlight to H_2 and O, H_2 lost to space, O left behind to accumulate),[2] the rise in oxygen was primarily due to biology. As Lovelock so vividly illustrated in his books on Gaia, and as was described in section 6.2.3, this enormous reservoir of oxygen (on average fully 21 out of every 100 molecules of air are O_2 molecules) arose and persisted on Earth through photosynthesis.

The puzzling nature of our oxygen-rich atmosphere, and the reason for its absence in other planetary atmospheres, derives from oxygen's extreme reactivity. Oxygen is volatile. It causes rusting and burning. Oxygen is continually lost from the atmosphere in order to "oxidize" (rust) iron[3] and other minerals in newly exposed rocks. It readily combines with many of the gases and minerals ejected from volcanoes, leading to a loss of free oxygen into the newly formed molecules. It reacts with volcanic hydrogen (H_2) to form water (H_2O) via the reaction $2H_2 + O_2 \rightarrow 2H_2O$, with volcanic hydrogen sulfide (H_2S) and sulfur dioxide (SO_2) to form sulfate (SO_4), and with carbon monoxide (CO) to form carbon dioxide (CO_2) via the reaction $2CO + O_2 \rightarrow 2CO_2$. Oxygen is used up following every entry of methane (CH_4) into the atmosphere, to convert it into carbon dioxide (CO_2).

7.1.1. Who Did It?

As explained in the last chapter, the great amounts of oxygen in the atmosphere in the face of large ongoing losses can only be understood if the oxygen is continually replenished by some large-scale process. That process is photosynthesis. Now I will consider how exactly oxygen first appeared in the atmosphere and ocean, and the changes this wrought on the living environment.

The first life on Earth was presumably chemolithotrophic.[4] That is to say, it must have obtained its energy from conducting chemical reactions between inorganic rather than organic chemical substrates. It is likely that such chemolithotrophs proliferated rapidly until all the profitable[5] chemicals in the environment were completely used up. Then, early in Earth history, microbes acquired the ability to capture sunlight energy followed by the ability to store the captured energy within biological molecules (for example, sugars) that they could keep for later use. Suddenly they had access to an inexhaustible energy supply. This was a change of great significance. When you look at a grain of sugar you can think of it as a piece of distilled sunlight energy in solid form—captured by a sugar cane plant. This crucial advance (taking energy directly from the sun) opened up a whole new means of subsistence.

The very first forms of photosynthesis may not have evolved oxygen. Still today, in hot springs around the world, purple and green sulfur bacteria use hydrogen sulfide (H_2S) and hydrogen (H_2) as electron donors during photosynthesis, and do not release oxygen as a by-product. In a surprising discovery, some of these non-oxygen-producing photosynthesizers even thrive on arsenic,[6] and in fact depend on it for their existence. However, evolutionary experimentation soon hit upon the fundamental chemical reaction[7]:

$$CO_2 + H_2O \xrightarrow{\text{sunlight}} CH_2O + O_2 \qquad \textit{(equation 7.1)}$$

as the most successful means of converting sunlight energy to chemical energy. From this point on, as long as life didn't die out, the Earth was probably destined to eventually acquire an oxygen-rich atmosphere.[8] Over time, as oxygen continued to be pumped out by the newly arrived photosynthesizers, all of the reduced substances on Earth (in the air, on land, and in the sea) became oxidized, after which oxygen was then free to build up in the atmosphere and ocean.

7.1.2. When Did It Happen?

The first[9] accumulation of oxygen in the atmosphere occurred so long ago (more than two billion years ago) that the geological evidence from that time is rather scarce and open to debate. There aren't all that many rocks on Earth that survive from so long ago, and those that do have mostly been subject to major alteration during the great duration of intervening time between then and now. Once rocks have been altered by immense heat and pressure (metamorphosed), they can no longer be fully trusted to paint a faithful picture of the ancient environment in which they were formed. Following metamorphism it becomes to some extent unclear which characteristics are original and which have been added by the metamorphism. For this reason, there is still some debate among geologists as to exactly how the "Great Oxidation Event" (GOE) unfolded. However, various lines of evidence[10] (including an abundance in the geological record of large deposits of "rusted" iron), puts a most likely date for this event at about 2.4 billion years ago. Thus by the time Earth was about half as old as it is now (4.5 billion years), oxygen had begun to build up in the atmosphere and in the surface ocean. After that, everything was different.

7.1.3. What Effect Did It Have on Planetary Habitability?

The GOE most probably had one highly beneficial impact: facilitation of the evolution of eukaryotes. Life on Earth can be divided into three domains: the archaea, the bacteria, and the eukaryotes. The bacteria and the archaea together make up the prokaryotes. Prokaryotes are more rudimentary than eukaryotes; all plants and animals (all complex multicellular life forms[11]) belong to the eukaryotic domain. The earliest life on Earth, the first to evolve, was prokaryotic. Prokaryotes are typically very small. The largest known prokaryotic cells, of the species *Thiomargarita namibiensis* and only fairly recently discovered in organic-rich sediments off the coast of Namibia, are an extraordinary 100–300 microns (0.1–0.3 millimeters) in diameter and therefore visible to the naked eye. However, these are giants among prokaryotes. These cells are absolutely unrepresentative of their domain. Most prokaryotic cells are much smaller, in fact invisibly small. The large majority are < 2 microns (0.002 millimeters) in diameter. Even eukaryotic single-celled organisms, by comparison, are much larger (2–100 microns or more) and more complex.

We are not certain when the eukaryotes, the most advanced domain of life, first evolved, although it appears to have been somewhere close in time to the GOE. Our uncertainty derives in part from the fragmentary and distorted nature of the fossil record from that long ago, in part from the microscopic size of algae and in part because it is not always clear which domain fossils of such antiquity belong to. The presence/absence of characteristic molecules (biomarkers), thought to be produced only by eukaryotes, can however be used. Such lines of indirect evidence suggest that eukaryotes evolved somewhere between about 2.7 and 2.1 billion years ago (Bya), but this evidence is again disputed.[12] Direct and unequivocal evidence (in the form of fossils of clear eukaryotic origin) is not available until about 1.5 Bya. Because most eukaryotes, in contrast to all prokaryotes, contain mitochondria and require oxygen, it seems unlikely that eukaryotes would have been numerically abundant prior to the great oxygenation event.[13] They are also less tolerant of ultraviolet (UV) radiation, and it is likely that their evolutionary radiation was aided by the formation of a planetary ozone (O_3) UV screen following the rise in oxygen.[14]

So, the rise in oxygen may have enabled the later evolution of eukaryotes. Looked at from the point of view of the organisms living at the time, however, the spread of oxygen throughout most of the biosphere was a disaster. Before oxygen accumulated in the environment, all organisms were of course anaerobic. They were well adapted to surviving without oxygen. Oxygen was in fact toxic to them, but up until that point it had posed no problem because there was very little of it around. But as oxygen levels built up, the anaerobes were poisoned.

As if this wasn't bad enough, oxygen also created a second and insuperable problem for the anaerobes. They became uncompetitive. Before the rise in oxygen, all organisms used chemical reactions that did not require oxygen. This was true across all the different "guilds" of life (ways of making a living), including photosynthesis, chemosynthesis, and organic matter breakdown. The problem is that the anaerobic reactions are inefficient compared to the ones that are possible when using oxygen. For instance, energy generation from organic matter yields significantly more energy if the decomposition reactions use oxygen than if they do not (table 7.1). The bioenergetic yield per unit food is greater when oxygen is involved in the breakdown process. This is critical. Aerobic (oxygen-using) organisms, with their inherent metabolic advantages, evolved and supplanted anaerobic microbes in all environments where oxygen penetrated.

The rise in oxygen was a great opportunity for the new organisms that evolved to take advantage of it, but not for the anaerobic microbes that first produced it. They forced themselves out of dominance in very many parts of the biosphere and banished themselves to live henceforth only in fringe environments where

TABLE 7.1.
Organic matter oxidation pathways and
corresponding energy yields

Reaction	Energy Yield (kJ per mole of CH_2O)
Oxic respiration	−475
Denitrification	−448
Mn-oxide reduction	−349
Fe-oxide reduction	−114
Sulfate reduction	−77
Methane production	−58

Note: All these reactions except the first take place
in oxygen-free or low-oxygen environments. Denitrification uses nitrate, which would also have been scarce
before the ocean was oxygenated (taken from Canfield
2005).

little or no oxygen is present, such as ocean and lake sediments, bogs and
marshes, animal guts, and deep underground.

As well as poisoning anaerobes, the rise in oxygen levels also had a transforming effect on climate. Oxygen-incompatible greenhouse gases such as methane
were removed from the atmosphere[15] and the Earth experienced what were possibly its first ever ice sheets, during the Huronian glaciation. This glaciation may
have been of Snowball Earth–type severity, with ice sheets in the tropics.[16]

7.1.4. Does the GOE Conform to the Gaian View?

While accepting caveats due to the considerable imperfection of the geological
record from so long ago, how does this example of internally driven environmental change look from this book's perspective of evaluating Gaia? From the
point of view of life in general, and if the evolution of more complex eukaryotic
life is taken to represent progress, then the rise in oxygen was certainly a good
thing. If the microbial ecosystems on Earth at that time could only have known
it (which of course they couldn't, being microbes), their actions contributed
eventually to the wonderful flowering of multicellular life on Earth.

For the anaerobes of the time, however, it was calamitous. In today's world
anaerobes occur only in the fringe environments just mentioned; they are odd-
ities, whereas in the early world they reigned supreme. The Great Oxidation
Event was the most fundamental of the internally induced revolutions to the
Earth environment. But does it agree or disagree with a Gaian interpretation of
planetary regulation? The verdict is equivocal, in my view; it was disastrous for

the dominant biota of that time, but, in the long run and with hindsight, rather beneficial for life as a whole.

7.2. Sucking Out CO_2—The Arrival of Land Plants

A second major revolution to the pattern of life on Earth was the greening of the continents. Early life on Earth was exclusively aquatic. Following the first evolution of life more than three billion years ago,[17] water most likely remained the sole medium for life over the next two billion years, apart from some green algae covering rocks along shorelines.

7.2.1. When and How Did It Take Place?

The colonization of the land surface by vegetation was a protracted process. It is likely that the first pioneers to penetrate inland were simple forms such as cyanobacteria and algae. By about 450 Mya plants similar to modern bryophytes (mosses and liverworts) had evolved.[18] The next wave of colonists included knee-high trees and shrubs. From a climate-perturbing point of view, the most important development was the evolution of rooted trees containing large amounts of carbon locked up in lignin-rich tree trunks. These evolved through competition to capture sunlight, which favors increasingly tall plants. Tall plants are able to intercept sunlight before it penetrates to plants below, outshading and thereby outcompeting their smaller neighbors. This selection pressure led over time to evolutionary innovations such as conducting vessels to pump water and nutrients against gravity (vascular plants), wood for structural strength and therefore height, and substantial roots capable of anchoring large trees against the enormous forces winds exert on them. The earliest known trees and forests that resemble those we have today are seen in the fossil record only after about 385–370 Mya, that is to say, during and after the Middle/Late Devonian[19] (fig. 7.1). Fossil tree stumps from the "earliest forest" were first discovered in the 1870s at Gilboa, New York State, and have more recently been supplemented by finds of additional fossils from a neighboring site (about 13 kilometers distant). This time the fossils were of the trunks and crowns, which could be joined to the fossil tree stumps to give a picture of the whole, showing that the stumps belonged to ~8 meters high "tree ferns" of the genus *Wattieza*.[20] Over the next twenty million years or so, the development of deeper root systems, improved means of water transport to the height of the canopy, leaves that intercepted large amounts of sunlight, and new reproductive strategies (seeds) also released plants from a restriction to moist habitats and allowed

Figure 7.1. How the first forests may have appeared. The four stages of the life of the *Archaeopteris* (the first "modern" tree) are illustrated. On the far left is a young, six meter *Archaeopteris*, in the center at twice that height is a medium-aged *Archaeopteris*, and on the right is a fully mature specimen. Farthest right is the collapsed and decaying trunk of a mature tree. © Walter Myers.

them to colonize drier inland habitats farther from standing water. Forests spread out over the land.

7.2.2. Impact on Atmospheric CO_2

Although trees were relative latecomers to Earth's evolutionary drama, upon their arrival they greatly altered the climate. At the same time that we see large amounts of plant debris beginning to appear as fossils in sedimentary rocks, atmospheric CO_2 seems to have begun a long decline (fig. 7.2). The evidence of a fall in atmospheric CO_2 is somewhat indirect because there is no ice that survives from that long ago, and therefore no bubbles in ice cores to give us a direct record of the atmospheric composition. However, several lines of less direct evidence support the conclusion of falling atmospheric CO_2: increases in the numbers of stomata (surface holes) in plant leaves to maintain the rate of "breathing in" of CO_2 despite it becoming increasingly scarce, changes to the carbon isotope ratios of calcium carbonate formed in the ocean (which reflect

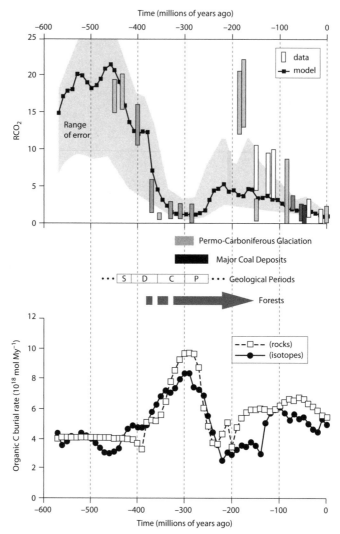

Figure 7.2. Carbon cycle effects associated with the first spread of trees across the land. (top) Atmospheric CO₂ versus time for the Phanerozoic (past 545 million years). The parameter RCO₂ is the ratio of the mass of CO₂ in the atmosphere at some time in the past to that at present (preindustrial times, 300 ppmv). The heavy line joining small squares represents the best estimate from a model, while the shaded area encloses the estimated range of uncertainty of the model estimates. Vertical bars represent independent estimates of CO₂ level based on the study of fossilized soils. Adapted from Berner (1997). (middle) The time of low RCO₂ (low atmospheric CO₂) comes some time after the first forests and corresponds to a time of coal formation and glaciation in the Late Carboniferous and Permian (the bulk of the world's black coal reserves, including deposits on all continents, dates from this time [Thomas 2002]). (lower) Rate of organic carbon burial calculated in two ways: firstly from δ¹³C marine carbonate isotope data and a model, and secondly derived independently from the abundance of organic carbon in different types of sedimentary rocks. Adapted from Berner (1998).

increasing burial of organic carbon), analyses of carbon isotopes in soil carbonates (which reflect declining atmospheric CO_2), and various evidence of associated climate cooling.[21]

7.2.3. Loss of Atmospheric CO_2 in Buried Plant Remains

It is likely that there were two principal reasons for the decline in atmospheric CO_2, the second of which will be discussed in section 7.2.4. The first reason is simply the burial of astounding quantities of plant organic matter. Living vegetation itself does not hold very much carbon in total (~600 Gt C) compared to the total amount in the atmosphere and ocean (~40,000 Gt C). However, burial of large amounts of organically derived carbon year after year provides the potential to greatly alter atmospheric CO_2 levels over time. When carbohydrates are formed via the photosynthetic reaction shown in equation 7.1 above, then CO_2 is removed from the air. Equation 7.1 is the most fundamental equation in the whole study of the Earth's climate system. The main point here is that the carbon in CO_2 is locked away into organic matter when the organic matter is formed. If that organic matter is subsequently broken back down to its base constituents in the very same place, for instance if the organic matter formation is rapidly followed by its decomposition by bacteria, then the CO_2 is released again to the same place from which it was extracted. When that happens there is no net, sustained, effect on atmospheric CO_2. But when the organic matter is incorporated into newly forming rocks, CO_2 is kept out of the atmosphere for many millions of years.

When massive quantities of organic matter were buried in the Middle and Late Devonian and also during the Carboniferous, the amounts of CO_2 drawn out of the atmosphere were quite staggering. It is estimated,[22] from assays of the organic carbon content of rocks dating from that time, that an additional amount of approximately 4,000,000 Gt C was buried between ~380 and 250 Mya (fig. 7.2), over and above the normal rate of deposition.[23] Comparing this number to those in the previous paragraph, it is clearly a very large amount. The locations in which this copious plant carbon was introduced into new rock were mostly under water. Some plant debris fell into swamps and then got turned into peat and then coal. Other land plant remains were swept down rivers before falling into the coastal seafloor mud of a river delta.

Intriguingly, the explanation for the long period of intense carbon burial may involve more than just one evolutionary advance. It may well owe something to an offset between two separate evolutionary advances. It is not possible to be sure from the geological evidence, but the evolution and spread of lignin (wood) as a successful building material for plant stems may have set a biochemical conundrum for fungi that took them many millions of years to solve. If there was a long hiatus between the advent of widespread tree trunks and the development of a widespread fungal capacity to degrade them (even today lignin is among the least digestible and slowest-to-degrade parts of plant matter), then for a long time plant stems and tree trunks would have littered the landscape with no effective means of decomposition. This idea[24] of a temporary failure in global recycling assumes a slow rate of evolution. It assumes that there was some barrier to the rapid evolution of enzymes that (1) broke down wood and (2) did so via chemical reactions yielding sufficient energy (or costing sufficiently little energy) to make their use by organisms worthwhile. Although hard to prove, this idea is tentatively supported by recent molecular clock evidence,[25] and is compatible with the clear and certain evidence for very high levels of organic carbon burial at that time.[26]

Further inspection of the fundamental biochemical reaction, equation 7.1, shows that not only is carbon dioxide consumed when plants photosynthesize, but oxygen is also released. As well as producing Earth's oxygen atmosphere in the first place, this reaction has special implications for oxygen at the time that the first large plants spread across the land surface. The astounding carbon burial must also, by the logic of the equation, have caused correspondingly large quantities of oxygen to be left behind in the atmosphere. That this actually happened is supported by the charcoal record, of interest here because fires require oxygen and produce charcoal. Charcoal is not found in many Devonian rocks but is present throughout the Carboniferous and Permian[27] when carbon dioxide was calculated to be low (fig. 7.2). Calculations suggest that oxygen made up more than 25% of the total atmosphere during the Carboniferous and Permian, compared to 21% today.[28] This extra oxygen has been posited as responsible for the various giant fossil insects discovered from Carboniferous times, including the famous half-meter wingspan dragonfly fossils discovered at Bolsover in England and Commentry in France, millipedes a meter long, and spiders with feet half a meter apart.[29] Insects do not actively transport oxygen using a single breathing tube (throat) and pair of lungs; they rely instead mainly on passive diffusion of oxygen from outside to inside through a network of hollow tubes. This means of oxygen supply is inefficient compared to the lungs of mammals and reptiles. Such an inefficient means of oxygen supply probably could not have supplied oxygen sufficiently rapidly to support the demands of large flying dragonflies unless the level of oxygen was higher than the current value of 21%.

Returning to the main topic of this book, the question of immediate interest here is: what relevance does this episode have to evaluating Gaia? To answer this question we need to establish firstly whether life played a strong role in the decline in atmospheric CO_2. Regardless of the intricacies of the arithmetic, it is clear that climatically significant quantities of carbon were buried at this time. Although not all details are known, the incorporation into new rocks of large amounts of wood (lignin), and other plant matter appears to have been responsible for a large part of the reduction in atmospheric CO_2.

7.2.4. Loss of Atmospheric CO_2 Due to Enhanced Silicate Weathering

The rise of plants also most likely had a second impact on atmospheric CO_2, again contributing to its decline. This impact came through plant roots, and their effect on the soil and rock into which they penetrated. Plant roots (together with their associated symbiotic fungi) release organic acids. Plants probably benefit from the production of the acid because it leads to enhanced leaching of nutrients from the surrounding soil and rock particles. But the enhanced acidity also leads to an associated secondary side effect, one of no great immediate benefit or direct relevance to the organisms responsible—the enhanced weathering of silicate rocks.[30] There are many different silicate-containing rocks, and silicate weathering as a whole therefore comprises many different chemical reactions, but the climatically important effect can be represented in simplified form by the reaction[31]:

$$CaSiO_3 + CO_2 \rightarrow CaCO_3 + SiO_2 \qquad equation\ (7.2)$$

Silicate weathering therefore extracts CO_2 from the atmosphere. The higher the rate of silicate weathering, all else being equal, then the greater the loss of CO_2 from the air. It seems fairly certain that the spread of plants across the land surface, and the more acidic soils they brought with them, led to a significant enhancement of global silicate weathering rates.[32] It is difficult to be sure about which of these two changes[33] (to burial of organic carbon or to silicate weathering) was more important in drawing down atmospheric CO_2.[34] But it doesn't greatly matter to the main argument: in either case, the CO_2 drawdown was driven by the evolutionary development of large land plants with tall and sturdy woody stems and with long roots.

7.2.5. Effect of Declining CO_2 on Planetary Habitability

In terms of Gaia it is not only relevant whether life caused these CO_2 changes, but also the effect they had on life. The answer is clear enough. There were severe and detrimental implications. Geological evidence shows clearly that repeated glaciations followed land colonization by plants. There was a pronounced fall in global atmospheric CO_2 and temperature between about 370 and 320 Mya (during the Late Devonian and Early Carboniferous). Reconstructions of atmospheric CO_2 suggest that it fell from about 4,500 parts per million down to maybe 500 parts per million (fig. 7.2a).

This new intervention in the climate system was not favorable for life. It did not lead to a better climate or to a more stable climate. It led to a much colder climate,[35] although clearly not as dire as the Snowball/Slushball Earth episodes to be discussed in the next chapter. During the Carboniferous there appear to have been periodic collapses into glacial conditions. Evidence for these glaciations comes from (1) rocks (called tillites) that appear to be compressed glacial moraines (moraines are the piles of rock debris deposited alongside and at the ends of glaciers); (2) rock faces with long, parallel scratches (glacial striations) scored in them, attributed to glaciers dragging other rocks across them; and (3) anomalous chunks of rock (dropstones) in mid-ocean seafloor sediments. It is thought that these misplaced large rock fragments could only have been transported away from the coast within icebergs that then melted out at sea, dropping the rock fragments into a foreign setting. Some of the rocks in which coal layers occur show evidence of unstable climates. Particularly in the later Carboniferous some coal deposits occur within "cyclothems"—rocks containing repeated cycles flipping back and forth between limestones and shales formed beneath the sea and coal beds formed in shallow swamps. The formation of these cyclothems is attributed to cyclical variations in global sea level due to waxing and waning of ice sheets.

The onset of icehouse conditions may well have been responsible for the Middle and Late Devonian biotic crisis, one of the "Big Five" mass extinctions that took place during the Phanerozoic. This extinction spanned an interval of about 20–25 million years, was made up of several separate and distinct events, and was particularly severe for the tropical marine biota.[36] About 40% of marine genera became extinct, although high-latitude marine faunas were relatively little affected. Tropical reef communities declined and were eventually replaced as temperatures plunged, perhaps by as much as 16°C. Ice sheets were present through most or all of the Carboniferous, usually restricted to high latitudes but descending to within almost 30° of the equator during the Late Carboniferous.[37]

The decline in atmospheric CO_2 must also to some extent have starved terrestrial plants of CO_2 (hence the more tightly packed stomata on leaf surfaces) and probably limited photosynthesis on land as a result, especially because the CO_2 decline was accompanied by a simultaneous rise in oxygen (O_2). The ubiquitous photosynthetic enzyme RuBisCO, possibly the most abundant enzyme on Earth, processes either CO_2 or O_2, depending on their relative availability; at high O_2 and low CO_2 its efficiency of carbon fixation is particularly reduced. On the other hand, there was also at least one redeeming aspect of the CO_2 scarcity. It is argued[38] that an indirect consequence of the drop in CO_2 was the stimulus to evolve large leaves. If CO_2 had stayed high and large leaves had never evolved, trees today would look very different.

7.2.6. Does the Greening of the Continents Conform to the Gaian View?

At the end of the last section I asked whether the Great Oxidation Event was compatible with a Gaian interpretation of planetary regulation, and I gave an equivocal answer. Although the first forests also produced both positive and negative consequences, the overall conclusion this time is I think far more straightforward. The CO_2 removal and climate cooling caused by plants invading the land was on balance overwhelmingly unfavorable for life.

7.3. Conclusions

Returning to the main argument, and taking the two case studies of this chapter together, life seems to mold the environment in rather haphazard ways, with neither rhyme nor reason to the eventual result. Indeed, for many major milestones of biological evolution (for example, the first multicellular organisms, the first animals) we are not as yet aware of any corresponding transformative impacts on the environment.[39] Where evolutionary developments have led to environmental impacts, then these have sometimes been beneficial and sometimes detrimental, with no obvious bias toward one or the other. But whatever the nature of the environmental modification, life has always changed to exploit and closely fit it, as it must because of evolution.

Oxygen-dependent photosynthesis and respiration evolved to take advantage of the appearance of oxygen in the biosphere, while anaerobes crashed from preeminence to relative obscurity. Land plants evolved to adapt to the low carbon dioxide and low temperatures that earlier generations had produced, and enormous insects thrived temporarily to take advantage of the transiently abundant oxygen, only to disappear again when oxygen levels subsequently subsided. New types of fungi eventually evolved to make use of the new food source.

The evidence just described is not particularly consistent with or supportive of Gaia. These internally generated (evolutionary) events produced some changes that generally improved the global environment for life, but also some that tended to spoil it. The two evolutionary innovations considered in this chapter produced a mixture of effects, in part beneficial, in part detrimental. By no means can it be said that these evolutionary breakthroughs unambiguously improved habitability. If anything the reverse is true, with one of the evolutionary breakthroughs having led to global poisoning (oxygen) and with both having produced major glaciations.

Chapter 8

A STABLE OR AN UNSTABLE WORLD?

A MAIN CLAIM of the Gaia hypothesis (assertion No. 3 of chapter 1) is that the Earth is a particularly stable habitat for life, has remained so ever since life became abundant on the planet, and is stable in part because of the actions of life:

> The most important property of Gaia is the tendency to keep constant conditions for all terrestrial life. (Lovelock 1979)
>
> Gaia theory predicts that the climate and chemical composition of the Earth are kept in homeostasis for long periods until some internal contradiction or external force causes a jump to a new stable state. (Lovelock 1988)
>
> The Gaia theory proposes that organisms contribute to self-regulating feedback mechanisms that have kept the Earth's surface stable and habitable for life. (Lenton 1998)

In this chapter we will carefully scrutinize this claim of stability by looking at data pertaining to past variability of the Earth environment. Earlier chapters looked at whether Earth conditions are optimal or at least particularly propitious for life. In the last chapter the question was whether key steps in biological evolution led to improved environments. Now the focus is different. The focus is no longer on how *favorable* a cradle the Earth is for life but instead is on how *stable* a cradle it has been. In the last chapter we learned that environmental change can be produced by evolutionary change—now we consider more generally how much and how rapidly the environment has changed in the past.

While the two events described in the last chapter show the possibility of one-off evolutionary changes causing enormous environmental alterations, it is also apparent that asteroid impacts and large-scale volcanism (eruption of flood basalts) can do the same. They have brought about mass extinctions and environmental changes—for instance, the asteroid impact that brought about the end of the Cretaceous. In this chapter we are interested in whether the general intrinsic pattern was one of stability or instability in the intervals between these externally or evolutionarily driven interruptions. A number of records of different environmental properties are shown in this chapter, and I consider what they reveal about Earth environment stability.

8.1. Reading the Rocks

8.1.1. By Eye

This very same question, stability or instability, is provoked by the direct visual appearance of the rocks. Rocks from past ages constitute a vast archive of information about past environments. In most places that archive is obscured; the land surface is covered by soil and vegetation and it is not obvious what sorts of rocks lie beneath. But in other places such as cliffs and escarpments and in road cuttings, in open-cast mines and on bare mountainsides, the archive is revealed and exposed to view. At these places we get a glimpse of what lies beneath the ground, and of the rock layers that have been laid down through the geological ages. The revealed strata prompt the same question as the subject of this chapter. This is because in some places we see very thick layers, and within each layer the rock is of a uniform and unchanging type, whereas in other places we see many thin layers of frequently changing rock type.

In some places whole mountainsides, hundreds of meters thick, are made up of rocks of similar composition and appearance (see for example fig. 8.1a–c). Whole grand vistas, of great dimension, are comprised of rocks of just one or a few types. Continuity of rock type in these places suggests that the same sort of material kept falling down to the seafloor over very many millennia. The nature of the rain of particles didn't vary, even over long ages of time. This could be taken as evidence that the normal state of affairs on Earth has been stability of the environment, with only occasional jumps to new types of environment.

We can look at one example in a bit more detail; a dramatic location for viewing a wide expanse of Earth history is the Grand Canyon (fig 8.1b). Here the rock sequence goes from more than 1,250 My old to less than 250 My old. That is to say, a billion years are on display between the deepest and shallowest rock layers. There are major gaps ("unconformities"), however, where rock layers are missing due to intervening periods of erosion rather than continued deposition. If we exclude these and look only at rocks above the unconformities, the cliffs in figure 8.1b are known to consist of nine primary bands (from Kaibab limestone down to Redwall limestone, with shales and sandstones sandwiched between), covering the age interval from about 245 to 360 Mya. The average duration between visible changes in rock type is therefore about 14 million years, still an extremely long stretch of time.

Similarly, it has been calculated that the chalk deposits that make up Beachy Head (fig. 8.1a), the White Cliffs of Dover, and other chalk cliffs along the English south coast, hundreds of meters thick in places, were laid down at a rate of maybe half a millimeter per year, although it is difficult to know the exact rate with any accuracy.[1] At that rate it would have taken hundreds of thousands to millions of years to accumulate the whole of the chalk sequence.

From these first examples we get an impression of general constancy of the Earth environment over long sweeps of time. However, this is not the whole story. In other places the appearance of the rocks is very different. In other places we see finely interleaved layers of, for example, limestones, shale, and coal. The color and nature of the rock can alternate every few centimeters (fig. 8.1d–f; remember also the cyclothems described in section 7.2.5). These other locations present a very different visual picture: they suggest environmental variability rather than environmental stability.

8.1.2. By Proxy

Simply looking at the rocks like this is interesting but can only take us so far. Visual appearance by itself is a naive way of trying to read the archive. Beautiful though the pictures in figure 8.1 are, a more subtle approach is needed because continuity of rock color may hide considerable environmental variability.[2] What do cleverer and more robust records tell us? What do we find when deciphering the rock record using more sophisticated techniques? In later sections of this chapter, some of these more sophisticated records will be presented.

That we are able to examine Earth stability in ever-greater detail and with more advanced techniques is thanks to the patient and resourceful work of paleoclimatologists. Their work is providing us with a growing number of climate and other environmental records that are increasingly finely resolved through time. Because temperature and other environmental parameters are not usually recorded directly, they have to be deduced from characteristics of rocks that changed in tandem with the environmental parameter of interest. Such indirect records are known as proxies. The number and the diversity of the different proxy records[3] is a testament to the ingenuity of the scientists involved. Bubbles in ancient ice are carefully analyzed with different instruments. Tree-ring widths are cataloged from fossil trees as well as from the oldest living trees. Stalactites in long-undisturbed caves are sectioned and then the variable nature of the annual incremental additions is probed to see what they can tell us about past climates. The calcium carbonate shell casings of foraminifera, that rained down to the seafloor throughout prehuman millennia, are recovered from cores drilled through thick ocean sediments, washed, and then their chemical compositions measured with exquisite precision. The numbers of pores (stomata) in fossilized leaves are carefully counted, and their density per unit leaf area calculated. This is just the tip of the iceberg—the imagination of paleoscientists has led to a whole armada of different techniques for reconstructing ancient envi-

a

b

c

d

e

f

Figure 8.1. Earth history writ large. (*a*) Chalk cliffs of Beachy Head in England, part of a thick chalk bed that reaches the surface in several locations in the United Kingdom and northwest Europe. (*b*) The upper sequence of layers in the Grand Canyon, taken at the south rim. (*c*) Eocene and Cretaceous sediments in the Badlands National Park, South Dakota, USA. (*d*) The Blue Lias sequence near Lyme Regis (UK) consists of a cyclical pattern of marls, shales, and limestones. (*e*) Dating of the layers (which are apparent if you look closely enough) in the Punta di Maiata sequence in Sicily shows them to be synchronized with periodic changes to the Earth's orbit around the Sun (Milankovitch cycles). (*f*) Periodical fluctuations in rock type are not restricted to the last few million years—this photo shows alternating rock layers from 350–390 Mya (Devonian age) from eastern Greenland. (*a*) © Can Stock Photo Inc. / Vinga; (*b*) Sally Scott / Shutterstock.com; (*c*) © Can Stock Photo Inc. / magmarcz; (*d*) Ian West; (*e*) Andrew Roberts; (*f*) John Murray, from T. R. Astin et al. (2010). In *Geological Society Special Publications*, 339: 93–109.

ronments. Some proxy techniques are later proven unreliable (not providing a faithful record of past environmental conditions, or being influenced by multiple different environmental factors so that isolating the history of the factor of interest is not possible), and some are later found to need further refinement. However, by extracting ever more data, by gradually becoming increasingly knowledgeable about the strengths and weaknesses of each record, and by adding more and yet more different records and intercomparing them, paleoscientists are slowly building up an increasingly reliable and detailed picture of the history of climate and environment here on Earth.

8.1.3. Perspective Matters

The immense duration of the history of life on Earth can, figuratively speaking, be viewed at different magnifications. If we zoom out to the very lowest magnification then we can see the whole scope in view, spanning 2–3 billion years or more but all in low resolution. Or we can zoom in to savor "snapshots" of only a few millions of years, or even less, at higher magnification. This degree of magnification is pertinent to an issue that recurs throughout this chapter. How do we decide whether to call a particular record variable or stable? All things are relative, and a record that from one perspective appears fluctuating, from another appears hardly to change. Each record looks different according to how much we are zoomed in or out on the time axis. They can also look vastly different depending on the scale on the other axis. If, during the time period of interest, the range of a particular parameter is between +991 and +996, then the parameter will look to vary greatly on a y-axis ranging from 990 to 1,000, but look to be actively stabilized on a y-axis ranging from 0 to 1,000. Even the apparent smoothness of a sheet of writing paper or of a plain cotton shirt looks rough and corrugated under an electron microscope.

For each of the records in this chapter, we therefore need to judge not only whether it looks flat or jagged on the plot, but also how it would look different if the time interval or vertical scale was altered. More precisely, we need to judge whether the amplitude of change is biologically significant.

8.2. OSCILLATING SEAWATER CHEMISTRY

Geologists have recently discovered new ways of getting information about the past chemical state of the oceans, which up until now has been largely a closed book to science. In particular, it has been found that when evaporation of seawater creates rocks made of sea salt (called "evaporites" or "halites"), little drops of ancient seawater get trapped (still in liquid form) in imperfections in the salt crystals. These liquid relics are starting to give us an archive of information about ancient seawater.

8.2.1. Fluid Inclusions

This is some of the most exciting geological work of recent years. In the same way that bubbles of air trapped between falling snowflakes have recorded the composition of long-since-disappeared atmospheres, so too little globules of seawater (known as "fluid inclusions") trapped in salt crystals record the composition of long-since-disappeared oceans. They are "time capsules" of ancient oceans.

Sea-salt formation occurs at the edges of oceans when the heat of the sun evaporates seawater to leave behind sea salt. As salt crystals grow in the more and more concentrated solution (evaporation driving off pure water to produce an increasingly salty brine) the fluid inclusions of partially evaporated seawater get trapped in imperfections in the growing salt crystal (fig. 8.2). These fluid inclusions are wonderful records of past oceans[4] but also suffer from two important disadvantages: (1) they are not of pristine, unaltered seawater, but rather are altered by evaporation (much as a rock pool exposed by the tide for several hours on a sunny day starts to become super-concentrated salt water); (2) they are mostly extremely small, smaller than sand grains, and so the task of measuring their chemical composition cannot be accomplished using conventional methods.

The first of these two problems prevents an easy and straightforward reconstruction of the chemical composition of past seawater from these fluid inclusions. However, much useful information can still be obtained. When present-day seawater is evaporated towards salt, the chemical composition of the ever-more concentrated seawater follows a well-known and repeatable trajectory, called the "evaporation pathway."[5]

The beauty of it is that, if one hundred samples of seawater taken from different oceans around the globe are all evaporated to solid salt, the evaporation pathways of each will be more or less identical. The sequence of chemical compositions that are passed through is the same in each case, give or take only very minor differences. And this is because the proportions of the most abundant elements in seawater (chlorine, sodium, magnesium, sulfur, calcium) vary little with location. They are almost identical in the Pacific and in the Atlantic Ocean, in the Arctic and in the Southern Ocean.

The second disadvantage has been overcome using a range of techniques including ion chromatography, inductively coupled plasma mass spectrometry, and x-ray dispersive systems. These advanced techniques allow analysis of incredibly small fluid inclusions. Some of the techniques are only applicable to

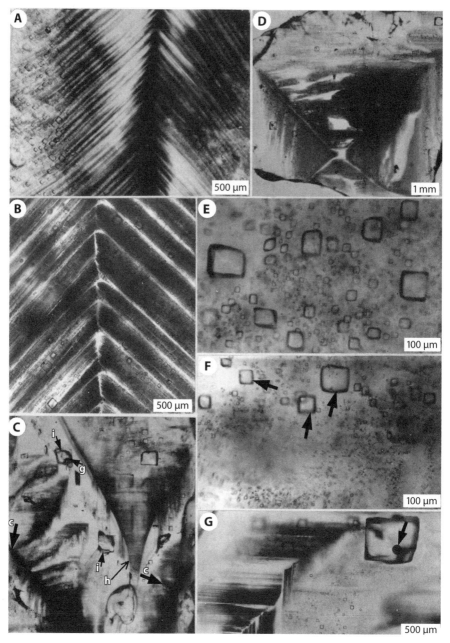

Figure 8.2. Photographs of fluid inclusions in fossil grains of salt from Slovakia, Ukraine, Poland, and Russia. Some grains have a chevron crystalline structure (*A* and *B*). Kovalevich, Peryt, and Petrichenko (1998). © 1998 by The University of Chicago. All rights reserved.

larger fluid inclusions (for example, > 200 μm), but others can analyze inclusions even as small as 20 μm in diameter, smaller than a grain of sand and small enough to fit within the width of a human hair.

The development of these techniques gave scientists the opportunity to peer inside these microscopic globules and measure their chemical make-up. The participants must have experienced an electrifying thrill of discovery when the first measurements of ancient fluid inclusions were made, once they realized that the measurements showed chemical compositions completely "off the beaten track" compared to the evaporation pathway of modern seawater. It is just not possible to arrive at the chemical composition of many of these ancient globules by progressively evaporating modern-day seawater. The ancient fluid inclusions have a chemistry that is distinct from those of any modern fluid inclusion.[6] This leads straightforwardly to an important conclusion: ancient seawater must have been very different. Whereas it has often previously been assumed that seawater chemical composition has been rather constant over geological time, at least as regards the major ions, these measurements have revealed previously hidden variability.[7] The earlier assumption of constancy was simply the application of Occam's razor. It was not based on any positive evidence of constancy but rather on a lack of any evidence of nonconstancy (in fact, on a lack of any evidence at all). Now that we have data, there is evidence of variability.

Although the use of fluid inclusions to reconstruct the chemical composition of pristine (unevaporated) ancient seawater is not without its difficulties,[8] much progress has been made. The key to the approach is that the original seawater, before evaporation, must have included the fluid inclusion chemical composition somewhere along its evaporation pathway. There are other questions about the reliability of individual fluid inclusions, for instance whether they have been subsequently altered by diagenesis (changes taking place during or after rock formation), or whether there were local peculiarities such as nearby freshwater springs providing inputs to the natural salt pans where they formed. However, these questions have been largely sidelined by the observation that fluid inclusions in rocks of the same age but from different parts of the world (for example, USA and western Australia, or Brazil, Congo, and Southeast Asia) all lie on the same evaporation pathway.[9]

8.2.2. Supporting Evidence for Variable Seawater (Mg/Ca)

The fluid inclusion measurements made so far have allowed estimation of past concentrations of a number of elements and ions, including sodium, chlorine, calcium, magnesium, sulfate, potassium, bromine, and lithium. For the rest of this section we will concern ourselves only with magnesium and calcium. This is in part because of biological relevance (the use of calcium by calcifers, in making their shells), and in part because the fluid inclusion history of these two elements has been corroborated by other, independent, measurements.

The fluid inclusion measurements have enabled the reconstruction of the ratio of seawater magnesium concentration to seawater calcium concentration (shown in fig. 8.3) as well as the individual concentrations (fig. 8.4). These reconstructions stretch back through most of the Phanerozoic, that is, most of the last 540 million years. Large changes (concentrations more than doubling or halving) occur between one interval and the next.

The dots in the upper panel of figure 8.3 are independent seawater (Mg/Ca) estimates from a very different method. All calcifying organisms (those that build shells from calcium carbonate) incorporate some fraction of Mg atoms in place of Ca atoms in the crystal lattice of their shells.[10] Some calcifying organisms apparently act to prevent this Mg substitution, presumably because the presence of Mg makes their shell more susceptible to dissolution. Other calcifying organisms, though, appear to incorporate Mg and Ca atoms with no selective screening against Mg atoms. In this latter case we can expect the ratio of Mg and Ca atoms that gets "locked into" the solid $CaCO_3$ to provide a faithful record of the (Mg/Ca) ratio in the seawater at that time. Echinoderms (a group that includes starfish, sea urchins, crinoids, and such) are one set of organisms whose $CaCO_3$ thus provides a record of past (Mg/Ca). The echinoderm-based estimates in the topmost panel of figure 8.3 substantiate the fluid inclusion–derived picture of changing Mg and Ca.[11]

The fossil echinoderm (Mg/Ca) ratios corroborate the fluid inclusion results, as do other independent data, from calcium carbonate veins within oceanic crust.[12] It is clear that there have been major swings in ocean chemical composition. As can be seen in the right-hand side of figure 8.3, the rate of change is particularly rapid approaching the present day.[13]

There is more. The recent work with fluid inclusions and fossil echinoderms confirms earlier suspicions that the world's oceans oscillated between "calcite seas" and "aragonite seas."[14] In synchrony with the oscillating seawater chemistry there has been an oscillation in the nature of the solid material precipitated inorganically from it. Calcite and aragonite are two different polymorphs of calcium carbonate (two forms of crystal sharing the same chemical formula but with different physical structures).[15] Calcite is easier to precipitate than aragonite at low seawater (Mg/Ca), whereas aragonite precipitates more easily at high (Mg/Ca), such as in the present-day ocean.

It was realized some time ago that there is an oscillation in the type (aragonite or calcite) of inorganic calcium carbonate reef cements found buried

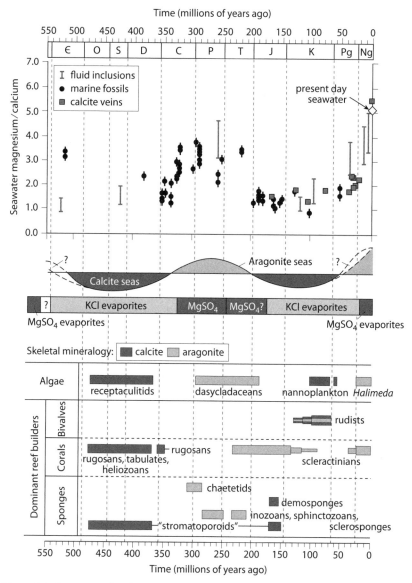

Figure 8.3. Summary of evidence supporting an oscillating (Mg/Ca) ratio (mol/mol) over the last 550 million years. In the upper panel, estimates from fluid inclusions are shown as vertical gray lines, estimates from echinoderm fossils as dots, and estimates from calcite veins as gray squares. Below that, the second panel shows how the chemical composition of evaporites (sea-salt deposits) has varied in synchrony with that of calcium carbonate fossils and calcite veins, leading to the concept of back and forth changes between "calcite seas" and "aragonite seas." The lower panel shows changes in the mineral form of calcium carbonate synthesized by the major sediment producers and reef builders in each period. Adapted from Montanez (2002) and Kerr (2002).

in the geological record. The fact that this record of reef cements agrees with the seawater chemistry reconstruction further strengthens confidence in the patterns.

There are also suggestions that this record of change in the inorganic $CaCO_3$ is mirrored by similar changes in the fossil record of biologically produced $CaCO_3$. The fate of different calcifying taxa may have been tied to the history of seawater (Mg/Ca). When (Mg/Ca) was low, then corals that built their skeletons out of calcite dominated. On the other hand, during aragonite seas such as today (high [Mg/Ca]), natural selection has presumably favored the success of aragonitic corals.[16] Furthermore, it has been proposed that the histories of a whole range of organisms (calcifying phytoplankton, calcifying seaweeds, corals, bivalves, sponges, and others) can all potentially be linked to oscillating seawater (Mg/Ca). As seawater chemistry varied, some $CaCO_3$-shelled creatures found themselves synthesizing the "wrong" form of $CaCO_3$ and became scarce or went extinct (if they were unable to adapt), to be replaced by others using the more appropriate form of $CaCO_3$.[17] Some have proposed that the fortunes of whole taxa have waxed or waned in phase with the changing chemistry of the sea, although a detailed statistical analysis casts doubt on this interpretation.[18] It should also be noted that these changes play out extremely slowly (fig. 8.3), allowing plenty of time for evolution to adapt organisms to cope with them.

8.2.3. Variable Seawater Chemistry and Gaia

Taken together this evidence strongly suggests that the major ion chemistry of the oceans has not been constant, as previously assumed. A "best-guess" reconstruction of past seawater chemistry from the fluid inclusion data calculates fluctuations of up to 300% in calcium (Ca^{2+}) concentration (threefold difference between minimum and maximum concentrations), up to 200% in magnesium (Mg^{2+}) concentration, and up to 300% in sulfate (SO_4^{2-}) concentration. Potassium (K^+), on the other hand, seems to have varied little over the last 550 million years (fig. 8.4). Although the exact amplitudes of variation are somewhat approximate, and although the cause of them is not well understood, it is clear that major fluctuations occurred in some ions.

Therefore, the ocean has not in all respects been a constant and stable environment for the life inhabiting it. While it has remained favorable for calcification of one sort or another,[19] which is obviously good for calcifying organisms as a whole, there have been great swings in chemical composition, to the detriment of some groups and the benefit of others. This evidence does not support the Gaian suggestion of environmental stability.

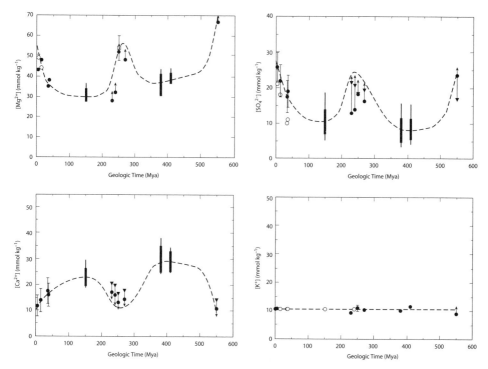

Figure 8.4. Reconstructions of major ion abundances in seawater as they have changed over the course of the Phanerozoic (last 545 Mya). Whereas the concentration of potassium (K^+) seems to have remained fairly constant over time, concentrations of magnesium (Mg^{2+}), calcium (Ca^{2+}), and sulfate (SO_4^{2-}) appear to have been much more variable. The data come mostly from fluid inclusions; the dashed line in each case represents a best estimate of the time history of the concentration. Reprinted from Juske Horita, Heide Zimmermann, and Heinrich D. Holland. Chemical evolution of seawater during the Phanerozoic: Implications from the record of marine evaporites. *Geochimica et Cosmochimica Acta*: 21: 3733–3756,© 2002, with permission from Elsevier.

8.3. Colder and Colder

The next record we will look at is one of temperature. The long-term temperature history of the Earth is not completely known.[20] However, there is now reliable evidence of a long-term overall cooling over the last 100 million years or so. Jim Zachos and colleagues[21] were the first to piece together many individual shorter-term records of $\delta^{18}O$-derived temperature to construct a record spanning the last 65 million years. The graph they produced has now become iconic; the original paper, published in 2001, has now been cited more than 2,500 times. Earlier data has subsequently been joined to yet more records, including some from even further back in Earth history, from the latter part of the Cretaceous.[22] The picture that is revealed is one of a generally cooling climate across the whole of the last 100 million years (fig. 8.5). Although there are some potential sources of error associated with $\delta^{18}O$-derived temperature, it is in fact one of the most robust and longest-used paleo proxies,[23] and the general trend the graph shows also agrees with other proxies for cooling climate and declining atmospheric CO_2 over this interval.[24] We can therefore be confident that the general trend, although not necessarily each and every minor fluctuation in it, is correct. This graph documents the cooling trend that has slowly changed the polar regions from being covered by lush forests (including tropical breadfruit trees) during the Cretaceous (section 5.3.1) to being covered by ice sheets today. As discussed in chapters 4 and 5, the shift to a cooler Earth has had a profound effect on life on Earth.

8.4. Ups and Downs of Greenhouse Gases

The record in this section is also of temperature, but at higher resolution. One of the most outstanding and revealing of scientific advances in Earth system science has been the revolution in the information available to understand glacial-interglacial cycles, with the discovery that bubbles trapped in ancient ice tell us about the composition of ancient atmospheres. The first reliable record of past atmospheric CO_2 from these ice cores was published in 1980, some time after the first Gaia publications in the 1970s. By drilling down through ice sheets and then analyzing the air bubbles at different depths (ages) in the cores, it has been possible to produce a wonderfully direct reconstruction of the history of the Earth's atmosphere over the last 400,000 years or so.[25] These wonderfully detailed records are very relevant to this chapter's quest of evaluating the Gaian claim of environmental stability over time.

If we look at the histories of carbon dioxide and methane (fig. 8.6), two of the most important greenhouse gases after water vapor, we can see that they march together in lockstep. They also march together with the oxygen isotope composition of ancient shells (the $\delta^{18}O$ of foraminifera), which indicates both ancient sea level and Earth surface temperature. Times of high carbon dioxide (CO_2) and methane (CH_4) correspond with high sea levels of warm interglacial times, such as at present; times of low CO_2 and CH_4 correspond with cold times when sea level was low, that is, the ice ages.

These treasure troves of information have stimulated many detailed hypotheses about the workings of the Earth system and continue to yield many insights. Here I will consider only what ice core records say about stability. Both temperature and greenhouse gases undergo large variations in synchrony with glacial-interglacial cycles. Carbon dioxide varies between about 180 (full gla-

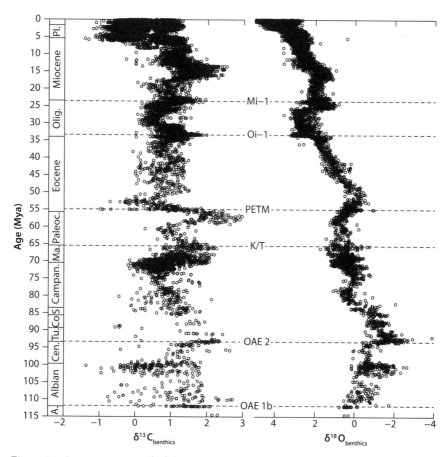

Figure 8.5. A composite record of climate change over the last 110 million years. The record was generated by joining together multiple sedimentary records obtained by drilling down through the seafloor at many different locations. The oxygen isotope composition of the shells of seafloor-inhabiting foraminifera, $\delta^{18}O_{benthics}$, correlates inversely with temperature, both because of a direct effect of water temperature on $\delta^{18}O$ and through an indirect effect associated with the total amount of ice on the planet. The increase over time of $\delta^{18}O_{benthics}$ points to a long-term cooling toward the present day. Mi-1: Miocene glaciation event; Oi-1: Oligocene glaciation event; PETM: Paleocene-Eocene Thermal Maximum; K/T: Creta-ceous/Tertiary boundary; OAE: Oceanic Anoxic Event. Reprinted from Friedrich, Norris, and Erbacher. *Geology*, Vol. 40, No. 2 (2012).

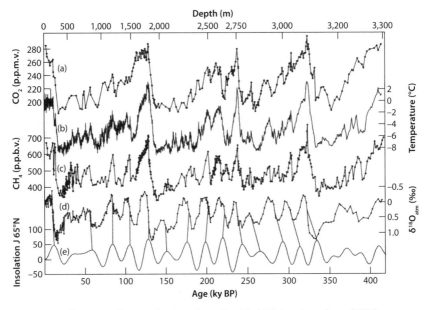

Figure 8.6. Climate and atmospheric carbon dioxide (CO_2) and methane (CH_4) concentrations from a 420,000-year-long ice core collected at Vostok in East Antarctica. Reprinted by permission from Macmillan Publishers Ltd.: JR Petit et al. *Nature*, 399: 429–436. ©1999.

cial) and 280 parts per million (interglacials), methane between about 350 and 700 parts per billion; there is therefore about twice as much methane in the interglacial as in the ice age atmosphere, and about 50% more carbon dioxide.[26] The period of the glacial-interglacial cycles (about 100,000 years) is about 1,000 times shorter than the timescales over which Ca and Mg concentrations change in the ocean (about 100 million years); the reason for this will become apparent in chapter 9.

The first ice-core records, incredible though they were, were of rather coarse temporal resolution. New cores and techniques have led to climate and atmosphere reconstructions of much finer detail. The picture that has emerged is one of erratic, one might almost say capricious, climate system lurches. Whereas the early picture was one of rather smooth descents into glacials and then more rapid climbs out of them, as the frequency of sampling has increased so too has the "spikiness" of the records. There is no "stately progression" into and out of ice ages—instead, climate seems to jump about in a less controlled fashion. The higher-resolution records have allowed us to zoom in on apparently smooth transitions, or on periods of apparently stable climate. This has revealed that the general trends are often punctuated by rapid but temporary shifts to very different climates.[27]

8.5. Erratic Climate Behavior: The Younger Dryas

We will consider just one example from the geologically recent past. About 13,000 years ago the world was emerging from the severe cold of the last ice age. The biota in general, Mesolithic humans included, were enjoying the benefits of a move to warmer times.[28] However, this warming was punctuated by the Younger Dryas period, at which time the Earth (certainly Europe and North America) was pitched back into full glacial cold for more than a thousand years, before climate once more resumed its warming trend. The existence of this frigid period was first identified in the 1930s from the pollen of *Dryas octopetala*, a hardy Arctic tundra flower that flourished at that time in Sweden, according to measurements of pollen abundance in lake sediments. Just as the terrible cold of the ice ages must have been fading to a distant racial memory, it suddenly reappeared to heap suffering on early humans. It can't have been a pleasant time to be alive. At the beginning of the Younger Dryas, temperatures dropped precipitously, by as much as 10°C within less than a century.[29] At the end of the period, following about 1,300 years of extreme cold, temperatures suddenly shot up again by 10°C or so within a few decades.[30] As suddenly and surprisingly as it began,[31] so too it ended. Climate then resumed its gradual trend toward the warmer world we enjoy today.

A temperature shift of 10°C is no small thing. Its severity can best be understood by comparing the impacts of much smaller climate blips of more recent times. During the Medieval Warm Period (circa AD 900 to 1300) Vikings colonized Iceland, Greenland, and North America. But this was followed by the Little Ice Age (circa AD 1500 to 1900), when temperatures dropped by only about 1–2°C. During the transition to this latter, only slightly cooler interval, the Viking settlements in Greenland foundered. During the Little Ice Age itself glaciers advanced in Norway, sea ice became more frequent around Iceland, and ice-skating on Dutch canals became a national pastime.[32] Global warming because of fossil fuel emissions is not predicted to greatly exceed 5°C within the next century, with very large predicted environmental effects. By extension, the impact of the 10°C drop during the Younger Dryas must indeed have been terrible.

It is most likely impossible for us to imagine the full hardships experienced by those of our hunter-gatherer forebears living in mid-latitudes at that time. They were nomadic or lived in caves. They had to glean the means for their subsistence from the natural environment. They had fires but no walled homes, that is, no proper means of escape from the cold. From our perspective within an unusually stable warm interglacial period, and cosseted by all the trappings of an advanced society, it is difficult for us to conceive the misery that this climate shift must have produced for those humans living at the time. Vegetation must have shriveled, food availability dwindled, cold winds pierced them, and many lives must have been blighted or brought to an early end. Numerous whole tribes, hitherto probably thriving and expanding in population, must then have found themselves unwittingly in the wrong place to survive a climate deterioration. They must have fought a desperate but ultimately in many instances an unwinnable battle against vanishing resources and the (literally) perishing cold. The Younger Dryas is one more example of a capricious climate swing, one that must have affected the whole of the biota, not just Mesolithic humans.

Closer inspection of ice cores in recent years has revealed that the climate record is littered with similar, albeit not always so extreme, short-lived climate excursions,[33] leading some ice-core researchers to the conclusion that "change—large, rapid, and global—is more characteristic of the Earth's climate than is stasis."[34]

8.6. Ocean Acidity

Out of the various environmental parameters relevant to life, Lovelock specifically proposed that temperature, atmospheric oxygen, ocean acidity (pH), and ocean salinity have all been kept at hospitable values for life. As discussed in chapter 3, the habitable pH range is extremely broad: for microbes it is about 0 to 12, and for eukaryotes at least 2 to 9.[35] The average pH of the present-day surface ocean is about 8.1. Although the total habitable pH range is very broad, certain classes of organisms are of course much less tolerant. In particular there is considerable concern at the present time about the consequences of ocean acidification (due to increasing CO_2) on organisms that use calcium carbonate as a building material for their shells and skeletons. Perhaps one-fifth of the world's coral reefs have already disappeared due to a combination of factors, including a reduction in surface ocean pH of ~0.1 units so far, with much greater coral losses predicted for the next few centuries if ocean pH drops even further. It seems probable that a fall in pH of ~0.3 units will make seawater too corrosive for corals. Therefore radical action in curbing CO_2 emissions may be required to save corals, even though the maximum predicted effects on pH are much less than a decline of 1.0 pH units, that is, well within the habitable range for the biota as a whole.

Paul Pearson and my Southampton colleague Martin Palmer used a novel and complex technique to try to reconstruct the history of ocean pH. This technique involves measuring the boron chemistry of ancient foraminifera shells. It

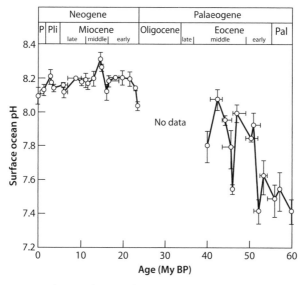

Figure 8.7. Estimate of sea surface pH change over the last 60 million years. Reprinted by permission from Macmillan Publishers Ltd.: P.N. Pearson and M.R. Palmer, *Nature*, 406: 695–699. ©2000.

makes use of the fact that the boron chemistry of seawater (and hence also of foraminifera shells) is heavily influenced by pH.[36] Boron measurements in fossil shells have therefore been used to infer a history of ocean pH (fig. 8.7). This technique is somewhat indirect and depends on a number of assumptions, some of which have been disputed.[37] Nevertheless, it may provide a window into how pH has varied over time, and its broad, long-term trend (increasing over time from lower values 60 Mya) agrees with that derived from an independent line of reasoning.[38] Their results suggest that the hydrogen ion concentration decreased approximately fivefold (pH went up from 7.5 to 8.2)[39] over the last 60 million years. However, their results also suggest that there has been a limit to the variation in ocean pH over this time period—hydrogen ion concentration has not varied by more than a factor of 10 (pH has not varied by more than one unit) during the last 60 million years.

8.7. SNOWBALL OR SLUSHBALL EARTH?

The final example of environmental variability in this chapter is also the most severe. If Lovelock were to have conceived the idea of Gaia today, rather than thirty years ago, would he still have developed it? Or would he have given it up, after due consideration of all that we understand today but didn't then? One area of subsequent research that might have given Lovelock pause for thought is the topic known as "Snowball Earth."[40] During these Snowball Earth events, which took place long ago, between 700 and 580 Mya, it is thought that Earth's thermostat failed several times. On these occasions the Earth apparently entered a planetwide deep freeze from which it didn't recover for many millions of years. This hypothesis has generated its fair share of dispute and rancor, giving rise to one of the more entertaining openings to a piece of scientific writing, by Euan Nisbet: "Snowballs can be fun. But when they have rocks in them, the arguments start. . ."[41]

Did the Earth really resemble a solid-centered snowball back in the past? Did temperature regulation really fail that badly? Was the planet completely covered in ice, pole to pole, including both sea and land at the equator? If so, then how did life survive? The debate still rages on some of these questions, although on some aspects a consensus is emerging.

8.7.1. Glaciation

A first question is whether there was any temperature plunge at all. The evidence for this is, however, pretty clear. Marine sediments typically consist almost exclusively of fine-grained particles, partly made up of the corpses of microscopic phytoplankton that had recently lived and died in the sunlit sea surface above, mixed in with fine-grained sand and silt swept off the land. Today's seafloor is, in most places, a muddy ooze. Rivers do not generally carry pebbles any distance out into the sea, because the pebbles rapidly sink down to the riverbed as soon as the river speed slows to anything more placid than a torrent. Likewise, when sandstorms develop and then (given the right wind conditions) sweep large quantities of dust out to sea, pebbles are not made airborne. Given the seeming impossibility of transporting such pebbles and stones far out to sea, how then can one explain the widespread appearance of large pebbles and stones interspersed in marine sediments from the Late Neoproterozoic (fig. 8.8a, c, d). While volcanic eruptions can throw rocks out to sea, they only do it on a local scale. The only known mechanism for transporting significant quantities of pebbles and rocks far out to sea involves ice. Rocks and pebbles become frozen into the undersurfaces of glaciers as they plow their way across the landscape. Those glaciers that peter out on land (without ever reaching the sea) form a "terminal moraine" as they deposit their rock load in a line across the end of the glacier. Those glaciers that reach the sea, on the other hand, eventually break up to form icebergs that float away into the surrounding ocean. These icebergs slowly melt and, depending on where the icebergs float

Figure 8.8. Evidence for Snowball or Slushball Earth conditions between 700 and 580 Mya: (*a*) diamictite (a rock made up of a heterogeneous mixture of assorted particles and rock fragments of various sizes, including many > 2 mm) above a striated (scratched) rock surface (Bigganjargga, Varangerfjord, north Norway); (*b*) faceted stone with surface striations (Adrar, Mauritania); (*c*) dropstone; and (*d*) diamictite containing dropstones (both from east Fransfontein, northwest Namibia). Reprinted by permission from John Wiley and Sons, from: P.F. Hoffmann and D.P. Schrag, *Terra Nova* 14: 129–155. (2002)

to, may release their rocky debris some time later to fall to the remote seafloor. This phenomenon is known as ice rafting, and the consequent anomalous pebbles in marine sediments are known as dropstones.

The predominant evidence for glaciations in the Late Neoproterozoic is these dropstones (fig. 8.8a, c, d). They are found in Late Neoproterozoic marine sediments in many locations around the world: Australia, China, USA, Africa, and Svalbard in the Arctic.[42] In some cases the amount of material deposited is quite surprising: in Death Valley, for instance, the layer of glacial diamictite[43] (rock fragment debris) is more than two vertical miles high. Other evidence of glaciation includes "striations" (scoring of lines into underlying rock as stones embedded in the bottom of glaciers are dragged across them—figure 8.8a, b—

and fractured rocks. We can draw a confident conclusion that there was severe glaciation at this time.

8.7.2. Extent of Glaciation

A second question is the extent of the glaciation. Was it restricted to higher latitudes, as during the recent ice ages, or was it truly global? Were the Neoproterozoic glaciations really something special, or "just another glaciation"? Did the appearance of the Earth from space closely resemble a flawless peppermint drop or were there instead areas of open ocean and ice-free land throughout. Essentially, was the Earth at that time a Snowball or a Slushball?

The evidence on this point is more equivocal and uncertain. Evidence from the magnetism in rocks suggests that dropstones were being deposited into seas near to the equator. The magnetic field of the Earth is oriented vertically at the poles (the magnetic field "enters" and "leaves" the Earth near to the north and south poles) but horizontally at the equator. As magnetized microscopic mineral grains fall down to settle on the seafloor, they tend to align themselves with the ambient magnetic field. The remnant magnetism of dropstone-containing Neoproterozoic rocks is horizontal in some locations, strongly suggesting they must have formed near to the equator.[44]

You might then ask why, given this evidence, there is any dispute. Why is there any doubt about a panglobal scale to the glaciation? A first reason is that one hypothesis[45] suggests that the orientation of the Earth with respect to the Sun could have been different from its orientation today. Just like Uranus, the Earth may have lain on its side, with its spin axis more nearly parallel to the plane formed by the Earth's orbit around the Sun, rather than more nearly perpendicular, as today. However, recent work[46] suggests that this scenario, interesting though it is, is almost certainly not what actually happened.

Secondly, the same magnetic evidence that suggests that the dropstones were dropped near the equator also shows that these dropstone-containing sediments were laid down over rather long periods of time. They were not laid down instantaneously. The Earth's magnetic field flips (reverses) occasionally, with north becoming south and south north. The frequency of the flips averages about once every few hundred thousand years. Dropstone-containing sediments in Australia record up to seven magnetic reversals, suggesting that the glacial conditions lasted for hundreds of thousands to several million years. On the other hand, however, if the Earth was in a full Snowball all of this time then it is hard to understand how glaciers could keep flowing out to sea and dropping stones into marine sediments. Once the whole ocean became covered in ice, then evaporation would cease, snow would cease to fall on land, and glaciers would cease to flow. The motive power for glaciers comes from the weight of new snow falling on top. No new snow means no more glacial flow means no more drop-stones. It is hard to understand how the geological evidence can be reconciled with oceans that remained completely frozen over.

Finally, if the surface ocean froze over completely then how did photosynthetic organisms survive the period? If the Earth suffered a full Snowball it is not obvious why algae wouldn't have died out completely. It is perhaps possible to conceive of one or two cold-tolerant species surviving on top of the ice, as ice algae do today in Antarctica, although conditions would have been harsh indeed. Microbial communities also thrive today within the concentrated brines contained in the interstices of sea ice. In addition to hardy ice dwellers, it is also theoretically possible that a few nonphotosynthetic strains of life survived in isolated refugia such as at hydrothermal vents or in marine sediments deep below the seafloor[47] and that photosynthesizers and other branches of life reradiated once conditions improved, with the rate of evolution potentially dramatically accelerated by a subsequent supergreenhouse. It is conceivable that all of life went through a narrow bottleneck and then spread out again. The available evidence, however, supports neither of these possibilities; instead, it appears more likely that sunlight-based life was able to continue in some shallow marine environments all the way through these events.[48]

These and many other questions are currently exercising the minds of geologists and paleoclimatologists. Most now agree that the Earth underwent some sort of prolonged Slushball or Snowball, although they might disagree about which of the two. Much of the debate now is not about whether the Earth underwent a cold shock of awesome proportions, but rather about just how extraordinary the shock really was. The consensus is that the glaciations at this time were special, potentially rivaled in severity only by those associated with the Great Oxidation Event (note 16 to chapter 7).

8.7.3. Impact on Evolution

Ironically, even though the Snowball/Slushball Earth events may have taken life perilously close to the brink, there are also suggestions[49] of an energizing effect on evolution. Evolution is sometimes thought to proceed only with difficulty during periods when the environment is rather stable. All ecological niches are occupied and presumably hard to invade. There are good reasons to believe that the ends of snowball events were followed by extreme hothouses due to build-ups of volcanic carbon dioxide in the atmosphere.[50] It is conjectured that once the snowball melted, average conditions jerked suddenly from about $-15°C$ to $+40°C$ in a matter of years.[51] If so, life was really put through the wringer, subjected to extremities of both heat and cold. The icehouse would have wiped out most organisms (clearing most niches), and then the extreme temperature variability may have forced the pace of evolution (note 26 to chapter 4) at the same time as organisms had to adapt to rapidly changing environments. It is possible that the catastrophic conditions had ultimately an invigorating effect on evolution, leading to rapid radiations of new forms.

All this is largely speculation, but what is not in doubt is that the earliest definitive fossil evidence of multicellular animal life dates from about 580 Mya, during the second half of the Ediacaran period, some millions of years after the end of the last Snowball Earth event (some earlier Australian rocks, predating the last Snowball Earth glaciation, also contain possible fossil animals, but the evidence is less clear cut[52]). If it is true that these extreme cold episodes were in fact beneficial to evolution, the benefit was fortuitous and coincidental. If Snowball Earths did indeed contribute to the radiation of complex animal life, it was an accidental by-product of collapses in Earth temperature regulation that may have come close to sterilizing the Earth.

8.8. Conclusions

The goal of this chapter has been to scrutinize the Gaian claim that conditions for life remained fairly constant through Earth's history, ever since life got going. In light of the knowledge gained during the last few decades, this claim now looks hard to defend. Our knowledge of Earth history has undergone several exciting revolutions in the last thirty years, some of which the scientific community is still in the process of assimilating. These revolutions are highlighting instability and nonconstancy of the Earth's environment.

There is no refuting the ice-core evidence that has given us such a detailed picture of how the serene and stately progression (as once believed) of ice ages and interglacials is in fact rudely and frequently interrupted by large and sudden swings in temperature, CO_2, and methane. In the last few decades, researchers have had to adjust their views to incorporate new evidence of surprisingly abrupt climate change such as took place at the Younger Dryas. Detailed, slice-by-slice analysis of these ice cores is making us frighteningly aware of past episodes of rapid climate change on Earth, of which we were hitherto ignorant. It seems that every time the development of a new method allows analysis of environmental change at a finer resolution, we find evidence of more rapid change. We can only expect more variability to be exposed in the future as more data are collected and as techniques and temporal resolution improve.

These advances in our understanding over the last thirty years contribute to a much more long-term transformation in the geological consensus. Earlier views emphasized relative stasis (slow change) even more strongly than does

Gaia. For instance, in Darwin's time views of general environmental constancy held sway among scientists; geological processes were assumed to all take place in a slow and gradual fashion, influenced by the concept of uniformitarianism of James Hutton and Charles Lyell. Following Lyell's publication of *Principles of Geology* in the 1830s, absolutely all geological phenomena were assumed to be explained by the ages-long application of the same processes as occur in the present-day world. Catastrophism was frowned upon. Explanations invoking eons of waves lapping on shores were fine; explanations invoking mega-tsunamis were considered unacceptable. Louis Agassiz and others thus faced great initial opposition when they first promoted the idea that ice sheets had once swept across large parts of Europe and North America during earlier ice ages. But eventually the overwhelming accumulation of evidence convinced even the doubters. There was a similar struggle for acceptance of the idea that an asteroid collision ended the Cretaceous and killed off the dinosaurs, with initial resistance eventually overcome by two key discoveries: the ubiquity of asteroid-derived iridium in sediments from that date, and the discovery of the impact crater on the Yucatán Peninsula. Nowadays, geologists still prefer uniformitarianism as the default explanation of phenomena, but there is increased openness to explanations invoking catastrophic events.

The Gaia hypothesis proposes that life has had a hand on the tiller of Earth climate, ensuring stable equable climates throughout Earth history. The picture revealed in this chapter is by contrast rather different. The data do not point to a constant environment, or to a cozy and hospitable one. In addition to the overall trend toward ever-icier climates over the last 100 million years, there is also compelling evidence that the tiller has on other occasions allowed climate to drift into dangerous states threatening to completely extinguish all life on Earth. That some life survived through Snowball/Slushball Earth episodes may owe as much to volcanic CO_2 and to the difficulty of freezing the entire deep ocean solid as to any biological control providing a steadying hand on the tiller.

The more we discover about our planetary history, the more fascinating it becomes. Previous notions of a quiescent ocean, with a constant chemistry immune to the large swings seen for atmospheric composition, are currently also in the process of being overturned. There have been large, albeit very much slower, fluctuations in ocean chemistry. The acidity of the ocean has also most likely varied considerably over the last 60 million years.

All this adds interest to scientific investigations of the past. If all were completely constant then the work would rapidly become tedious. But at the same time it weakens some of the support for Gaia. We can safely conclude that the assertion that Gaia has helped to keep the Earth environment stable is not supported, because the environment has not been all that stable.

All variability is relative, however, and while conditions have fluctuated it is also important to recognize that they never fluctuated so far as to transgress the habitability constraints for life as a whole. Despite the fluctuations life has, somehow, remarkably, survived through all of the geological ages. Therefore, while remaining cognizant of all the environmental variability, we must also appreciate this continued habitability. Throughout an immensity of geological time beyond our capacity to truly conceive of, throughout more than two billion years, not once have all the nooks and crannies on Earth become simultaneously uninhabitable. As described in the next chapter, this is even more surprising because fatal shifts in climate could have occurred quite rapidly. Molecular phylogenies and the dates of speciation events deduced from molecular clocks suggest that life has continued in an unbroken thread across all this vast expanse of time.[53] Possible explanations for this intriguing and important fact will be explored in the next chapter. Whereas this chapter has focused heavily on data, the next will be more theoretical, looking at how we can extract a correct interpretation from these two apparently discordant facts: long persistence of life despite a highly variable environment.

Chapter 9

THE PUZZLE OF LIFE'S LONG PERSISTENCE

THE PREVIOUS CHAPTER examined the claim that the planetary environment has been particularly stable over time. Many instances of contrary evidence were found. But life has nonetheless survived, even if it may have occasionally (for example, Snowball Earths) been a close-run thing. In this chapter I will now describe why this persistence of life over such an immensity of time is, from one point of view, extremely puzzling, but from another, no surprise at all.

As part of describing one view, I will explain why atmospheric CO_2 is susceptible to rapid change. As a consequence, the Earth could, in theory, quite easily have shifted to a state of freezing cold or of boiling heat within only a short interval of geologic time. In either case, if the shift was extreme enough, we might expect life to have been completely and irrevocably extinguished. That such a shift never took place, that all life never perished even once over such an immensity of time, seems an improbable outcome.

Climate failure could easily have happened but never did, leading to interest in the possibility of stabilizing mechanisms that prevented any fatal shifts in climate. According to the Gaia hypothesis, life's involvement contributed to stabilization and kept conditions comfortable. But this is not the only way to explain the lack of any climate failure. It may not be quite so simple.

The second view is based on the so-called anthropic principle. In contrast, it cautions that, given that we are here, life must necessarily have survived. Essentially, if we take as our starting point that intelligent life did evolve on this planet, then the absence of preceding fatal climate swings becomes totally unsurprising.

In this chapter these two separate viewpoints will be explained, then used as a foundation for considering what the fact of long-term life persistence means for the evaluation of Gaia.

9.1. RESIDENCE TIMES AND STABILITY

In order to discuss how easily atmospheric CO_2 can vary, we need first to become acquainted with two important preliminary concepts: reservoir sizes[1] and residence times.[2] The relevance of these concepts will become clear as the chapter progresses.

9.1.1. Long Residence Times

Some "reservoirs" have very long characteristic residence times, and as a result the associated environmental properties are resistant to rapid change. One example is the amount of heat stored in the ocean, with the associated property being water temperature. Water has a high heat capacity (the volumetric heat capacity of seawater is about 4 joules cm^{-3} K^{-1}, compared to about 0.001 joules cm^{-3} K^{-1} for air). As a result, more than three orders of magnitude more energy input is required to raise the temperature of a cubic meter of seawater than a cubic meter of air. Furthermore, there is a lot of water in the ocean. This leads to the total heat capacity of the whole ocean being also about 1,000 times that of the whole atmosphere. And this in turn leads to an important property of the Earth system with regard to how it responds to global warming—namely, that this thermal inertia of the oceans induces a time lag such that even if anthropogenic CO_2 emissions are reduced sufficiently to stop atmospheric CO_2 from rising any further, the oceans and Earth as a whole will still carry on warming for several decades.[3] The increase in air temperature that has taken place so far would have been greater if not for ocean thermal inertia. More than 80% of the extra heat retained up to now by the stronger greenhouse has gone toward warming up the ocean and currently resides there. As ocean and air temperatures slowly catch up with CO_2 change, only then will the full extent of the warming due to fossil fuel CO_2 be felt.

A second example of large reservoirs (in the sense that the word is used here) leading to impossibility of rapid change is atmospheric oxygen. There are about 37×10^{18} moles of oxygen (O_2) in the planet's atmosphere.[4] Just less than 21% of the atmosphere is oxygen (the rest being mostly dinitrogen, 78%, and argon, 1%). As discussed earlier in chapter 6, every year about 0.02% of that atmospheric oxygen (8.4×10^{15} moles O_2) is lost due to a combination of (1) respiration by animals and plants (by far the largest flux), (2) oxidation of reduced gases emitted by volcanoes, (3) oxidation of biogenic methane, and (4) oxidation of reduced rocks that become exposed at the Earth's surface. Given these losses, how much can we depend on the continued availability of oxygen that we need for our survival? Is there any worry that we will so damage the planet as to imperil our own survival through oxygen depletion? A frequently raised concern is that by cutting down rain forests such as the Amazon ("the lungs of the planet"?) we will asphyxiate ourselves. Fortunately the reality is much less alarming.

The oxygen losses are normally counteracted by the large source of oxygen from photosynthesis. However, even if all plant life were somehow to die off, simple calculation predicts that it would still take at least 4,400 years (37×10^{18} / 8.4×10^{15}) to remove all oxygen, or at least 440 years to remove even one-tenth of the current amount of oxygen in the atmosphere. Because there is such a large quantity of oxygen in the atmosphere, it is simply not possible for us to greatly alter its concentration in just a few centuries.[5] Large volcanic eruptions likewise make insignificant dents in atmospheric oxygen. If you've ever worried about whether there will continue to be enough oxygen to breathe if we destroy the rain forests, then you can set your mind at rest. Whatever we do to the planet, even under the most extreme scenarios, there will be no palpable shortage of oxygen for us, for our grandchildren, or for our grandchildren's grandchildren.

Calculations such as these show us that, for some variables, we should not be surprised to see stability over timescales of thousands or even millions of years. In fact, in some cases instability is impossible, certainly on timescales anywhere near that of a human lifetime, because no fluxes exist that are capable of significantly affecting these tremendous reservoirs in such a short time. Another example would be the average altitude of a continent, which we confidently expect not to change greatly within human lifetimes, because of the very slow action of all processes tending to denude continents or build up mountain ranges.

9.1.2. Short Residence Times

On the other hand, for other variables stability is less easily accounted for. Some environmental variables, such as water level in a river for instance, have short residence times (in this case because input/output fluxes, such as the rate of rainfall across the catchment, can be very large relative to the volume of the "reservoir," the river) and hence are susceptible to rapid changes such as floods.

Even some global environmental variables can be susceptible to changes within years or decades, although not usually over days or weeks as for river levels. The Earth's "natural" (preindustrial, that is, before ~AD 1750) atmosphere contained about 0.028% CO_2 and 21% O_2. There was thus approximately one thousand times more oxygen than carbon dioxide in the natural atmosphere. If all else is equal, it is much easier to perturb a small than a large reservoir (think of how long it takes to fill a small cup versus a large bathtub from the same tap). And in this case the major sources and sinks affecting both CO_2 and O_2 (photosynthesis, respiration, fossil fuel burning) are similar in magnitude, because they have more or less equal and opposite effects on CO_2 and O_2 (effects are in a ~1:−1 ratio). Therefore the residence time of atmospheric CO_2 is about 1,000 times shorter than that of atmospheric O_2. It is possible to double or totally deplete carbon dioxide in about 1/1000th of the time that it takes for oxygen. According to this simplistic calculation the residence times of oxygen and carbon dioxide in the atmosphere, with respect to photosynthesis and respiration, are about 4,000 years and 4 years, respectively.

Stability because of enormous inertia would seem therefore not to be a feasible explanation for atmospheric CO_2. This picture is a simplification of a more complex reality,[6] and in actuality residence times for both CO_2 and O_2 are a great deal longer than just calculated, in part because of continuous gas exchange between the ocean and atmosphere such that they are not really separate reservoirs. Nevertheless, the overall point remains: residence times give us insights into susceptibility to change, and atmospheric CO_2 is much more susceptible to rapid change than is atmospheric O_2.

We have already seen in the previous chapter how surface temperature on Greenland changed by about 10°C within only a few years, at the beginning and end of the Younger Dryas. Although atmospheric CO_2 can change rapidly, it cannot change quite as rapidly as that. It appears that the Younger Dryas cooling was driven by changes in the ocean circulation rather than by global atmospheric CO_2 change. Research described in the previous chapter (section 8.5) suggests that atmospheric CO_2 was indeed somewhat lower during most of the Younger Dryas, but that the triggers for the large Greenland temperature changes at the beginning and end may have had more to do with changes in northward heat flow in the Atlantic Ocean. The 10°C change was most likely regional in nature and caused by heat redistribution instead of global in nature caused by atmospheric CO_2 decline.

The preceding discussion gives an example of the sorts of calculations that Earth system scientists carry out all the time as part of their daily work. Residence times are an important tool in Earth system calculations. They also give an insight into the expected stability of Earth's climate.

9.2. A Guardian of Global Climate?

Equipped with the concepts of reservoirs and residence times, we can now more thoroughly consider the stability of atmospheric CO_2. This leads us to the first of the two views mentioned at the start of the chapter, that continued habitability requires stabilizing feedbacks. In this section I will just describe this viewpoint, saving any evaluation until later.

In the 1990s two well-known biogeochemists, Robert Berner (Yale University) and Ken Caldeira (Stanford University), used residence time calculations to argue for the presence of some negative feedback process that regulates atmospheric CO_2. Without it, they argued, it was simply too improbable that the Earth could have remained continuously habitable for life (temperatures on Earth never less than freezing or hotter than boiling everywhere) over many millions and even billions of years.[7] The residence time of CO_2 in the (ocean + atmosphere) system is ~400,000 years (as I said, it is very much longer than the four years mentioned earlier), and so a 25% shortfall of influxes compared to outfluxes would completely empty the (ocean + atmosphere) system of carbon in ~1.6 million years. There were thus about six chances for CO_2 to become completely exhausted during the last ten million years alone. If each outcome is independent and randomly determined by a coin flip (assuming 50% odds each time of such an imbalance occurring) then the chances of a satisfactory outcome on each and every occasion six times in a row are quite low (about one in sixty-four). And they rapidly get worse as the time period is extended further, decreasing to only one in 6×10^{18} over 100 million years. Because CO_2 exhaustion has evidently not happened in the last ~600 million years (no evidence of later snowballs[8]), Berner and Caldeira argue that some process must have kept kicking in at low CO_2 to safeguard against its exhaustion, and at high CO_2 to prevent accumulation to excess.

Berner and Caldeira take their argument one step further. They argue that the absence, over such a long period of time, of any excursions into fatal climates implies the presence of a failsafe protective mechanism such that a corrective change was bound to kick in every time CO_2 levels started falling too low or rising too high. So not only did restorative changes happen to occur every time, but they were bound to. This may seem like a subtle distinction, but the two arguments are in fact rather different. It is all about how much is attributed to luck and how much to mechanism.

We have seen that long-uninterrupted habitability can't all be attributed to inertia of a large reservoir because the inertia of the CO_2 reservoir is in fact not all that great. So this leaves us to choose between luck and stabilizing feedbacks, or some combination of the two. Berner and Caldeira put it all down to stabilizing feedbacks and attribute none of it to luck. Their argument allocates no role to coincidence because, they believe, the coincidences involved would have to have been just too large.

We can make an analogy with the end-of-month balance of a bank account, one that is observed to stay fairly constant over many decades despite large incomings and outgoings every month. Without knowing anything about the account, we might briefly entertain the idea that this could just be down to luck. Much more likely of course, in this example, is that it is down to the account holder spending more when the money is available and taking corrective actions (for example, trimming outgoings or working overtime) once the balance gets too low.

9.3. Or Just Plain Lucky? The Anthropic Principle

We now come to the second of the two viewpoints mentioned at the start of the chapter. It is possible to take a radically different perspective on continued habitability. The change in perspective does not involve questioning the numbers or the math Berner and Caldeira employed, but does involve a very different philosophical approach. This philosophical approach is one that has featured prominently in discussions of how the universe came to be the way it is,[9] but has not featured so prominently, up to now, in discussions of how the planet came to remain habitable for so long.

This alternative perspective relates to the fact that the planet in question (Earth) is special and distinctive in one particular regard—it has given rise to intelligent life. It is far from being a randomly selected planet. Our ability to talk knowledgeably about how planets in general behave after they have gained life is severely limited because we only know about a small sample size of such planets—to be precise, just one. And it so happens that our sole specimen is bound, a priori, because it is of course our own planet, to be a planet that has been habitable for a considerable length of time. Therefore even our sole data point is biased in the sense of not having been selected at random. Even this one data point is statistically compromised.

This second viewpoint is based on the Weak Anthropic Principle. It is given special attention in this chapter because many of you reading this book will be unfamiliar with it, and also because it is potentially of great importance both to the subject of this book and also more generally to our understanding of Earth system history. I will now explain it and review arguments that have been advanced both for and against it.

9.3.1. The Weak Anthropic Principle

There is a special class of facts that, logically, it is impossible for us ever to observe. We cannot observe that which refutes our existence. If such a fact is truly observed then the reasoning based on it must be mistaken (so that it does not actually preclude our existence). Either that or else the observation itself must be mistaken. This special class of facts is therefore distinct from all others. For example, if you see a death certificate stating that your great-great-grandmother died at age five, you can be quite sure that either the document is erroneous or else the person in question wasn't actually your great-great-grandmother. This line of reasoning is known as the weak anthropic principle, also stated as "conditions that are observed in the universe must allow the observer to exist." This is a simple enough concept, but one with extremely wide-ranging ramifications. The connection to this book arises because any fact that would lead incontrovertibly to the conclusion that "the Earth was recently sterile" belongs to this special class of unobservable facts.

As Andy Watson[10] has previously argued, our ability to consider the question of Gaia, and to contemplate the reasons why our planet has remained continuously habitable for life, depends on the evolution of intelligent life. Microbes cannot ponder planets, and so discussions of planetary habitability can only follow after the evolution of more complex life capable of rational inquiry. Such an evolutionary process takes time. It presumably requires many millions or billions of years of incremental evolutionary advances, each building on top of what has come before. This ongoing progression culminating in intelligent life is possible only if the planet remains continuously habitable.[11] So when it comes to questions about planetary habitability, we are not independent observers[12]; instead, we are observers whose very existence predetermines the answer to the question we are asking.

According to this argument, continued suitability of Earth for life (extremely long durations of consistently equable temperatures and other factors) may well have been a largely fortuitous rather than a regulated affair. Perhaps the chances of conditions on a planet like Earth remaining habitable for so long are 1,000,000,000 to 1 against. But if so, then logically we can only have arisen on the 1, and not on any of the other 999,999,999, and therefore we can hardly judge the odds objectively from analyzing just the history of our own planet.

9.3.2. Blind Cyclists and Other Self-Selecting Samples

One thought experiment that is used to explain the weak anthropic principle is that of 10,000 cyclists each asked to cross the city during rush hour, and each wearing a blindfold. Imagine that an amiable organizer pats each on the back before they start, helps them on with their blindfold, and reassuringly tells them not to worry: "It'll be fine." Nevertheless, the carnage is terrible and only one of them makes it through unharmed. The organizer comes up to the survivor afterward and congratulates him: "There, I told you it wouldn't be so bad, didn't I!" If the single survivor knew nothing of the 9,999 other participants (supposing him/herself to have been the only participant, or possibly one out of many that were successful) then they would most likely develop a skewed perspective of the ease of negotiating rush-hour traffic while blindfolded. Has the environmental history of Earth been more like that of the lucky blindfolded cyclist, or was it always likely to be a trouble-free journey?

If the argument from the weak anthropic principle is correct, then our position as privileged end products of a long and uninterrupted evolutionary process argues against a simple acceptance of the Berner and Caldeira argument. We know that our planet survived, but it necessarily must have done so in order for us to be contemplating these questions. What we don't know is how many failed. For all we know, the universe is littered with millions upon millions of failed planets that were all once equable and habitable but are now no longer. For instance, in our own solar system Venus is now too hot to be habitable, but its climate may have been similar to Earth's early on, before it succumbed to a runaway greenhouse effect. And there is much evidence suggesting that liquid water once flowed on Mars's surface, before it lost its atmosphere and became too cold. If some sort of similar climate disaster had befallen Earth (if the Earth had, even just once during the last 500 million years, been sterilized completely), then we would not be here to ponder these questions.

As with trying to evaluate whether it is a good idea to do the lottery by interviewing only past winners, we are not able to perform an unbiased analysis. The key to this issue is the ability to get access to an unbiased dataset from which to draw conclusions. Historians have to cope with just such a problem, encapsulated by the phrase "history is written by the winners." Whenever one country or nation or tribe or sect conquers another, it is typically the conquerors who control how the conflict is subsequently portrayed. The Bayeux Tapestry is a notable example. Just as presidents and kings can have problems getting untainted advice (becoming surrounded with sycophants who don't usually imperil their own livelihoods by giving brutally honest advice), so too there is a problem here with a self-selecting sample leading to a biased perspective. In the same way that completely candid advisors may not be employed for long, so too planets that become uninhabitable don't go on to generate intelligent and curious observers (fig. 9.1).

9.3.3. Family Trees, Luck, and Health

A close parallel lies in the incredible improbability of our own individual existence. If we stop and reflect upon it, it is apparent that each and every one of us has a "perfect" set of ancestors.[13] "Perfect" here is meant in the sense that all of our ancestors managed both to survive to adulthood and to subsequently succeed in reproducing. And this is all the more remarkable given that, as was noted in chapter 2, many more young are produced in all species than can normally survive to adulthood. When we look at the natural world, including at primitive

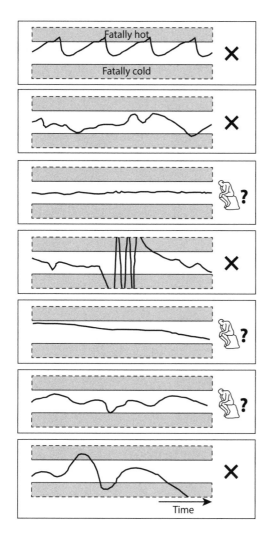

Figure 9.1. The weak anthropic principle. Due to the time required for evolution, the home planet of an intelligent observer cannot recently have been sterile. Therefore, intelligent observers cannot look back at unsuccessful planetary regulation.

human tribes not using modern medicine, large proportions of the very young die before reaching sexual maturity. Despite this, no such failures occurred anywhere in our bloodlines. Not even once. Every single one of our forebears was a winner, in the reproductive sense. This is necessarily true for every single one of our ancestors, including all of our human and protohuman ancestors, all of our mammalian ancestors, and, stretching even further back in evolutionary time, every one of our premammal ancestors. We are products of a long and undiluted story of flawless success, stretching back in an unbroken chain of millions of generations to the dawn of life. The owners of thoroughbred racehorses and pedigree dogs recite their animals' ancestries with pride. But in fact we need look no further than the ancestries of each and every person alive for incredible records of bloodlines stocked full of successful progenitors.

Among our ancestors not a single one was barren and infertile (up until the invention of IVF at least) or a life-long virgin. Not one was exclusively homosexual. Excluding cases of rape, all were able to attract mates and none were so disinclined as to never give sex a try. When epidemics of typhoid, smallpox, influenza, and plague ripped through ancient populations, our ancestors were always among the (sometimes small) band of privileged survivors, or at least only succumbed after having already produced children. We are descended exclusively from the most hardy and the most disease resistant. Our bloodlines contain only survivors and never victims.

When starvation decimated populations, our ancestors made it through to sexual maturity even at times when hardly any of their fellow tribe members, villagers, or townsfolk did. When temperatures plunged during the Younger Dryas and babies and children in some regions were exposed to terrible extremes of cold, our ancestors were among the fortunate ones possessing the robust constitutions needed to survive. On occasions when one tribe massacred another during prehistoric times, our ancestors may have been among the murderers but were most certainly not among the murdered. World Wars I and II carried off millions of men, but not one of those was our direct ancestor unless by that time in their lives they had already produced the next generation in their (and our) family trees.[14] Not a single one of our ancestors succumbed during youth to the 1918 flu epidemic.

The a priori odds against our bloodlines being such paragons of unbroken success are astronomical.[15] Each person alive is the upshot of a fantastically improbable sequence of good genes and good luck. But of course it cannot be any other way, because each and every evolutionary dead end (each and every animal and person that died before puberty, for instance) gave rise to no descendants and therefore no one today can be descended from them.

I have pursued this digression at considerable length due to the importance of this alternative perspective offered by the weak anthropic principle. The preceding discussion helpfully illuminates the different perspectives on probability that arise from (1) looking forward through time to an unknown future, compared to (2) looking backward through time from a known present. If we imagine looking forward from the time at which life first began on Earth, to calculate the odds that each and every single organism through time that turned out to be an ancestor of one particular person, let's say Charles Darwin, would in fact survive and reproduce successfully, then of course the odds are ridiculously low. But if we adopt the different perspective of looking backward, secure this time in the knowledge that Charles Darwin did in fact exist, to calculate the odds that so too did all his ancestors, then in this case the probabilities are all of course 100%.

It can be proposed that the logic is identical for the development of planet Earth. No matter how appallingly poor the odds of continued habitability might be when looking forward, from a backward perspective of course the odds are completely different. Given our privileged viewpoint, we cannot hope for an unbiased perspective. The probability of an observer's home planet having been habitable for a long time is 1. Our sense of wonder at the continued survival of life throughout the extraordinarily long reaches of geological time should, according to this line of reasoning, be tempered by the realization that, given that it is our own planet we are talking about, it could not possibly have been otherwise.

9.3.4. Suggested Shortcomings of the Anthropic Principle

The weak anthropic principle is considered by many to be useful in explaining why our planet, solar system, galaxy, and universe are all currently suitable for life, and why our planet has remained so for billions of years. Many think it highly pertinent. But by no means everyone. The anthropic principle has certainly not met with universal favor in the past, and is by no means universally favored now. Several major criticisms have been leveled at it, which I will now describe.

Firstly, it has been pointed out with good reason that the name itself is rather misleading. The term *anthropic principle* is clearly a misnomer. It is not restricted to humans, as "anthro-" implies, but instead, hypothetically at least, applies to any intelligent beings capable of pondering the mystery of their existence. The principle is concerned equally with all intelligent observers, whether human or not. A better name would be "observer selection effects."

Secondly, it has been attacked for its associations with Intelligent Design. Stephen Jay Gould appeared to attack the anthropic principle in his essay "Mind and Supermind,"[16] but in fact it is not the anthropic principle per se that he disliked. What Gould actually took objection to is (what we now know as) Intelligent Design, and the use of the anthropic principle as an argument for this. It is this one specific use of the anthropic principle that he challenged; that is to say, he disliked the use of the observation that the Earth and the universe appear to be "just right" for producing sentient life (more of which below) as a justification for arguing that there must therefore be an intelligent designer behind it all. Gould improperly conflated two arguments (the anthropic principle and Intelligent Design) and thereby condemned the anthropic principle by association. But the anthropic principle stands alone as a separate argument that need not be associated with Intelligent Design. For instance, it is used here without any such connection. In this case, therefore, Gould's criticisms do not apply.

More contentious versions of the anthropic principle, some unsurprisingly labeled the Strong Anthropic Principle, have also been proposed. Various criticisms have been raised against these stronger variants of the anthropic principle. Again, however, because only the less controversial weak anthropic principle is being employed here, these criticisms are not relevant.

Finally, and most seriously, the anthropic principle has often been criticized for its lack of testability.[17] Some have even called it nonscientific. Explaining a phenomenon in this way has been labeled a cop-out, a "just-so" explanation that immediately puts the argument beyond the reach of any further scientific investigation. Attributing the cause of a particular phenomenon to pure luck rather than to mechanism is all very well, but such an assertion is impossible or at least very difficult to test, so the criticism goes. For instance, anthropic explanations for the suitability of the physical constants in the universe attract criticism because they are obviously untestable (we cannot observe universes with different values of the fundamental constants, and we do not even know if any such exist).

This general criticism of the anthropic principle obviously has some foundation. On the other hand, however, one can question whether the difficulty in testing a hypothesis has any bearing on the likelihood of its being correct. As a general point, it is obviously always a mistake to confuse what we might like to be true with a dispassionate assessment of what actually is true. We should try to keep our personal preferences and desires strictly separate from our attempted objective assessments. It should always raise suspicion when the outcome that someone argues for, ostensibly on a rational basis, in fact coincides rather precisely with the outcome that is most desirable and convenient for them. Likewise, it also pays to be a little suspicious of your own reasoning when the upshot dovetails neatly with what you want to do anyway. It is common knowledge that when trying to be truly objective and impartial in making a judgment about the truth of a proposition, wants and desires and any considerations of expediency should always be put firmly to one side.

And it seems to me that it is no different in this case. If an explanation involving the anthropic principle is inconvenient, I suggest that should be neither here nor there in terms of assessing whether it is admissible as the actual explanation of the phenomenon. Just because it does not yield easily testable predictions, that should not rule it out. It is clearly possible (the aforementioned lottery winners, for example), for luck to be the true and correct explanation of a phenomenon. We should therefore concern ourselves only with whether or not a given hypothesis is a plausible explanation of the facts.

By analogy, if doctors adopted a similar standpoint, they would forever bar themselves from making diagnoses of depression, because there are typically no clear physical symptoms or manifestations that can be searched for in order to prove the accuracy of such a diagnosis. A diagnosis of depression is not completely testable, but nowadays we nevertheless accept that it is in many cases the correct diagnosis. Similarly, the difficulty in proving or disproving an anthropic explanation does not by itself make it an inadmissible explanation.

9.4. PROSPECTS FOR TESTING THE ANTHROPIC PRINCIPLE AS IT RELATES TO CONTINUED EARTH HABITABILITY

Although there is a general difficulty in testing explanations based on the anthropic principle, with regard to Earth history it is somewhat easier than it is for anthropic explanations of other phenomena. Whereas we are completely unable to inspect alternative universes, if indeed there are any, we do have a limited capacity to inspect other planets. We also have information about Earth's own history. It is encouraging that, in stark contrast to most uses of the anthropic principle, an anthropic explanation for continued Earth habitability is not doomed to be forever beyond the scope of testing. Some ways in which it might eventually be tested are as follows.

9.4.1. Remote Sensing of Extrasolar Planets

The answer to the question of whether planets, once inhabited, necessarily develop stabilizing mechanisms (whether "once inhabited" implies "always inhabited," because life is bound to take over the controls of the environment) would be most satisfactorily answered by comparison of the fates of a large number of planets that had at some time harbored life. It would be wonderful to know how long life had persisted on a large sample of planets, how many times regulation had eventually failed, how stable the environment had been in the interim, how the stabilities of large collections of life-bearing and uninhabited planets compared, and so on. We could then easily assess whether the advent of life tends to endow planets with unusual environmental stability.

Unfortunately, however, such information (the biological and geological histories of many planets) is far beyond our reach for the foreseeable future. The distances to even neighboring star systems (the closest, Alpha Centauri, is 4.37 light years away) exclude for the time being any possibility of sending even an unmanned spacecraft and getting a reply back. Although directly visiting planets outside the solar system is no more than a distant hope, our ability to find and inspect them remotely is rapidly improving.

Astronomers have in recent years made great strides in using telescopes to detect the presence of planets circling other suns (so-called extrasolar planets). At the time of writing, this is only just becoming possible for planets the size of Earth. But better instruments, including on the recently launched Kepler observatory, will in the near future allow Earth-sized planets to be detectable in greater number.[18] What's more, we can look forward to a time, probably not very many years distant, when telescopes will be capable of sensing the atmospheres of Earth-sized extrasolar planets, as Lovelock suggested years ago for Mars (chapter 1). From the precise composition of the spectrum of radiation emitted from those planets, or from observing how the color of starlight changes when viewed through their intervening atmospheres, we will gain insight into the chemistries of their atmospheres.[19] It should become possible to detect the presence or absence of oxygen and other gases. We will therefore be able to pick up clues as to whether or not they support oxygenic life,[20] even while unable to visit them in person or with probes. As instruments continue to improve we can hope that this approach will eventually provide much interesting information. And the information collected will allow us to test the plausibility of an anthropic explanation for continued habitability.

For example, it would be hard to continue to credit the anthropic principle if evidence is generated to show that out of a large number of Earth-sized planets within the Goldilocks Zones[21] of their parent stars, a reasonable percentage contain abundant oxygen in their atmospheres, which has not got there through the hydrogen escape mechanism discussed in chapter 7. After the Great Oxidation Event, it took nearly another two billion years before atmospheric oxygen rose all the way to its current high level. High oxygen levels on another planet might therefore be taken as evidence that life has long survived there. Conversely, if abundant oxygen is not sensed on any of a large number of Earth-sized planets, then this would strengthen the argument for the anthropic principle being the correct explanation. It would also be interesting if significant quantities of both oxygen and methane were detected to be simultaneously present in a distant atmosphere (chapter 6). As we learn more about other planets beyond the solar system, we will be better placed to judge if Earth is special.

9.4.2. Mars

Even without studying planets outside the solar system, we are not in all respects limited to our sample of just one. It is interesting that research strongly

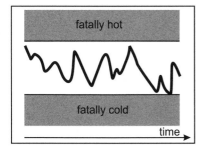 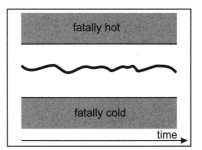

Figure 9.2. Cartoon showing contrasting expected histories of Earth's temperature, depending on whether the continued habitability was mainly due to: (*a*) blind luck (anthropic principle), or (*b*) habitability maintenance, i.e., active and effective stabilization to keep temperature at comfortable values. Axes are deliberately devoid of units. The range of habitable temperatures is shown here as constant through time. However, for sentient life to develop there must have been some progression in biological complexity. This implies that the environmental requirements became more stringent toward the present day, or in other words that the permissible temperature range narrowed.

suggests that Mars once contained water.[22] If we eventually discover that it once also contained life, but then later became uninhabitable, then our sample size of planets on which life arose will have been increased to two, one successful and one unsuccessful. For this reason alone the missions to Mars are of the greatest scientific interest. Such a discovery would start us on the path toward a more objective perspective. Proof that Mars changed from being once living to now sterile would demonstrate in a concrete fashion that Earth's history may not be typical, because of anthropic bias. It would help us conceive more readily that the universe may be littered with failed planetary habitats.

9.4.3. Earth's Own History

Although we can't yet judge between anthropic and nonanthropic hypotheses based on any large sample of planetary histories, Earth's own history may conceivably also shed some light. Earth's own history should look different depending on whether it has arisen by chance or through active stabilization. The schematic in figure 9.2 shows in cartoon form two types of contrasting planetary temperature histories that we might expect according to the two scenarios. Which of the two does Earth's history most resemble?

Although we do not possess such a detailed record before about 200 Mya (the age of the oldest surviving oceanic crust), there is nevertheless strong evidence (presented in section 8.7) that the Earth experienced severe glaciations in the Late Neoproterozoic between about 700 and 580 Mya. The Earth veered several times into extreme cold. The degree of cold may have closely approached the lower limit to habitability of figure 9.2. Another near-fatal excursion may have taken place even earlier in Earth history, at the time of the Great Oxidation Event (note 16 to chapter 7). As discussed in the last chapter, these severe glaciations appear to have been occasions when the climate of Earth wandered perilously close to being so cold as to completely sterilize the planet. Although there have been other severe glaciations on Earth, there is no suggestion that they also led to thick ice near the equator or that life was nearly extinguished in its totality.[23] We do not know of any occasions within the Phanerozoic (last 540 million years) when life as a whole nearly perished due to cold.

We are also not aware of any severe greenhouses that came close to sterilizing the planet through excess warmth (as on Venus), although we would perhaps be unable to tell if any such events had taken place, given that severe heat does not leave behind the same clues in the rocks as does severe cold. In my view the currently available data about Earth's climate evolution are not conclusive in terms of revealing whether the planet's temperature has been actively maintained or not. However, the Snowball Earth episodes argue strongly against any very tight control.

9.5. Large Coincidences and the Size of the Universe

Before weighing up the two viewpoints and seeing where that leaves us, we should return briefly to the Berner and Caldeira argument of section 9.2. If you remember, their argument was that it would be too implausible a coincidence to have avoided fatal icehouses and greenhouses solely by chance. But just how much luck is demanded, and is such a degree of luck reasonable? At what stage does the proposed fluke become so impossibly large that it is no longer credible?

Obviously if the calculated chances of over two billion years of never-interrupted habitability are X:1 against, and X is much larger than the number of planets in the universe (let's call this Y), then an anthropic explanation is implausible. But what are the values of X and Y? First let's consider Y, the total number of planets in the universe.

Over the last few decades there has been a fundamental shift in our conception of the universe, altering our perspective on the size of Y. It used to be thought that we could in principle see to the farthest extent of the universe, using telescopes. Therefore, in order to estimate Y, all that was needed was to scan around the sky with powerful telescopes and count up everything that could be seen (or at least count up all the galaxies, and then multiply by an average number of stars per galaxy and planets per star). When this exercise was carried out it led to an answer of about 10^{22} planets.[24]

However, another view has now taken over, and it is now thought that we can see only a fraction of the total universe. The universe is now thought to be much larger than just that part we can see. The reason for this belief is related to the distance light (or any other electromagnetic radiation) can travel in 13.7 billion years, the time since the Big Bang. The speed of light sets a constraint. This would not present a problem if the universe was smaller than the distance light can have traveled since its beginning. But it does present a problem if it is bigger. It now looks as if most of the universe is too remote for light from it ever to have reached Earth.[25]

This puts an immediate obstacle in the way of estimating Y. Because we are unable to observe most of the universe by any means, we are unable to gauge its size.

Several discoveries[26] have led to this overturning of the previous dominant interpretation (that all of the universe is in view) and its replacement by the new interpretation. Together these discoveries revealed that the part of the universe that can be seen from Earth is rather homogeneous and isotropic. In other words, it does not vary from place to place, or with direction of view. Although stars group into galaxies, and galaxies into superclusters, the density of these does not appear to thin out (or, conversely, to increase) toward the edge of the observable universe. Instead, at these larger scales matter is observed to be spread uniformly throughout the part of the universe that we can see. There is no indication that the edge of our visible universe is any different from those parts closer to hand.

This is not what was predicted by the earlier understanding. If the whole of the universe were in view, we would have expected to see changes toward the edge of the outward expansion that followed the Big Bang, and nothing beyond the outer edge. Instead, no such changes are seen and no outer edge can be discerned.

As a result, there is no reason to believe that the edge of the visible universe constitutes the edge of the universe as a whole. The universe could be vastly greater. Observations do not even rule out a universe that is infinite in extent. So, when it comes to estimating the value of Y, the total number of planets in the universe, we can only be secure about the lower bound, about 10^{22}. We do not have an upper bound. It can only be said with confidence that the actual number of stars lies somewhere between 10^{22} and infinity.

Since Y could, as far as we can tell from observations, be infinitely large, it must be possible for it to be larger than X, however large X is. Or, in other words, in a universe that is infinitely large, even large flukes can occur somewhere.

The end result of this short analysis is that an anthropic explanation for long habitability remains plausible. Even if, in the absence of a controlling mechanism, continued habitability for two billion years would require an immense fluke, such a possibility cannot be dismissed. We may recoil from such an explanation, given the unattractive appeal to coincidence. We may instinctively wish to be able to refute it. However, on these grounds at least, we cannot.

9.6. What Sort of Planet Should We Expect to Be On?

So far in this chapter I have described and discussed two contrasting viewpoints on life's long survival. Either Earth has been actively kept habitable by feedbacks or it has fortuitously stayed habitable. Either it was due to mechanism, or it was due to chance. However, we are not logically restricted to either/or. Now it is time to consider a third possibility, that it could have been due to some mixture of the two. The true explanation could instead be a combination of the two viewpoints articulated above. A more reasonable explanation may be to ascribe life's continued survival to a mixture of both luck and environment-stabilizing mechanisms, albeit mechanisms that don't work all that well.

This also relates to our thinking about what sort of planet we should expect to find ourselves on. More specifically, should we expect to observe on this planet: (a) only negative feedbacks; (b) some random mixture of both negative and positive feedbacks; or (c) some mixture, with negative (stabilizing) feedbacks predominating?

According to the Gaia hypothesis, the Earth system contains a homeostatic set of controls keeping important environmental parameters at life-comfortable values, analogous to the tight physiological homeostasis taking place within animal bodies. In contrast, an anthropic perspective could be taken to suggest a very different picture, more akin to "anything goes." If continued survival is all down to luck, then the system of feedbacks could in theory be of any degree of effectiveness for planetary regulation. It could be perfectly well organized and robust, or horribly awkward and fragile, or indeed anywhere in between. Regardless of the inherent suitability of the planet's feedbacks for stabilization, luck could have carried us through. That is one way of looking at things from an anthropic principle perspective.

Andy Watson notes, however, that the consequences of the anthropic principle might work out in a slightly different way. An alternative and more nuanced interpretation is perhaps a more likely explanation:

> Planets on which negative feedback mechanisms operate to stabilize the environment are more likely to sustain long-lived biospheres. Thus most planets on which observers arise are likely to have Gaia-like properties. . . . In this sense, the

Gaia hypothesis could be regarded as a consequence of observer self-selection: biospheres that are not self-regulatory may simply tend to expire before they evolve observers. (Watson 2004)

Andy Watson's point is that a survey of many planets, all of which were initially habitable, would most likely have found that they differed widely in terms of their inherent probabilities of remaining habitable across vast stretches of time. For instance, each one would have started with a different initial amount of water, a different chemical composition of its ocean and atmosphere and rock, a different intensity of volcanic activity, and even different sorts of life that first evolved on them. Some of those planets would have been inherently more likely to stay habitable from that point on, whereas others would always have been more likely to suffer runaway feedbacks toward uninhabitable conditions. Out of all the planets, complex life (and observers) would be most likely to evolve on those planets that were at the beginning naturally more predisposed to remain habitable.

This idea has similarities with the biblical Parable of the Seeds. The parable is as follows: an amount of seed was scattered randomly, and out of that seed, some fell on good ground, some on stony ground, some among choking thorns, and some on the path where the seeds were easily visible to birds. Although each and every seed has a finite chance of surviving, the chances are much better for those seeds sown on good ground. Most of the survivors several months later on are likely to have been those sown on good ground. If there was only one seed that survived to an adult plant then it would most likely have been one of those sown on the good ground.

Watson's line of argument seems reasonable. However, despite the sound reasoning, that does not mean that we should necessarily expect anything like Gaia, or at least not in the sense that Gaia is typically understood. If it were possible in the first place for planets to develop tight physiological regulation resembling that operating in an animal's body, then, yes, we should indeed expect to find ourselves on a Gaia-like planet. But planets, unlike animals, are not products of evolution. Therefore we are entitled to be highly skeptical (or even outright dismissive) about whether to expect something akin to a "superorganism," even on the most propitious of planets.

If we use the word *Gaia* in a weaker sense, it is still unclear just how well suited we can expect even the most initially favored planets to be. What is clear is that if we accept this reasoning, we should not expect a completely random assortment of negative and positive feedbacks. Instead we should expect a bias toward self-stabilizing tendencies. Certainly the most significant and critical feedbacks should be broadly favorable for maintaining life-favorable conditions.

In terms of our single known example of a planet on which observers evolved, Earth, while we can list individual stabilizing mechanisms that do operate on this planet,[27] we also know that they are accompanied by many destabilizing mechanisms.[28]

Andy Watson also considered other evidence when pondering the relative roles of innate suitability and luck in fostering life's long survival. Survival in the face of a warming sun (increase in solar output through Earth history, see note 20 to chapter 8) is, Watson suggests, additional evidence for a Gaian mechanism, given the need for compensation. However, it is also fully possible for some sort of imperfect compensation (for instance overcompensation, if the planet actually cooled over time despite the warming sun) to have taken place by chance. Watson considers the evidence for self-regulation in our knowledge of the history of Earth's climate, and then compares the evidence for what he calls "Innate Gaia" versus "Lucky Gaia." He concludes:

> The biosphere appears to have a sometimes efficient, but frequently erratic, thermostat. Though the thermostat has extended the lifetime of the biosphere to several billion years, it is hard to make the case that it is sufficiently reliable to virtually guarantee that outcome. . . . "Lucky Gaia" seems a simpler and more defensible hypothesis than "Innate Gaia." (Watson 2004)

He looks forward, as do I, to further information about Snowball Earth and early life on Mars.

9.7. Taking Stock

This chapter has already covered a lot of ground, and several different ways of accounting for Earth's long habitability have been described. It is worth pausing briefly to recap where we have got to.

The early parts of this chapter laid the foundations in terms of introducing various concepts: reservoirs, residence times, the potential for instability of atmospheric CO_2, and the weak anthropic principle. Two very different explanations have been proposed for habitability maintenance: control mechanisms and blind luck. A third possibility, a combination of the two, has just been introduced. In addition, two great facts are central to the debate. The first of these is the evidence that life on this planet was not interrupted even once throughout an enormity of time. This great fact is the source of this whole debate in the first place. The second great fact is that climate and environment on Earth have been highly variable, as laid out in the previous chapter.

This second fact argues against accepting the proposition that all-powerful stabilizing mechanisms have been solely responsible for keeping the Earth comfortable for life. A strong and robust temperature controller would not have allowed Snowball Earth events, for instance. Given the turbulent nature of Earth's past, including near-fatal dips in temperature, it no longer seems credible to believe in a fail-safe, strongly stabilizing mechanism (or set of mechanisms). On the contrary, the evidence points toward stabilization having been imperfect. It seems more likely to me that crude regulatory mechanisms on Earth increased the chances of continued survival from vanishingly small to just very small, but that luck also played a part.

9.8. Implications for Our Understanding of Silicate Weathering

This leads us neatly into a discussion of silicate weathering, which has long been proposed as an overriding regulator of climate. This section is a short digression reappraising silicate weathering. The understanding just developed allows us to look at silicate weathering with fresh eyes.

Silicate weathering could, it has often been suggested, act as a non-Gaian climate controller.[29] In theory, it could control climate without any involvement of life, through purely inorganic processes, though life may strengthen the feedback. It has been put forward as an abiotic alternative to Gaia in accounting for the otherwise puzzling long persistence of life-friendly conditions.

Earlier in the chapter (section 9.2), I described how Berner and Caldeira assert that something must have kept Earth climate in check. They are by no means alone. Wally Broecker and Abhijit Sanyal made a similar argument,[30] concluding that silicate weathering consumes CO_2 and potentially stabilizes its concentration. Broecker and Sanyal argued that atmospheric CO_2 must police the rate of silicate weathering, because in the absence of such an effective feedback control "the CO_2 content of the Earth's atmosphere would have wandered over a large range threatening life either by overheating or by carbon dioxide starvation."

But, as we have seen, matters may not be so simple. According to the anthropic principle, this lack of fatal wandering could have been down to luck. It is conceivable that silicate weathering did not and does not exert a powerful and reliable thermostat-like control over Earth temperature. From this view, perhaps there never was such an all-powerful controller.

Alternatively, if we prefer the combined explanation, the operation of silicate weathering could have been an important feature of this planet that made it from the outset more likely to play host to the evolution of intelligent life. It could have initially biased this planet toward climate stability. To return to the Parable of the Seeds analogy, perhaps the operation of silicate weathering makes this planet equivalent to "good ground." From either perspective, however, given that Snowball Earths happened, silicate weathering must have executed at best a somewhat weak and imperfect stabilization of climate.[31] Much was still down to chance. But this is not how silicate weathering is normally viewed.

To give one example, Lee Kump (at Pennsylvania State University) and colleagues reviewed the effectiveness of silicate weathering as a negative, stabilizing feedback on climate. In a thorough review[32] they discussed some of the reasons behind the ongoing debate associated with silicate weathering. Ceding some ground to contrary evidence and the unarguable gradual decline in global temperatures over the last 50–100 million years (which silicate weathering, if a fully effective climate stabilizer, should have acted to oppose), Kump and co-workers concluded that "numerical models and interpretation of the geological record reveal that chemical weathering has played a substantial role in both maintaining climate stability over the eons as well as driving climatic swings in response to tectonic and paleogeographic factors."

But what is the specific evidence for its role in maintaining climate stability? In large part it seems to come down to arguments based on the assumption that *something* must have stabilized climate. To quote again from Kump and colleagues: "Numerical models highlight the importance of feedback in the carbon cycle to prevent excessive fluctuation in atmospheric pCO_2 and climate, and the most likely source of feedback is the climate sensitivity of chemical weathering and erosion . . . ," and "the corollary of this solution of the Faint Sun Paradox is that some mechanism must have regulated the drawdown of these gases to prevent permanent global glaciation or extreme warmth." And once more: "Nevertheless, the planet has remained well within habitability limits of higher organisms for at least the last few billion years, which suggests the presence of a strong regulatory feedback mechanism much like the climate-weathering feedback."

As just discussed, the weak anthropic principle presents an alternative view to this line of argument and allows attribution of long-term habitability to luck rather than to mechanism, or to a combination of the two. From this perspective there is no requirement for silicate weathering or any other process to work as an all-powerful climate-stabilizing feedback. Hopefully future investigations of silicate weathering can evaluate it without the prior assumption that some all-powerful thermostat (probably silicate weathering) has continually guarded Earth's climate. And, as said above, Snowball Earth episodes in any case make it questionable whether any such controller exists. The many oscillations and long-term trends in paleorecords (chapter 8) also demonstrate that regulation has been far from perfect, as does the present excessive coldness of Earth. In my opinion it is more likely that silicate weathering is at best a partially effective thermostat that may have contributed to continued habitability, but has not ensured it.

9.9. IMPLICATIONS FOR OUR UNDERSTANDING OF EARTH SYSTEM HISTORY

As well as providing a different perspective on silicate weathering, the weak anthropic principle suggests that we should modify how we think about Earth system history more generally. If our planet has remained habitable partly due to luck, then we need to revise how we think about its history.

Rather than searching for perfect thermostats, we should instead expect that the assemblage of Earth's regulatory mechanisms is more of a random mishmash, a product of happenstance. I suggest that we should be more interested in contingent factors, idiosyncrasies of global carbon cycling, fallible and partial feedbacks (such as feedbacks that protect only against excursions to very cold but not to very hot climates), and time-varying forcings[33] than in perfect control mechanisms that do not exist. We can expect some predisposition toward climate stability, but we should not expect it to be perfectly arranged for life's particular convenience or so as to guarantee life's continued sustenance. Although our planet does possess stabilizing feedbacks, it seems likely that they are not so robust that Earth was guaranteed to remain habitable and stable once life had started.

Richard Dawkins has also considered the role of the anthropic principle, but as it relates to the chances of life originating on a planet, rather than persisting on a planet:

> ... It has been estimated that there are between 1 billion and 30 billion planets in our galaxy, and about 100 billion galaxies in the universe. Knocking a few noughts off for reasons of ordinary prudence, a billion billion is a conservative estimate of the number of available planets in the universe. Now, suppose the origin of life ... was so improbable as to occur on only one in a billion planets ... even with such absurdly long odds, life will still have arisen on a billion planets—of which Earth of course, is one.
>
> This conclusion is so surprising, I'll say it again. If the odds of life originating spontaneously on a planet were a billion to one against, nevertheless that stupefyingly improbable event would still happen on a billion planets. The chance of

finding any one of those billion life-bearing planets recalls the proverbial needle in a haystack. But we don't have to go out of our way to find a needle because (back to the anthropic principle) any beings capable of looking must necessarily be sitting on one of those prodigiously rare needles before they even start the search. (Dawkins 2006)

Equivalent logic applies when we consider this other phenomenon that may also have been stupefyingly improbable: the chances of the "flame of life" not once sputtering out through the course of billions of years of Earth's history. Now the even more prodigiously rare needles consist not only of planets on which life once arose, but also those on which it subsequently persisted. When we look up at the night sky and wonder at the multitude of stars (noting that the true number is now thought to be much greater than a billion billion—section 9.5), we should expect that at most only a few of those stars are likely to harbor life in their vicinity, given the doubled rarity coming from the need for both life origination and life persistence.

If luck indeed played a large role, then on those infrequent planets on which life once starts it is quite possible that life usually dwindles away long before it ever gives rise to complex life. Perhaps that is the more normal fate for life. If true then of course it makes it more probable that Earth with its evolved intelligent life, far from being something commonplace in the universe, is instead exceedingly scarce.

This leads to a profound question: is the Earth an incredibly rare jewel in the universe, or is it instead something much more mundane? If we follow the anthropic principle then it points us toward the first answer and suggests that the Earth is special, not only because it hosted life in the first place, but also because it managed to continue to host life across vast expanses of time, through geological ages.

In this section we are interested in how our understanding of Earth history is affected by including the anthropic principle in our thinking. To illustrate how it might alter our thinking, it is interesting to draw a parallel to the life cycles of different ocean wanderers (fig. 9.3). At one end of the spectrum of chanciness, ocean sunfishes release many millions of eggs in each spawning event (one adult female was observed to contain 300 million eggs). Each of these, after spawning, is subsequently left to fend for itself without, as far as we know, any parental care whatsoever. For the ocean not to be full of sunfish, the prospects for each newly released egg must be extraordinarily bleak; their chances of surviving predators and other dangers in order to reach adulthood must be much less than a million to one. Compare this to the assured and pampered upbringing of a wandering albatross chick. Each monogamous adult couple

Figure 9.3. Three ocean wanderers that produce very different numbers of eggs and whose hatchlings face very different survival prospects: (*a*) ocean sunfish (*Mola mola*), (*b*) green sea turtle (*Chelonia mydas*), and (*c*) wandering albatross (*Diomedea exulans*). (*b*) IrinaK / Shutterstock.com; (*c*) www.photo.antarctica.ac.uk.

produces just a single egg every two years. The egg is incubated in a large nest for two to three months, and once hatched the fledgling is subsequently fed and nurtured by both parents for another nine months. Under this tender care, each egg has about a 30% chance of surviving to become an adult bird. The wandering albatross therefore sits at the other end of the spectrum of chanciness. In between, at intermediate levels of chanciness, are many other creatures. A well-known example, sticking with ocean wanderers, is that of sea turtles, which every few years visit beaches to bury and then abandon clutches of about one hundred eggs.

What we would dearly like to know is how good a bet the Earth was at the outset, in terms of the chances of the environment remaining sufficiently propitious for long enough to allow life to develop to intelligence. Were the initial prospects closest to those of a sunfish egg or a turtle egg or an albatross egg? It is rather remarkable that we know so little about where on this spectrum of chanciness the early Earth sat.

Another interesting implication is the link that the anthropic principle makes between the geological history of Earth as a crucible for life and the physical nature of the planet and the universe. Arguably there have been uncanny coincidences in all three. The explosive force of the Big Bang at the beginning of the universe seems to have been quite finely tuned compared to gravitational attraction. If the early expansion had been too fast then the rate of separation of matter would have been so rapid that galaxies and stars would have been unable to coalesce; if too slow (or gravity too strong) then the Big Bang would have been followed shortly thereafter by the Big Crunch. Likewise, Earth is "luckily" large enough in size to have retained its atmosphere, and "fortunately" situated at the right distance from the Sun (within what is known as the Habitable Zone or the Goldilocks Zone), so that it is neither too hot nor too cold for life.[34] Through the anthropic principle we can understand all of these apparent coincidences. But the point of this paragraph is that the anthropic principle is not only relevant to astronomers and cosmologists. It is also relevant to geologists. It can also help us understand why none of the five major mass extinction events of the last 500 million years killed 100% of all species, even though each killed > 50% and one, the Late Permian, killed > 95%. It can also help us understand sustained habitability. The anthropic principle potentially explains in part both life's "miraculous" survival through Snowball Earths, and also the "surprisingly" long duration of a climate congenial for life.[35]

In geology and Earth system science the anthropic principle is at present only infrequently considered (it is simply not mentioned at all in standard textbooks). It should instead be at the forefront of thinking whenever the topic of discussion relates to Earth habitability. Although it remains to be determined exactly how large a role luck has played in Earth's uninterrupted habitability, a significant contribution seems highly probable. The weak anthropic principle is thereby relevant not only to the physical characteristics and location of Earth but also to the history of how Earth's environment has developed over time.

9.10. Conclusions

This chapter has been a wide-ranging one, and implications beyond Gaia have been considered. Returning to the main focus, how has the discussion in the last two chapters contributed to our evaluation of the Gaia hypothesis?

The previous chapter reviewed evidence to show that if indeed there has been a Gaian hand on the tiller to keep the environment stable, it must have been a rather shaky one. But, nevertheless, despite the environmental instability, life survived. There has been a continuous, uninterrupted thread of life persisting from the infancy of the habitable planet ($>$ 2 Bya) right through to the present day.

This chapter has considered whether or not this is a strong argument for Gaia. An explanation is surely required for why this thread has remained intact for so long, despite various planetary catastrophic events such as asteroid impacts, and despite the rather short residence time (susceptibility to instability) of atmospheric CO_2. The potential volatility of atmospheric CO_2 implies that fatal icehouses or hothouses could easily have occurred. One proposed solution has been the Gaia hypothesis, whereby life is assumed to have intervened in the controls to keep the Earth environment equable. Another proposed solution has been that non-Gaian stabilizers exist, including a thermostat such as silicate weathering based entirely or mainly on inorganic processes. But in either case it is assumed that there must have been a strong, protective thermostat. Surely the only possible explanation for such improbable persistence over such an immense duration of time is that the Earth possesses a giant thermostat ensuring tolerable temperatures are maintained.

Not necessarily. As discussed at length in this chapter, the anthropic principle leads us to question or at least qualify such a conclusion. There are thermostat-like processes that operate on the Earth, although their efficacy is not yet in all cases well established. What we can say with confidence, however, is that the fact of long continued habitability does not prove the existence of a fail-safe climate protection mechanism on Earth. We may well have also been lucky. It may well be the case that like the rush hour traffic to blindfolded cyclists, planets are in fact distinctly perilous habitats for life, on average and when considered over billions-of-years timescales. Perhaps life becomes extinguished on the vast majority of planets on which it ever gets started. Perhaps complete extinction is by far the normal outcome. Since this question can only ever be contemplated by survivors such as ourselves, an unbiased answer is not to be expected, or at least not until we can progress beyond studying only the history of our own planet. Because of observer selection, Earth cannot be typical.

The subject of this chapter is a fascinating one. It is to me an intriguing question whether the history of life on Earth has been one of almost guaranteed, built-in success, or one of extreme good fortune, or somewhere in between. What role has hazard played in life continuity? Are we here primarily due to "destiny or to chance,"[36] or to some mixture of the two? It may take a long time to better elucidate the relative importances of luck and mechanism, but the search should be immensely interesting.

The question at the start of this chapter was: "what does the fact of long-term life persistence mean for our evaluation of Gaia?" At the end of this chapter we are left, in my view, with a somewhat unsatisfying answer. Although Gaia doubtless could, if true, account for extended life persistence, evidence of environmental variability (most notably Snowball Earths) argues against any tight and powerful stabilization. Moreover, the weak anthropic principle questions the need for Gaia or any other explanation invoking completely effective negative feedbacks, because continued habitability could have been in large part down to chance. For these reasons it can be concluded that the long and uninterrupted duration of life-tolerant conditions does not prove the existence of an all-powerful thermostat, and does not prove the existence of Gaia.

Chapter 10

CONCLUSIONS

THE GAIA HYPOTHESIS is one of the best known of all modern scientific hypotheses. Many scientists have enthusiastically adopted it, while others have rejected it out of hand. It has attracted generous measures of both opprobrium and acclamation, bouquets as well as brickbats.[1] But which judgment is correct? In this book I have looked at the reasoning advanced in support of the Gaia hypothesis and have examined it to see if it withstands detailed examination and if it makes a strong case for Gaia. Does the hypothesis still seem plausible when subjected to close inspection and probing scrutiny?

In this book Gaia has been broken down into its separate subcomponents and assertions, each of which has been carefully examined in turn. Each aspect of the hypothesis has been compared against a large amount of evidence, including much that has been generated in the time since the hypothesis was first formulated. Now in this chapter I weave together the individual strands of analysis into an overarching evaluation of Gaia. I come to a conclusion as to whether the evidence as a whole tends to support or to contradict Gaia. You will have realized from the preceding chapters that it is going to be a mostly unfavorable conclusion.

Later in the chapter I also discuss why the answer we get is important. I explain why we should care, deeply, about whether or not we have a proper conception of how the Earth system works. This last chapter is therefore split up into two main parts: what the overall assessment is, and why it matters.

10.1. REACHING A VERDICT ON GAIA

Before coming to the overall assessment, it is useful to remind ourselves of two things: firstly, what we expect from a successful hypothesis, and secondly, the ground that has been covered in this book.

10.1.1. What Makes for a Successful Hypothesis?

For us to embrace a hypothesis, or even just tentatively accept it, we expect several things. We require of course that it be able to explain satisfactorily a phenomenon or a class of facts. We give it more credence if it can explain several phenomena or classes of facts simultaneously; we respect it further still when it can explain things that other hypotheses cannot.

The greatest hypotheses wonderfully bring whole sweeps of knowledge into order, enabling us at a stroke to make sense of a large variety of different trusted facts that previously seemed immensely puzzling or incompatible with one another. This was the case, for instance, with Darwin's suggestion about how natural selection and evolution worked. *The Origin of Species* is a long discourse showing how just one hypothesis can reconcile a vast and diverse array of observations about organisms living on Earth; it explains their behavior, their biogeographical distributions, patterns of fossil change through geological time, embryological development, artificial selection under domestication, and many other phenomena.

A hypothesis is also preferred when it has predictive power: ideally it will be able to predict correctly the answers to experiments or observations that have not yet been carried out. The more times a hypothesis survives fresh attempts to test or challenge it, the more confident we become in its veracity. We also require, obviously, that there should be no facts or observations that are incompatible with the hypothesis. Or in other words, there should be nothing that falsifies it; it should be consistent with all that we know.

Finally, there is also simplicity of explanation. The more concisely a hypothesis, theory, or law can be stated, the more intellectually pleasing it is. The reason that $E = mc^2$ holds such a high place in our esteem is not just because of its power and the clarity of understanding that it brings, although these are of course of primary importance, but also because of its majestic elegance and brevity. Likewise, there is an austere simplicity to $rb > c$, the basis of kin selection (chapter 2). Again, the inverse square law of gravity, $F = Gm_1m_2/r^2$, has a beautiful simplicity to accompany its enormous predictive power (for instance its ability to predict orbits of planets centuries into the future, when used in conjunction with Newton's laws of motion).

10.1.2. The Road Traveled . . .

This book has been an attempt to assess whether or not Gaia satisfies these criteria for a successful hypothesis. The principal method of analysis has been to examine the validity of Lovelock's three assertions (chapter 1). In order to do this, a wide variety of pertinent evidence and reasoning has been presented, and compared to expectations if the assertions are correct. The major points that have been covered are as follows.

The first chapter introduced the Gaia hypothesis, described the three supporting assertions, and also described two competing hypotheses to Gaia: coevolution of the environment and life, and a "geological" hypothesis in which life is considered as a passive passenger on a planet over which it exerts no influence. I described how the rest of the book would include a comparison of those three hypotheses against the evidence, to see which, if any, are consistent with all the facts available to us.

In chapter 2 the possibility of a mechanistic basis for Gaia was considered, but none was found. The question of how Gaia might arise naturally out of ecological and/or evolutionary dynamics was considered. It was seen how environmental regulation (homeostasis) is readily achieved in nature by collectives of related genes, whether those genes are housed within the bodies of individual multicellular organisms, within colonies of eusocial insects, or within families of higher animals such as beavers. But there was found to be no suggestion, from the available evidence, of tight environmental homeostasis or control being produced by collectives of unrelated or only moderately related organisms, such as herds, flocks, species, ecosystems (including coral reef communities), or the global biota. Planets do not mate, reproduce, inherit characteristics from other planets, or undergo any struggle for existence in competition with other planets. It is preposterous therefore to invoke evolution acting at the planetary level, and no serious suggestion of this sort has ever been put forward. The occasional suggestion (by Lovelock and others) that the planetary biota or biosphere is a planetary-scale superorganism is not, on this reasoning, at all true in an evolutionary sense.

While the superorganism idea can be clearly rejected, it is not, however, completely inconceivable that Gaia could be produced by evolution. It could be some sort of emergent property that arises in some as-yet-unappreciated way out of the extraordinarily complex web of ecosystem and evolutionary interactions. That said, it has not so far been possible to identify, even in outline, any plausible reason why such an emergent property would be likely to arise. We are left at the present time without any mechanistic explanation or bottom-up understanding of Gaia. Equally, however, we also have no categorical proof that something like Gaia could not possibly emerge.

Chapters 3–5 scrutinized the first of Lovelock's assertions in favor of Gaia—namely, that the Earth environment is well suited to life. Extremophiles living in constantly extreme environments become wonderfully well adapted to those environments and, interestingly, over time lose the ability to survive in less extreme environments. Because of this process of "habitat commitment," tepid water (which we would think of as a mild environment) kills outright many cryophiles and also many thermophiles. Tepid conditions, benign to most organisms, are insufferable to the specialist extremophiles. The marvelous plasticity endowed by evolution forces organisms to come to fit their own environments snugly, however objectively good or bad they might be, but this can come at the expense of losing the ability to tolerate different environments.

Habitat commitment indirectly gives rise to the impression that all environments have been made to be uncannily well suited to their inhabitants, even though the causal arrow actually points in the opposite direction. A lengthy examination, taking the organism-molding properties of evolution into account, led to the conclusion that the global environment is far from optimal for life. Two facets were considered in more detail: (1) The planet is too cold. An Earth that was warmer than at present over many millions of years would come to host a greater biodiversity (more species) and a greater total biomass. Most species would probably evolve more rapidly under a warmer climate, and individual organisms would mostly "live faster," because temperature governs the pace of life. (2) Widespread nitrogen starvation is not necessary. There is no a priori need for the curious phenomenon of widespread nitrogen starvation within a nitrogen-rich biosphere. A Gaian Earth would not be organized this way.

Chapter 6 considered the second observation that Lovelock advanced in support of Gaia: that life has altered the Earth. The facts are clear that this has indeed been the case. Several important properties of the planetary environment have been shifted to different values than would be found on a lifeless Earth, and life continues to maintain the offset. Thus Lovelock was correct on this issue, both in terms of biological control over aspects of seawater chemistry, and also in terms of the composition of gases in the atmosphere.

If Gaia is an accurate characterization of how the natural world has worked, and if Gaia arises in some way from evolutionary dynamics, then we might expect that major evolutionary innovations (for example, the evolution of large trees leading to the first proper forests and their spreading out over the land) would have been beneficial to the global environment and have aided stability. However, the impacts of two such innovations were examined in chapter 7, and the evidence found to be mostly to the contrary. The arrival of oxygen-producing algae and the first forests both had highly deleterious as well as some favorable impacts.

Chapter 8 examined the third of Lovelock's assertions, that the Earth has been a stable environment for life. From the Younger Dryas to ice ages to long-term cooling trends to Snowball Earths, multiple lines of evidence now point to a lack of environmental constancy over multiple timescales. We now possess an abundance of evidence of environmental variability, but it is nonetheless apparent also that the Earth environment has remained continuously habitable over billions of years. Some use this by itself to infer the need for Gaia or some other environment-sustaining mechanism. But is the logic justified? Is it correct to deduce from the immensely long persistence of life-suitable conditions that there must therefore be an all-powerful planetary protection system that has continually headed off any tendencies for conditions to become uninhabitable?

It seems at first thought that a stabilizing mechanism must be required, if we are not to allocate too large a role to coincidence. The stabilizing feedbacks need not necessarily be Gaian in nature, but they must exist, it would seem. However, as just discussed in chapter 9, if the anthropic principle is accepted then it leads to a modified perspective. There are an enormous number of planets in the observable universe alone, and likely vastly more beyond. We are also exceptionally ignorant of how frequently once-habitable planets wander off to sterility. The anthropic principle cautions us to remember that we can only ever find ourselves on a long-inhabited planet, however rare such an occurrence might be.

The choice of possible explanations for long habitability is, however, not limited to either mechanism or chance. As just considered in chapter 9, the actual explanation could involve a combination, a mixture of good fortune and stabilizing feedbacks. Whether by itself or in combination, the anthropic principle is probably part of the explanation for why our planet has remained continuously habitable for so long.

Such has been the structure of this book, which has consisted of detailed scrutiny of each of the separate assertions and components in turn. This diverse information is now synthesized into an overall understanding.

10.1.3. Weaving the Strands Together

The most important evidence bearing on Lovelock's various assertions is summarized in table 10.1:

TABLE 10.1.
The most pertinent evidence relating to Lovelock's three assertions

Assertion	Important Evidence	Conclusion
1: Environment is well-suited for life	"habitat commitment" in extremophiles, impoverished terrestrial biota during ice ages, "unnecessary" nitrogen starvation	Not supported: environment is adequate but is far from optimal
2: Biota has shaped the environment	coexistence of O_2 and CH_4 in the atmosphere, strong depletion of biologically utilized elements in seawater	Supported: the biota has shaped the environment
3: Global environment has been stable	ice core atmospheric CO_2 records, $\delta^{18}O$ temperature/ice volume records, fluid inclusion ocean chemistry record, low-latitude dropstones during Snowball Earth	Not supported: the environment has not been stable, although it has remained habitable

Upon examination, the second of the three assertions was found to be abundantly supported by data, but the other two were found to not be compatible with the available facts. Far from being benign, the planet is too cold to be optimal for life, and this has particularly been the case during the inhospitable ice ages that have dominated the recent history of our planet (occupying > 75% of the last few million years). As more and more data come in, and increasingly illuminate the history of the Earth, the lack of climate constancy has become more apparent. From the dramatic glacial cycles of climate and greenhouse gases, to the Cenozoic long-term cooling, to the brinksmanship of the Snowball Earth events, a more and more detailed picture is being built up of a dynamic and eventful climate history. The actual, complex, climate history is far more interesting than earlier views of a predominantly stable Earth.

On the basis of this evaluation it is clear that the Gaia hypothesis is built on suspect foundations. Two out of three of the underpinning assertions are seen to be unsafe when carefully compared to all of the available data, although much of this data has only been generated in the years since Gaia was first proposed.

The lack of any established bottom-up mechanism that can explain how Gaia is produced also weighs against it.[2] No one has been able to explain convincingly how Gaia could emerge out of evolutionary or ecological dynamics. It is therefore perhaps not surprising that major evolutionary advances, such as the evolution of oxygenic photosynthesis, or the first foresting of the land by sizeable trees (the evolution of lignin), have been associated with environmental catastrophes (oxygen poisoning of the preexisting anaerobes and glaciations).

We can now take this synthesis another step further, and compare not only Gaia but also the two competing hypotheses (introduced in chapter 1) to several of the major observations and classes of facts that have been reviewed in this book (table 10.2).

One conclusion that can be drawn from this book as a whole, and that is highlighted in tables 10.1 and 10.2, is that there is no particular strong and compelling evidence that demands us to accept Gaia. There is no single body of facts or line of unimpeachable reasoning that sways the debate conclusively in favor of Gaia. This is shown in table 10.2 by the absence of any row with a tick for Gaia and a cross for the other two hypotheses; there is no observation that only the Gaia hypothesis is compatible with. To put it another way, no key piece of evidence has been identified in its favor; there is a lack of any clinching fact or argument that points exclusively to Gaia as the only plausible hypothesis for how Earth and life influence each other.

That said, the Gaia hypothesis does have one distinct advantage in explanatory power, which the other two at first sight appear unable to match. Because

TABLE 10.2.
Comparison of the three competing hypotheses to major facts

	Hypothesis		
Observation	Gaia	Geological	Coevolution
Suboptimal planetary environment	✗	✓	✓
Biological role in shaping the environment	✓	✗	✓
Highly variable environment over time	✗	✓	✓
Sustained planetary habitability over several billion years	✓	✓ (with anthropic principle)	✓ (with anthropic principle)
Evolutionary innovations with detrimental environmental consequences	✗	✗	✓

✓ denotes compatibility of a hypothesis with an observation or class of facts; ✗ denotes incompatibility

Gaia posits a network of stabilizing mechanisms it can thereby account for the continued habitability of the Earth over immense stretches of geological time (probably more than three billion years), even though the residence time of CO_2 is very much shorter (chapter 9). Although CO_2 has fluctuated considerably over time, the induced climatic fluctuations have never been sufficiently large to wipe out all life, although the Snowball Earth episodes may have come fairly close. Many scientists believe that Earth's CO_2 and climate must have been actively maintained over this enormous span of time, in the same way that a thermostat connected to a heating system maintains the temperature of a room at a near-constant value.

While Gaia would indeed be capable of explaining this key fact of long-sustained habitability, the fact can also be explained in other ways. A mostly inorganic feedback, such as silicate weathering (section 9.8), could have been responsible. In addition, reasoning based on the weak anthropic principle offers an alternative explanation that is different in a more fundamental way.

Logically our very existence as intelligent contemplators of our planetary history is contingent on a long and uninterrupted evolutionary process. On planets where the evolutionary process has been interrupted, such questions never get asked, because there are no intelligent beings there to ask them. Given that we are considering this question about the planet on which we evolved, a long history of successful evolutionary development is a logical prerequisite. From this perspective continued habitability is something that must have happened, rather than a curious fact that needs explaining. The anthropic principle removes the need for Gaia or any other thermostat, or at least cautions that while it is reasonable to expect the Earth to possess a more than usually favorable set of feedbacks, we need not expect anything like a perfect stabilizing system (section 9.6).

Therefore this key fact of long-sustained habitability does not point exclusively to Gaia as the only plausible hypothesis. The anthropic principle removes the need for any overriding mechanism making sure that the planet has been kept fit for life. Instead it suggests that hazard and happenstance are part of the explanation for the prolonged adequacy of this planet for life.

10.1.3. Comparing the Three Hypotheses

We can consider how the multifarious topics and facts discussed in this book allow discrimination between the three competing hypotheses, and whether they allow us to decide which sits best with the evidence.

In light of what we know about the planet as it is today, and its history over time as revealed by various pieces of geological data, I consider that the only viable hypothesis is the coevolutionary one. There are no natural phenomena that either Gaia or the geological hypothesis is uniquely able to explain. More broadly, there is no great class of facts that either Gaia or the geological hypothesis can account for but that coevolution can't. The reverse is, however, true. As just listed, Gaia is inconsistent with several important facts. The geological hypothesis also runs counter to some important evidence: it is ruled out by the clear intervention of life in the setting of some global properties of the atmosphere and ocean. Coevolution, on the other hand, is fully compatible with what we know. Thus, out of the three hypotheses initially considered, only coevolution remains plausible.

The coevolutionary hypothesis is a less daring and a less grand vision than Gaia. It does not commit itself to a bold conclusion in the same way that the Gaia hypothesis does. It doesn't stick its neck out. As a consequence of its relative reticence, however, it has the benefit of not contravening our observations of how the Earth works and how it has changed over time. Coevolution limits itself to stating that life has had significant impacts on the environment, and, which is obvious, that the environment has also had a strong influence on the evolution of life. But whereas Gaia goes further to invoke an emergent property arising out of this two-way interaction, coevolution makes no such claim. Coevolution does not, unlike Gaia, specify any outcome or direction to the back-and-forth interactions between life and environment. It does not specify that the interactions lead to a modulation of the environment toward a more hospitable and stable state for the organisms inhabiting it. Coevolution limits itself to declaring that the interaction is bidirectional and leaves it at that. Coevolution does not of course rule out the possibility of negative feedbacks and stabilization of certain environmental parameters, but equally does not require it.

Coevolution is therefore a more modest and a more circumspect proposal. In one sense one could even call it more tedious, because of its relative lack of ambition and grandeur compared to Gaia. However, in another sense it is much more appealing, for the simple reason that it seems to be true. It presents an accurate picture of how life and Earth exert influence over each other. It may be a rather reserved hypothesis in terms of not being willing to commit itself very far, but it has the distinct advantage of being compatible with the known evidence of how the Earth and its biota behave today, and are thought to have behaved throughout the vast ages of geological time.

10.1.4. Kirchner's Gaia Variants

As described earlier (section 1.2), James Kirchner realized that many different authors had used the phrase Gaia hypothesis to mean different things, and he sought to clarify the debate by distinguishing several different variants of the hypothesis, which he referred to as "a taxonomy of the Gaia hypotheses."[3] These ranged in strength from what he called "weak" to "strong," where these terms referred to the extremity (degree of novelty and potential controversy) of the hypotheses: (1) *influential Gaia*, which asserts only that biology affects the physical and chemical environment to some degree; (2) *coevolutionary Gaia*, which limits itself to stating that the biota and environment are somehow coupled (equivalent to the coevolution hypothesis of this book); (3) *homeostatic Gaia*, which emphasizes the stabilizing effect of the biota; (4) *teleological Gaia*, which implies that the biosphere is a contrivance specifically arranged for the benefit of the biota; and (5) *optimizing Gaia*, which suggests that the biosphere is optimized in favor of the biota. Although usage is indeed variable, Gaia is most often used in one of the latter senses (3, 4, or 5), which I have focused on in this book. Over time Lovelock has modified the Gaia hypothesis away from (5) toward (4) and (3). In table 10.3 I give a view on which of Kirchner's flavors or variants of Gaia seem to be compatible with the facts and conclusions presented in this book.

The three strongest forms of Gaia, according to Kirchner's taxonomy, are at odds with some of the facts and deductions laid out in this book. The two weakest definitions are easy to accept but are not substantially different from what has been proposed elsewhere under different names. Or, in other words, it would seem that those variants that are credible (1, 2) are not particularly novel, and those that are novel (3, 4, 5) are not credible.

Table 10.3.
Evaluation of various alternative forms of the Gaia hypothesis

Gaia Variant	Description	Accept?	Reason to Reject
Influential Gaia	biota influences the abiotic biosphere	✓	none
Coevolution-ary Gaia	biota and biosphere influence each other	✓	none = coevolution
Homeostatic Gaia	biotic influence on biosphere imparts stability	✗	environmental instability; biota can impart instability (e.g., first forests)
Teleological Gaia	regulation of the biosphere by, and for, the biota	✗	environment too cold, evolutionary innovations sometimes make it worse
Optimizing Gaia	biota manipulates biosphere to make it optimal	✗	environment too cold, evolutionary innovations sometimes make it worse

✓ denotes compatibility of a hypothesis with an observation or class of facts; ✗ denotes incompatibility

Table 10.3 might make it seem that the analysis in this book leads to rejection of only some of the variants of Gaia and not others, because some variants are found to be viable. However, Kirchner's taxonomy is misleading in this respect. In practice Gaia is never, or only very rarely, taken to imply either (1) or (2). These two weakest definitions are either identical to, or even weaker than, the coevolutionary hypothesis Schneider and Londer put forward, and they do not correspond to normal usage of the word Gaia. Referring to these much weaker hypotheses as Gaia in fact only generates semantic confusion. As documented in the early chapters of this book, although Lovelock has moved away from (5) over time, he continues to define the Gaia hypothesis in terms corresponding to (3) and (4). The term *Gaia hypothesis* has therefore been used in this book to refer only to the stronger variants.

10.1.5. Concluding Comments on Gaia

To conclude then, in my view Gaia is a fascinating but a flawed hypothesis. It is not a correct characterization of planetary maintenance and life's role therein. Some of Lovelock's claims (such as ice ages being more favorable for life) are seen to be dubious when probed more deeply. Some of the key lines of argument advanced in support of Gaia are insecure, or else give support in equal measure to other hypotheses as well as to Gaia. There is nothing that can only be explained by Gaia.

New life forms have evolved and have had enormous impacts on the environment, sometimes beneficial, sometimes not. Some periods of Earth's environmental history are characterized by favorable conditions for life as a whole, others by Snowball conditions. There is no obvious consistent pattern to it. Such a disorderly and haphazard aspect to planetary history is completely compatible with what we might expect according to coevolution. But, I suggest, it does not resemble the more orderly pattern we might expect according to Gaia. Although the mass effect of the activities of whole groups of organisms is sometimes in a direction helpful to sustaining planetary habitability, equally it is frequently aligned in a detrimental direction.[4] Given this, it is not surprising that the Earth's environment is far from optimal for life.

10.1.6. Where Next?

This concludes my assessment of the Gaia hypothesis. I believe Gaia is a dead end. Its study has, however, generated many new and thought-provoking questions. While rejecting Gaia, we can at the same time appreciate Lovelock's originality and breadth of vision, and recognize that his audacious concept has helped to stimulate many new ideas about the Earth, and to champion a holistic approach to studying it.

In the process of evaluating Gaia in this book, new questions have come up, each one an intriguing major research topic in its own right:

1. If we could design a planet from scratch so as to be best suited for life (taking into account the tendency of evolution to fit life to environment), what would that planet look like?

2. As part of this, what planetary temperature would be best for life in the long run?

3. Environmental change on Earth is not at an end point but is ongoing. Where is it headed?

4. Why has the Earth been getting progressively colder over the last 50–100 million years?

5. Out of chance and mechanism, which of the two has been most important in shaping Earth history?

6. How many thermostat-like processes (such as silicate weathering) have helped keep Earth's temperature on track?

There is certainly no shortage of interesting questions for future research.

10.2. Why It Matters Whether the Gaia Hypothesis Is Correct or Not

I now move on to the second part of this chapter, to consider the implications of this assessment against Gaia and in favor of coevolution. Or, to put it another way, why should it matter whether we believe Gaia or not? What difference does it make?

For a start there is the great intrinsic interest in understanding why our planet remained fit for life, and how it works as a system. This alone makes the subject of great interest. But in addition to this contemplation of how our planet has been shaped and modified over the geological ages, there is also a second, more pragmatic, reason for wanting to understand how our planet works. The question of Gaia's plausibility is especially important at the present time when we need to learn to be good stewards of our planet.

10.2.1. Planetary Management Requires Solid Understanding

It has so far proved extremely difficult to limit global emissions of CO_2. However, the predicted consequences of these emissions are so serious that many strategies (some rather outlandish) to counteract their effects are now being considered. These include (a) direct injection of CO_2 into rock formations to lock it away from the atmosphere, (b) ocean fertilization with iron, (c) dedication of large tracts of land to growing biofuels or forests, and (d) placement of mirrors in space to reflect sunlight away from Earth. Because our own activities are already having a great impact on the natural world and show no signs of slowing, it seems increasingly unlikely that we can just leave the Earth to its own devices and simply assume that a "hands-off" approach is always best. Doing nothing to offset or counterbalance our overwhelmingly large effect on the planetary environment and ecosystems is in itself a decision in favor of planetary modification, albeit inadvertent modification. Such a decision will still lead to change, because this is happening in any case, although the changes in this case would be limited to the unplanned ones. It looks increasingly likely that we as a society are slowly moving away from a belief that we should leave the biosphere to its own devices and instead, at some point in the not so distant future, will embark on some degree of active intervention and management. Despite the dangers of tinkering with a system that we do not fully understand, it may well be decided that the best course for trying to protect the planetary environment is through proactive stewardship.

If we accept that we are going to make additional conscious interventions with the aim of counteracting our original unconscious ones, then those decisions will have to be based on the best available scientific understanding of how the Earth's climate system works. If so, then it is critically important that we rapidly increase the depth of that understanding.

10.2.2. Planetary Scientists and Plane Mechanics

Safely managing a planet is, in some ways, analogous to safely managing an airplane. The safe operation of the vehicle is obviously of great importance to the passengers. For the same reason that airplane passengers are indebted to aircraft mechanics for their careful maintenance and scrutiny of the airplanes, so too our global society has to hope that it can cultivate clear-thinking Earth system scientists. Aircraft mechanics should not be excessive optimists, or else they will clear even the most fault-ridden plane to fly and will put passengers' lives at risk. Nor should they be unduly pessimistic, otherwise airplanes will be continually grounded to general frustration, as took place following the eruption of the volcano at Eyjafjallajökull on Iceland in 2010. The mechanics have to take a considered view of the risks of flying with different faults when they make each and every decision as to whether a plane is deemed airworthy.

As for aircraft, so too for planets, to some extent, except that—needless to say—we can't emergency land or disembark from our planet if it gets into trouble. If our planet starts malfunctioning then we can't get off. Earth system scientists, like aircraft mechanics, not only have to strike the right balance between pessimism and optimism but must also aim to be as knowledgeable as possible about how our planet works so that they can come to an informed opinion about what the risks of different courses of action are. Collectively they must be incisively intelligent in evaluating the risks of different modifications to planetary functioning. They have to try and be aware of possible dangerous effects on the Earth system. It isn't helpful to be unduly pessimistic if the evidence doesn't warrant it, but likewise it is imperative to sound the alarm loudly if a careful assessment of the risks suggests a significant cause for concern. Unlike aircraft mechanics, Earth system scientists cannot veto a potential course of action if they deem it too dangerous; all they can do is to spell out the dangers and then hope that politicians and society as a whole will listen and take appropriate action.

This brings us to why it is important to evaluate Gaia correctly. For reasons about to be described, I believe that a "Gaia mindset" unconsciously predisposes toward undue optimism. As far as planetary stewardship is concerned there could be serious adverse consequences of adopting an incorrect and overly optimistic paradigm view of how the Earth operates as a system.

10.2.3. Gaia Imbues Undue Optimism

Accepting or rejecting Gaia will have consequences for how we understand and therefore how we decide to manage the Earth system, because Gaia posits a framework for how the system operates and self-stabilizes without human interference. A major feature of the Gaia hypothesis is the focus on stabilizing, thermostat-like controls, and the expectation that most aspects of the environment are kept in check by negative feedbacks, or at least were until humans started interfering with the system. Rejecting Gaia should also entail a rejection of any such complacent assumptions. It should introduce a new wariness into our conception of how the Earth behaves.

Although the large majority of Gaia adherents to whom I have talked over the years are also keen proponents of action to protect Earth's environment, nevertheless Gaia, by the very nature of the hypothesis, inculcates a predisposition to suspect natural feedbacks to be stabilizing. It can inspire a false sense of security. And this Gaian worldview has spread more widely, as can be illustrated with a few examples. A prominent climate skeptic blog (there is also an identically named book) is titled *The Resilient Earth*, with the title giving away the preconception of the authors. An "Evangelical Declaration on Global Warming," available on the web, includes the statements: "We believe Earth and its ecosystems . . . are robust, resilient, self-regulating, and self-correcting . . . Earth's climate system is no exception," and "We deny that Earth and its ecosystems are fragile and unstable products of chance." A third example, the Amsterdam Declaration, was mentioned in chapter 1.

Unfortunately, the actual workings of the planet appear to be considerably more complicated than the rather simple picture initially suggested by Gaia. While this makes the topic more interesting from an intellectual point of view, it also makes it more worrying, especially in light of the Earth being the only available home for mankind. A realization that the world does not at all resemble Gaia in its workings will hopefully lead to an increased awareness of the possible fallibility and capriciousness of Earth's natural climate system. In the previous section I mentioned the need for Earth system scientists to strike a correct balance between pessimism and optimism. Gaia has the potential to interfere with this balance because it predisposes toward more optimism than is warranted.

10.2.4. The Need for an Unbiased Worldview

We are currently driving the Earth outside the envelope of its recent history. During the last 800,000 years (the current limit of the ice core record), atmospheric CO_2 has never made up more than about 0.03% (300 parts per million) of the atmosphere. In contrast, at the time of writing we have already caused it to rise to nearly 400 parts per million, and the rate of increase is still accelerating. The speed at which we are adding carbon dioxide to the atmosphere is probably unprecedented during the last 50 million years or more.[5] We are currently stressing the Earth environment in unprecedented ways. Who knows what will happen?

As we carry on playing Russian roulette with the only habitable planet available to us, it is to be hoped that environmental and climate scientists can come to understand the system sufficiently thoroughly, and sufficiently rapidly, and can convey their findings sufficiently lucidly to politicians and the general public. If so, then perhaps the worst dangers of the future can be predicted and the most disastrous courses of action avoided. Without improved scientific insight, however, we will continue to happen upon potential environmental catastrophes unforewarned, and the next one hidden in the future could be even worse than those we are already aware of. In the history of our dealings with our planet we have already accidentally and unintentionally given it a hole in its ozone layer, and, as will be described below, it is only thanks to a minor quirk of fate that this wasn't very much worse. The next discovery could turn out to be even more serious than the ozone hole and global warming. With each of these environmental problems scientific understanding has thankfully been able to provide society with at least a little advance warning, even if easy solutions are not always available. But there is no guarantee that this will always be the case. The next time we discover a problem, the damage might already have been done and it could be too late to repair.

There is a need either to cease strongly modifying the Earth (unlikely, given the size of the global population, which is still growing) or else to work out how to do so in safe ways. It is therefore vital that we ascertain which human-associated environmental modifications are likely to be acceptable in terms of their future impacts, and which are not. In this struggle to discern a safe course for the future it is going to be vitally important to be equally open to the possibility of positive (destabilizing) as well as negative (stabilizing) feedbacks, and to the possibility of sudden jumps, thresholds, points-of-no-return, and regions in which runaway feedbacks dominate, as well as to smooth gradual changes. It is essential that we stay equally aware of potentially harmful or catastrophic feedbacks,[6] as we are of potentially helpful and stabilizing ones.[7] It is important that we adopt a correct paradigm view of how the natural Earth system operates (how life and environment interact with each other), as a basis for understanding how to manage the planet effectively.

The Gaia hypothesis can, if accepted uncritically, lead us to attribute any harmful consequences purely to anthropogenic influences, because of its implicit assumption that the natural system, if left to its own devices, is highly stable and self-regulating toward stability. As discussed in chapter 9, however, the long-term persistence of life on Earth could have had as much to do with luck as in-built safety mechanisms in the Earth system. If this view is correct, then it demands that we should adopt a more suspicious and less complacent approach in our expectations of the future behavior of the natural system. There is a tendency, exacerbated by Gaian thinking, to conceive of "Mother Earth" as always behaving helpfully and that any environmental problems are solely down to our pollution.[8] Such a confidence in Earth's natural resilience is emotionally comforting, but is most likely to be misplaced and could be dangerously deceiving. Sharp changes in climate prior to human intervention were described in chapter 8. We need to rid ourselves of the notion that the natural world is normally a stable, automatically self-regulated system. As I'll shortly describe, anthropogenic influences can potentially combine with unfortunate quirks of the climate system to yield points of weakness leading to the potential for disastrous consequences. It's not always only the pollution that is the problem, but sometimes also Earth's surprising susceptibility to it. The story of CFCs is a cautionary tale that illustrates this point.

10.2.5. CFCs and the Ozone Hole

Ozone (O_3) is a molecule comprising three oxygen atoms. The "ozone layer" is a layer high up in the Earth's atmosphere (in the lower part of the stratosphere[9]) where most of Earth's ozone resides. It intercepts $> 90\%$ of the potentially harmful ultraviolet radiation emitted by the Sun before it reaches the Earth's surface. This matters to us because exposure to too much UV radiation increases the risk of cancer.

Chlorine is transported through the atmosphere within chlorofluorocarbon (CFC) molecules, and when these reach the stratosphere sunlight breaks them apart, releasing the chlorine. The formation of the ozone hole was brought about by the tendency of each chlorine atom that reaches the stratosphere to cleave many thousands of ozone molecules before it too gets removed through combination with methane or some other gas. In most chemical reactions the substances that are required to be present for a reaction to start all get used up as they are converted into new substances by the chemical reaction. But in some reactions one of the necessary substances acts as a *catalyst*; without its presence the reaction would not take place, but it does not itself get consumed in the reaction. Chlorine turns out to be a catalyst in the chemical reaction that destroys ozone in the stratosphere. After instigating each ozone-degrading reaction, the chlorine remains in place to repeat the job over again. It also turns out that the natural system is very poor at mopping up excess chlorine atoms once they reach the stratosphere.

These two unfortunate properties of the natural system were responsible for making it vulnerable to the accelerating industrial production of chlorofluoro-carbons from the 1930s onward.

We all owe a great debt of gratitude to atmospheric scientists such as Sherry Rowland and Mario Molina. Firstly, that they had the perspicacity to identify the problem with CFCs in the first place. And secondly, that they and others had the perseverance and sense of public duty to raise their concerns in the public arena, and then to hold to their positions even in the face of corporate-financed attempts to discredit them.[10]

Paul Crutzen also contributed fundamental research that helped society to recognize the serious problem with ozone. In his Nobel Prize acceptance speech,[11] he pointed out an important but typically overlooked aspect of the problem. Ozone depletion could easily have been much, much worse. If not for a fortunate circumstance, we would probably have largely destroyed the ozone layer before we even had any realization of what we were doing.

In the initial research into which chemical compound should be used in spray cans and refrigerators, both chlorofluorocarbons and bromofluorocar-bons were considered. But it so happened that CFCs worked better and came to be adopted as the standard for these purposes. This was rather fortunate. The ozone-destroying properties of the two molecules were completely unknown and unsuspected at the time the choice was made, but it turns out that each bromofluorocarbon molecule in the stratosphere catalyzes the destruction of about fifty times as much ozone as does each chlorofluorocarbon molecule, before being itself removed. If bromofluorocarbons had happened to be chosen, the impact on the ozone layer would have been much more serious; large-scale ozone depletion would have occurred before we had even appreciated that there was a problem. As Crutzen put it:

> This brings up the nightmarish thought that if the chemical industry had devel-oped organobromine compounds instead of the CFCs—or alternatively, if chlo-rine chemistry would have run more like that of bromine—then without any preparedness, we would have been faced with a catastrophic ozone hole and at all seasons during the 1970s, probably before the atmospheric chemists had de-veloped the necessary knowledge to identify the problem and the appropriate techniques for the necessary critical measurements. Noting that nobody had given any thought to the atmospheric consequences of the release of Cl or Br before 1974, I can only conclude that mankind has been extremely lucky. (Crut-zen 1995)

Every time we sit in the sun and enjoy its warmth, we should be thankful that we can still do so safely. The Earth's UV-protecting layer still exists, thankfully, and reasonable doses of sunlight are still overwhelmingly healthy rather than harmful.

This story[12] of CFCs and bromofluorocarbons should ring warning bells. We had a narrow escape. It would be nice to think that we survived the ozone scare due to intelligent foresight and prudent planning. But, regardless of how comforting such a view might be, it would be totally inaccurate. The truth instead is that CFC production initially forged ahead without any understanding of the consequences of those actions.

Lovelock himself played an important role in the debate over CFCs. After providing crucial scientific data at an early stage, he had a less helpful policy contribution later on. He was dismissive of early concerns about ozone depletion and came to an overly benign opinion about how the natural system would cope with CFCs.[13] The shock discovery of the ozone hole over Antarctica showed, at a stroke, just how wrong he had been. Perhaps, if Lovelock had at the time been developing a hypothesis based on coevolution, without the imputation of resilience to the natural system, he would have adopted a more cautious approach to the dangers of CFCs.

This discovery of ozone depletion happened more than thirty years ago now, but not so much has changed since. We are still ignorant about many aspects of how the Earth works as a system, and even so we are still imposing large changes upon it. And we are still getting surprises. We have been lucky so far in that not even one of those surprises has been catastrophic, even though, as just discussed with bromine, they easily could have been. The next time we stumble across the existence of such a problem, will it be civilization-threatening, and will it already be too late to counteract it?

10.2.6. More Russian Roulette Anyone?

The lesson from all of this is that we should be much more circumspect in how we think about the Earth climate system. It most likely contains other serious flaws and potential lines of weakness[14] that we have not yet even begun to appreciate. It is not a robustly self-maintaining, self-healing system; it is not a superorganism in the sense of being endowed with the tight physiological regulation and safety mechanisms that we expect of the bodies of organisms, which have been exquisitely tuned by millions of generations of natural selection, which in each generation ruthlessly culls failures and propagates successes. Because the Earth climate system has transpired, as opposed to evolved, there is no reason to expect it to be particularly robust or fail-safe.

James Lovelock in his Gaia writings equates the Earth system with a super-organism, with all the attendant connotations of tight physiological control, well-ordered functionality, and robust performance. But what if this analogy is in fact drastically incorrect and very far from the truth? What if it is not just wrong but misleading? As was discussed in chapter 2, we know that the Earth has not undergone biological evolution, and is not an organism. It has not evolved under natural selection and it has not been designed—it has just come to pass. A better analogy would be to compare the development of the Earth's environment with that of a language, where the historical development is haphazard and unplanned, determined mostly by accidental and arbitrary events. Likewise, when we look at maps of very old cities that have grown slowly through the ages by gradual accretion, the arrangement of streets seems bizarre and rather random, at least from the point of view of efficiency and ability to proceed swiftly from one location to another. Their arrangements differ greatly from those of modern planned cities, arranged on a grid. The point here is that the superorganism analogy is unwarranted and excessively optimistic. It is also rather unhelpful in the current situation. Given the lack of any evidence or logical basis for this view, it should be dropped from further use.

As just discussed, there is clear evidence that the system of climate and life on Earth is, from the perspective of maintaining favorable conditions for life, both flaw-ridden (extreme sensitivity to chloro- and especially bromofluoro-carbons) and fault-prone (Snowball Earth, Younger Dryas).[15] The main plank of evidence that would argue for all-powerful self-stabilization is the long-continued habitability of the Earth, but, as discussed earlier, the anthropic principle argues against such a straightforward conclusion. For these reasons I argue that Gaia is a potentially dangerous misrepresentation[16] of how the Earth system works.

Due to the very great complexity of the whole Earth system, and the very many interacting components that contribute to it, our understanding of it is far from perfect. It is still fairly rudimentary at the present time. We continue to be surprised by facets of the behavior of the system, such as our recent realization of potentially serious consequences of ocean acidification including loss of coral reefs, consequences that are being actively investigated at the time of writing. We know that the planet is not a superorganism and is far from completely resilient, but we do not understand all aspects of its fragility.

Echoing a point made by many others, I suggest that our incomplete knowledge also means that we should proceed with extreme caution rather than with recklessness in terms of our impacts on the Earth's environment. We should adopt the precautionary principle. We should be equally open to possible calamities ahead as to possible natural restorative behavior. We should be fully aware that even apparently innocuous human activities might cause calamities, where they happen to impinge on deficiencies or lines of weakness in the Earth system. A complacent belief in the comforting power of Earth to self-heal, which can come as unwanted baggage with the Gaia hypothesis, is neither merited nor helpful. We should instead adopt a less complacent and more cautious approach to dealing with the only habitable planet available to us.

––––––––––––

This book has been about evaluating whether or not the Gaia hypothesis is a reasonable picture of how Earth and life interact with each other. In my view the evidence piles up to say that it isn't. In this last chapter I have also endeavored to show that the answer to this debate is not only of academic interest. It matters greatly. We need a deep and accurate understanding of how our environment works in order to successfully negotiate the next few centuries of tremendous human-induced changes. Ensuring that the global environment remains propitious for life is up to us, and there is no Gaian safety net to come to the rescue if we mismanage it. On the contrary, we need to keep our eyes open for Achilles' heels in the Earth system that could make it particularly vulnerable to anthropogenic impacts. The workings of Earth's climate and environment are considerably more complicated than the simplistic view painted by Gaia, and we must struggle to understand them. This struggle will have a better chance of success if not fettered by the faulty paradigm view that is Gaia.

NOTES

Chapter 1. Gaia, the Grand Idea

1. (Lovelock 1972; Lovelock and Margulis 1974; Margulis and Lovelock 1974)

2. The theory of endosymbiosis has been supported by the discovery that mitochondria and chloroplasts have their own genetic means of replication, separate from those of the host cell. The nucleus of the host cell does not possess all of the genes needed to create its own endosymbionts, and if there are no endosymbionts already in the host cell, then the host cell is unable to manufacture any new ones from scratch. In fact, we now know that only mothers pass mitochondrial DNA to their children, and so changes in mitochondrial DNA over time are influenced only by mutation, and not by crossover. The analysis of mitochondrial DNA has been of great use in determining the structure of the tree of life, because, if mutation rates are assumed constant, the degree of difference between the mitochondrial DNA of two different species (the number of mutations that has accumulated) should be directly related to the time since the two species diverged from their common ancestor.

3. (Lovelock 1979)

4. Daisyworld (Watson and Lovelock 1983) served its original purpose admirably, which was to show that regulation could potentially arise out of population dynamics without any self-sacrifice on the part of the organisms, and without any conscious cooperation or intent. As discussed further in chapter 2, however, Daisyworld includes some unrealistic assumptions.

5. (Lovelock 1988; Lovelock 1991)

6. The proceedings of the first Chapman Conference on Gaia were published in a volume (Schneider and Boston 1991) containing forty-four papers.

7. The proceedings of the second Chapman Conference on Gaia were published in another volume (Schneider et al. 2004) containing thirty-one papers.

8. (Lenton 1998)

9. (Kerr 1988)

10. For instance, one recent book highly praises the Gaia hypothesis: "But the Earth System always strives for balance . . . [Lovelock's] observation that various planetary mechanisms act almost in an intentional way to maintain a temperature favorable to life is an accurate one" (Lynas 2007), and the dust jacket of a recent biography even acclaims Lovelock as "a prophet whose prophecies are now coming true" (Gribbin and Gribbin 2009). Not all share such a view, however. In another recent book Gaia is condemned as "suspended uncomfortably between tainted metaphor, fact and false science" (Beerling 2007). Another recent book (Flannery 2011) espouses Gaia, but is criticized for doing so by Michael Benton, in his review of the book in *Nature* (Benton 2011). Peter Ward's recent book (Ward 2009) rejects Gaia utterly and proposes in its place a directly contradictory hypothesis: that the impact of life on planetary habitability is more often biocidal than bio-favorable. This alternative hypothesis will be briefly evaluated in chapter 10.

11. See articles (Lenton 2002; Kleidon 2002; Kirchner 2002; Volk 2002) in volume 52 of *Climatic Change*, and follow-up articles (Lovelock 2003a; Volk 2003; Lenton and Wilkinson 2003; Kirchner 2003) in volumes 57 and 58 of *Climatic Change*.

12. (Gribbin and Gribbin 2009)

13. The Amsterdam Declaration on Global Change is at http://www.essp.org/index.php?id=41.

14. (Gillon 2000)

15. (Kirchner 1991)

16. The process of coevolution is one in which reciprocal evolutionary influences lead to a series of modifications and countermodifications between highly dependent species; the eventual outcome is a close match between the ecologically intimate species (Futuyma and Slatkin 1983; Nitecki 1983; Thompson 1982).

17. Coevolution can generate some wonderful patterns. Field ecologists studying purple-throated hummingbirds (*Eulampis jugularis*) and *Heliconia* flowers in the Caribbean discovered one such pattern (Temeles et al. 2000; Temeles and Kress 2003; see also the Perspectives article: Altshuler and Clark 2003). They found that the male and female hummingbirds, the only available pollinators for the *Heliconia*, have different beak lengths and shapes: the beaks of the males are short and relatively straight, those of the females are longer and hooked. These match exquisitely to the two different species of *Heliconia* on the islands of Saint Lucia and Dominica, and so the males feed on one species and the females on the other. Not only that, but where one of the *Heliconia* species is absent, the other has evolved two "morphs," one of which fits the male beak, the other the female beak. Another example of coevolved long tongues and long nectar tubes is found in the cloud forests of the Ecuadorean Andes, between the bat *Anoura fistulata* and the flowers of *Centropogon nigricans* (Muchhala 2006).

18. (Langmore, Hunt, and Kilner 2003)

19. (Futuyma and Slatkin 1983; Nitecki 1983; Thompson 1982)

20. (Schneider and Londer 1984; see also Ehrlich 1991)

CHAPTER 2. GOOD CITIZENS OR SELFISH GENES?

1. Overproduction of progeny leads to exponential population increase, as Malthus first realized long ago. This can never be sustained for very many generations because the numbers rapidly become ridiculously large. For instance, the progeny of just one single bacterium (10^{-12} g), doubling in number every half hour by binary fission, could, at such a rate of increase ($P_t = P_0 * 2^{(t/0.5\text{hours})}$), attain the weight of the solid Earth (6×10^{27} g) within fewer than three days, as long as there was no limitation by predation and finite food supply. Of course in reality there is such limitation. There is also a trade-off between "more" and "better" in the production of offspring, with many species having evolved toward producing fewer but more successful offspring. But even an Adam and Eve pair of elephants, to use Darwin's example of the slowest of breeders, would, if reproducing at the normal slow rate for elephants but with all descendants surviving to sexual maturity, multiply in number from two up to circa twenty million living descendants after 750 years. Even those species that do not "breed like rabbits" produce many more young than are necessary simply to replace losses and maintain constant population levels.

2. To give a further example, an adult oak tree may produce as many as 100,000 acorns per year, every year, over a lifespan of, say, 500 years; but on average only one out

of those fifty million acorns will survive to become a fully sized mature oak tree; the other 49,999,999 must fail or be eaten. Not all organisms produce large numbers of young, and some species regulate their reproduction according to environmental circumstances, producing fewer young when times are hard and more when circumstances are propitious; however, these are not exceptions but just matters of degree: without exception all species overproduce progeny.

3. The competition may not always be direct. For instance, in the famous example of the peppered moths that became predominantly black during and after the industrial revolution in Britain, this was because of indirect competition. Although there is presumably competition between individual pepper moths for food supply, experiments have shown that the agent of selection was not a difference in ability to obtain food, but was instead selective predation by birds on those moths that were less well camouflaged. As soot turned the bark of trees blacker, black peppered moths became more successful.

4. It is not a comfortable view because a necessary corollary of high birth rates has to be high infant or juvenile mortality. The alternative is exponential population increase, which leads inexorably, eventually, to resource exhaustion. This was demonstrated in an unintentional experiment on free-living feral donkeys (*Equus asinus*) brought to Australia many decades ago. Following escape from captivity these wild asses increased in number and are now numerous in the semidesert outback of the Northern Territories of Australia. Over the century or more in which they have roamed free there, the population of donkeys seems to have come to an equilibrium with its food supply. In a single part of the range only, more than 80% of the population was culled in 1982–83. The fate of this culled part of the population was subsequently compared to that of the unmanipulated remainder of the population, whose numbers appear to have been balanced at the carrying capacity of the environment. The number of donkey foals being born every year was not significantly different after the culling. Birth rate did not increase. But close monitoring of the population showed that the culled population was recovering at the rapid rate of 20% or more per year, in contrast to the unculled population, which did not increase in size. If births did not increase then how could population? Further detailed investigations showed that the quality and quantity of vegetation makes a significant difference to the healthiness and chances of survival to adulthood of the foals. A much smaller proportion of juveniles died after culling (21% mortality per year) compared to the unculled population (62% mortality per year). The killing of so many of the donkeys in 1982–83 set in motion a chain of events to eventually return the system to its former equilibrium. Reduced grazing pressure allowed recovery of the grasses and herbs that, it is presumed, the donkeys normally keep heavily grazed down. Although the increase in quality and quantity of forage did not affect the rate at which foals were being born each year, it strongly improved their chances of avoiding starvation to survive to adulthood. The resulting population spurt will have eventually returned the density of feral donkeys to carrying capacity (equilibrium levels), where increased malnutrition (death of most juveniles) once more prevented any further population increase (Freeland and Choquenot 1990; Choquenot 1991).

5. To emphasize a point made by many: there is no rational reason for taking the workings of the natural world as a recipe for our own morals. The two are unconnected, and there is no reason why one should inform the other. Evolution is an automatic process with no tendencies toward promoting justice, maximizing pleasure, or minimizing pain; it doesn't care in the slightest. Just because we are products of evolution it does not

in the slightest follow logically that we should adopt its principles as our moral code. Using nature for moral guidance would lead us to legalizing infanticide, rape, and murder. Thankfully we are capable of thinking for ourselves and rising above our natural state. The quirks of nature are certainly not to be taken as a prescription for how we should organize our own society.

6. Think of all the acorns. It is not an efficient way of maintaining forests if only about one in fifty million acorns ever completes the voyage from acorn to seedling to sapling to acorn producer. On a more somber note, think of the starving feral donkeys (note 4). Although some overproduction of progeny is necessary in order to allow population numbers to recover from occasional random dips, and to expand when conditions are favorable, the incessant massive overproduction is not efficient. We tend to think of nature as being extremely efficient in terms of recycling waste, but this is not wholly true; when organic waste products are not nutritious or energy rich they are not scavenged, they simply pile up in the environment (this is the case for calcium carbonate shells and skeletons, as explained in section 2.6.5, and may at one time have been the case for lignin-rich tree trunks, chapter 7).

7. Many such instances of siblicide and sibling conflict have been carefully documented (Mock 2004).

8. Even though they produce on average five sets of molars during their lifetimes, eventually they all wear out.

9. (Wynne-Edwards 1962)

10. For instance, male deer when fencing with antlers, or male elephants battling with tusks, only rather infrequently inflict serious injury on the opposing combatant. Most competitions are resolved following highly ritualistic displays without any recourse to physical contact. Even when the aggression becomes physical, the contests are often carried out in a way likely to keep injuries to a low level. Trials of strength are more likely than fights to the death, with the loser having the opportunity to retire unharmed. Even if an opponent's flank is exposed to attack, foul blows are usually avoided. Lorenz argued that such ritual displays have arisen because of the benefits to the species as a whole.

11. (Goodall 1986)

12. (Hamilton 1964; Williams 1966)

13. Kin selection is the idea that the process of natural selection selects for behaviors and body characteristics that favor the survival and reproductive success of close relatives, as well as of the individual itself. This is the same as saying that evolution acts to optimize an organism's *inclusive fitness* (the sum of its own personal reproductive success together with that of its relatives, multiplied in the latter case by the degree of genetic relatedness with each relative) rather than just the individual fitness. The selfish gene theory says that natural selection is a competition between genes rather than between organisms per se, with organisms just being vehicles for transporting genes from one generation to the next. Kin selection and the selfish gene (gene-centered) view make similar predictions.

14. (Maynard Smith 1964)

15. (Dawkins 1982)

16. As Wade (1976) described, experiments were performed on *Trilobium* beetles, which live in flour. Forty-eight separate colonies of beetles were deposited in separate containers of flour and left to breed for thirty-seven days, after which a new generation of forty-eight colonies was started, each containing sixteen beetles initially. Artificial

selection was imposed by disposing of those colonies that had grown fastest (highest population densities at the ends of the experiments) and choosing the new seed populations only from those colonies that had grown slowest (lowest final population densities). In this way an artificial selection pressure was created toward slow-breeding groups. As the experiment continued through multiple generations, it was found that the average final population density decreased gradually over time, due to the selection pressure. The experiments show the possibility that selection between groups can affect which individuals are successful, that is to say, that group selection can potentially be a factor in evolution.

17. While it is theoretically possible that this type of selection could operate in nature, in combination with selection between genes/individuals, the evidence suggests it is not very influential. If group selection dominated then we would expect to see certain characteristics in groups. For instance, we would expect to see restricted rates of reproduction so as not to exhaust local resources, whereas we mostly see organisms always reproducing as rapidly as possible. Likewise, group selection would predict only one or possibly a few males in each group, because each male can father an almost infinite number of young. The fitness of the group is diminished by the wasted resources expended on males; groups consisting almost exclusively of females would have higher fecundity and would therefore be more successful. If group selection is a strong force, then we would expect a sex ratio that is strongly biased toward females. Instead, a balanced sex ratio is almost universally the case in nature, as predicted if evolution hones individual (or gene) rather than group fitness. Likewise, if group selection is dominant then we would expect infanticide (killing of young group members) to be a vanishingly rare event, rather than the fairly common occurrence observed among meerkats (Clutton-Brock et al. 1998) and other species. Female meerkats, which usually stay in their home group for life, often kill whole litters of pups born to other females in their group, a practice likely to be detrimental to the group as a whole but beneficial to the survival of their own progeny (the infanticide usually occurs shortly before the killer herself gives birth, removing competition for food for her pups; group members share responsibilities for feeding of the young—more young means less food to go round). One reason for the seeming lesser importance of group selection may be that migration of individuals between groups prevents group-oriented behaviors from being uniquely transmitted to the next generation of group members.

18. (Krakauer 2005). Similarly, cooperative breeding in birds (where birds help raise their relative's young rather than their own) is common in species where breeding pairs are faithful, less common in more promiscuous species (Cornwallis et al. 2010).

19. (Mehdiabadi et al. 2006). There are many other cases where microorganism cooperation is found to be associated with close genetic relatedness (Buckling et al. 2009).

20. In animals with memories, cooperation is also observed to occur between unrelated individuals (non-kin), on a "you scratch my back, I'll scratch yours" basis. There are a number of examples from higher animals of such reciprocity, as Tim Clutton-Brock (2009) detailed. These include mutual grooming, alarm calls, and food sharing.

21. The expression (Hamilton's rule) that determines whether natural selection should favor an act of altruism from one organism to another is satisfyingly simple: $rb > c$. Where r is the degree of relatedness between two organisms, b is the individual fitness benefit that would accrue to the recipient of the altruism, and c is the individual fitness cost to the organism carrying out the altruistic act. As a famous quote puts it (from J.B.S.

Haldane): "I would lay down my life for two brothers or eight cousins" (the degree of relatedness is 0.5 for brothers, 0.125 for cousins). There is a concise elegance to this simple expression. Its simplicity belies its profound importance in helping us to understand evolution.

22. In fact, it turns out not to be solely just the degree of relatedness in a colony that controls whether reproduction is the preserve of only a few. In bee colonies, at least, it is also a function of the efficiency of the "policing" undertaken by the workers in honeybee colonies. If colony members other than the queen try to lay eggs, in colonies with effective policing these are usually detected and eaten.

23. Eusocial animals live in social groups typically containing sterile workers and a queen. Some other typical characteristics are that (a) brood care is carried out collectively within the nest, in contrast to each adult or adult pair caring for their own offspring; and (b) offspring growing up in the nest will often devote their lives to supporting the nest.

24. (Reeve et al. 1990)

25. There are still some instances that remain puzzling, however, and seem to resist easy reconciliation with kin selection. One example is wasp species such as *Agelaia angulata* and *Parachartergus colobopterus*, whose nests contain large numbers of queens with the result that nest members are on average much less closely related than in single-queen nests (Queller, Strassmann, and Hughes 1988). Because of the low average relatedness most female worker wasps are raising young that are not their own (although see note 49 for a possible explanation for this phenomenon). Another puzzling example is the aforementioned siblicide in birds' nests; it seems counter to kin selection to kill your own sibling. A tentative explanation is that, in environments in which there are not enough resources to enable the parents to raise two chicks, it may still pay to lay two eggs but have only the healthiest survive in order to minimize the chances of no chicks at all surviving because of infertile eggs or early death of the only chick.

26. These experiments are fascinating because, for the first time, they advance kin selection from being (merely?) an elegant explanation of a large set of facts, to being a theory that is amenable to direct experimental testing. The study with bacteria (Griffin, West, and Buckling 2004) grew populations comprising two different strains. Altruistic activity (after a total of forty-two bacterial generations, to allow evolution to take place) was assessed by the rate of production of an iron-binding protein that individuals synthesize and then release into the surrounding medium, but which any individuals can recapture and benefit from (to take up the iron). When each population was genetically distinct (the two strains kept apart), altruism was found to be greater than when each population was genetically diverse (the two strains mixed together). The bee study (Langer, Hogendoorn, and Keller 2004) was able to manipulate experimentally the degree of relatedness of breeding female bees. When a group of ten related bees was put in the same experimental plot containing several possible nest sites, they usually chose to nest together in pairs or threes. When groups of unrelated bees were released, however, they more commonly tended to nest alone, despite lower survival rates of solitary nests. It was demonstrated that the decision whether to cooperate or "go it alone" was affected by relatedness. The results of both of the experimental studies support kin selection, although other factors were also shown to modulate the degree to which altruism was favored.

27. Although Lovelock also mentions eels, it is impossible for them both to perform the same phosphorus-transporting function, since they perform the opposite migration.

Salmon spawn inland, after which the young salmon migrate to sea before returning as adults to spawn. Eels spawn out in the open ocean (the Sargasso Sea, famously) where the larvae mature into elvers, which then migrate up rivers to mature inland. Finally, after growing into adults, they return to the ocean to spawn. Eels are thus net *exporters* of phosphorus from the land (because of the extra body mass they put on while living in freshwater before returning to sea).

28. (Maynard Smith 1982)

29. The ESS concept is extremely useful, but it would nevertheless be a mistake to believe that the structure and behavior of organisms is always perfectly adapted. Evolution is always constrained to work with what has gone before. It is a blind process without any foresight, and so cannot go "back to the drawing board" and start again from scratch, even if that would eventually give better results. Nature is full of clumsy solutions such as the recurrent laryngeal nerve in the giraffe, or vestigial relics that are no longer useful, such as the appendix and wisdom teeth in humans (for other examples see Williams 1996).

30. In practice it is sometimes a little more complicated than sketched out here. For instance, the ESS is sometimes a variable strategy that depends on the actions of the opposing organism in an interaction, such as the "tit-for-tat" strategy. Or sometimes there are multiple ESSs, and stability is only predicted for a mixture of different strategies.

31. Needless to say any *single* salmon mutating to this strategy would have considerable difficulty in finding a mate.

32. The "you scratch my back, I'll scratch yours" type of cooperation (reciprocal mutualism) is an exception. (For a review discussing possible causes of cooperative behavior other than kin selection, see Clutton-Brock 2002.)

33. (Watson and Lovelock 1983). An animated introduction to Daisyworld can be viewed online at http://library.thinkquest.org/C003763/flash/gaia1.htm.

34. An explicit mathematical description of Daisyworld can be found in Watson and Lovelock (1983).

35. (See for example Weber and Robinson 2004; Nordstrom, Gupta, and Chase 2004; and Harding 1999)

36. Page 264 of Lovelock's autobiography (Lovelock 2001).

37. Greenhouse gases warmed the world, in preindustrial times prior to human influence, by about 33°C; on the early Earth, with much higher levels of atmospheric CO_2 and/or methane, the effect must have been much greater. More than this, the available evidence shows that Earth surface temperature and atmospheric CO_2 concentration have varied in close concert with each other across glacial-interglacial cycles of the last 800,000 years (see chapter 8), and perhaps also for much longer (Royer, Berner, and Park 2007). Changes in Earth albedo also have the potential to impart significant changes in planetary temperature, but are less important than greenhouse gases. For instance, if the overall Earth albedo (sum of sunlight reflected from land surfaces, oceans, and clouds) was to increase by 10%, up to ~40% from its current value of ~30%, then this would cool the planet by almost 10°C.

38. (Robertson and Robinson 1998)

39. (Wood et al. 2008)

40. A new approach is an improvement but also suffers from a lack of generality. A more sophisticated recent model generates global-scale environmental stabilization without Daisyworld's critical assumption that what is good for the individual is also good for

the global environment. Space does not allow a full description of this model here (for details see Williams and Lenton 2008), only a brief description. The planet is modeled as a ring of interconnected flasks containing microbes. Microbes can alter their local flask environment according to their genetic traits, and if they (as a flask population) alter their environment beneficially then the flask will be able to sustain a higher biomass. Flasks with a higher biomass export more microbes to adjacent flasks, along with their genetic traits, and in this way Gaian traits tend to be propagated. The model is imaginative and interesting but in my opinion is another "parable" rather than a realistic depiction of the natural world; there is again a built-in correspondence between what is good at the local level (this time the flask rather than the individual) and the global level. The model still seems unrealistic compared to, for instance, the well-mixed global atmosphere, which is not split into separate flasks that interact separately with different regions or continents or ocean basins. A universal explanation for how Gaia could emerge from natural selection remains elusive.

41. (Paasche 2001)

42. The term *homeostatic* is used when a system has a propensity to resist perturbations, or in other words has a tendency to return toward some "set-point" value if nudged away from it. The best example is that of a thermostat controlling room temperature. The "set-point" temperature is controlled by turning the dial to whatever temperature you want the room to be. If the sensed room temperature is below that set point then the thermostat turns on the radiators or furnace until the room warms up to the set point. If the sensed room temperature is above that set point then the thermostat turns off the radiators and may also turn on air-conditioning.

43. Richard Dawkins was the first, to my knowledge, to describe this concept of the influence of the genes spreading out beyond the boundaries of the individual bodies they are housed in (Dawkins 1982).

44. (Odling-Smee et al. 2003; see also Turner 2000)

45. (Dawkins 2004). Dawkins suggests a distinction between "niche changing" (inadvertent side effects on the environment that happen to take place, but not because of the consequences) and "niche construction" (beneficial effects on an organism's environment that occur because they are beneficial to the organism itself, and are therefore maintained and favored by natural selection). Dawkins also advocates caution in taking the notion of the "Extended Organism" (Turner 2000) too far.

46. Colonies of some species have multiple queens; in other species each colony has a single queen, but she mates with multiple males.

47. Haplodiploidy is a curious system of genetic inheritance and sex determination found in some bees, wasps, ants, thrips, and beetles. When sperm and egg combine and develop into a new individual, in haplodiploid species this individual will always grow up to be a female, and never a male. Males of these species develop only from eggs that are not fertilized by sperm. From this it can be seen that males have mothers but no fathers, and never have sons. Moreover, they have only one set of genes (inherited from their mothers) and must always pass this very same single set of genes to all of their daughters, who for this reason are genetically even more similar to one another than sisters normally are. The term *haplodiploidy* comes about because one of the sexes possesses only one set of chromosomes (i.e., is haploid), whereas the other sex possesses the more normal two sets of chromosomes (i.e., is diploid).

48. See the review paper by Crozier (2008). Although modeling studies do support the hypothesis that haplodiploidy favors the evolution of eusociality, it may be that its

importance has been overestimated in the past, although this subject is still being debated. Clearly haplodiploidy cannot be the only explanation, because by no means all eusocial species are haplodiploid (termites for instance are diploid). While haplodiploid sisters are unusually closely related (3/4) to one another, favoring increased cooperation, this is counteracted by a more distant than usual relationship to their brothers (1/4). The genetic similarity of these "supersisters" is such that they are considerably more similar to one another than they are to their own offspring; if faced with a choice between rearing eggs of the queen that will develop into future supersisters, or with rearing their own eggs, they should therefore choose the queen's eggs, all else being equal.

49. In a fascinating recent paper (Hughes et al. 2008), the mating habits of 267 species of eusocial bees, wasps, and ants were analyzed. This analysis suggests that the evolutionary ancestors of all living eusocial insects were monogamous at the time in their evolutionary history when eusociality evolved. That is to say, at that critical time each female (queen) mated with only one male, and there was only one queen per colony. It seems likely that multiple matings, and also multiple simultaneously breeding queens in a colony, both of which tend to reduce relatedness, only evolved later, and only after evolution had "locked in" eusociality by making some castes irrevocably sterile. The upshot of this study is to give strong support to the hypothesis that eusociality developed exclusively in species exhibiting high genetic relatedness within colonies. One group of scientists disagree however about the importance of relatedness in explaining eusociality, placing more emphasis on the production of a defensible common nest site as a precondition for the evolution of eusociality (Nowak, Tarnita, and Wilson 2010).

50. (Jacklyn 1992)

51. (Grigg 1973)

52. The level of support for such a view was recently made clear by the scale of the response to an attack on it. The article mentioned in note 49, which rejects kin selection as the explanation for eusociality (Nowak, Tarnita, and Wilson 2010), drew a fierce riposte from other evolutionists. There was an overwhelming and overwhelmingly negative response to their article. Five separate rebuttals to Nowak et al., featuring nearly two hundred authors altogether, were published in the *Nature* issue of March 24, 2011.

53. Except in cases of (1) mutually beneficial symbiosis, when it is not really altruism at all, and (2) when the favor may be repaid later, i.e., "you scratch my back, I'll scratch yours." It is worth repeating that this should not in any way be taken as a prescription for human morals—we of course can, and do, adopt more principled arrangements.

54. Bleaching events on coral reefs occur when the polyps expel their symbiotic algae (zooxanthellae), from which they derive food, and are so named because the corals look paler without their symbionts. Bleaching events are a stress response, for instance to high temperatures. If the stressful conditions abate then new symbionts are usually recruited. But if the stressful conditions persist then the bleaching will be fatal.

55. This is not to deny that forests and corals have very strong effects on their environments. They do, but the question is whether those effects are incidental (happenstance) or whether they are effectively coordinated toward homeostatic regulation. Trees have strong effects on the forest floor on which they stand, by dropping needles, leaves, and twigs that go to form soils, and by sending down roots into the soil and even down toward the bedrock to mine for nutrients. They also influence the temperature by affecting the local albedo, sometimes rather strongly as can be seen by the contrast at high latitudes between bright snow and dark green coniferous forests, in which each branch is inclined downward to encourage snow to slide off. Coniferous forests that blanket

northern Canada and Scandinavia absorb significantly more sunlight than the adjacent snowfields. Rainforest trees significantly impact the local water cycles by sucking up vast amounts of water and then evaporating it through their leaves. There is no doubt that trees play a large role in constructing their own niches. But there seems little evidence to suggest that there is any cooperative venture between all the species in a forest. There is no counterpart at the scale of woods or forests to the tight control over temperature or other variables that is seen in individual bodies or in the nests of some eusocial insects. When the inside of a bees' nest gets too hot or too cold then the worker bees collectively adopt behavioral strategies (section 2.6.3) that bring the temperature back to normal. No parallel response can be seen in forests.

56. For more on the importance of waste accumulation in determining the properties of the Earth system, see Tyler Volk's articles (2003; 2004) and Wilkinson (2004).

57. The stony corals that build reefs depend on sunlight, and cannot grow at depths deeper than about fifty meters below the water surface (less where the water is turbid). Not all corals require sunlight, however. Some corals do not possess symbiotic algae. Deep-sea corals have been discovered in many places in recent decades, living on the seafloor hundreds of meters down, in darkness to which virtually no sunlight penetrates. These, unsurprisingly given the lack of light, possess no algal symbionts; they live instead by filter feeding. In general it appears that these cold-water corals secrete limestone more slowly and do not rival the architectural wonders of their surface counterparts. However, the finding that the legs of some oil rigs, deep underwater, have already been colonized by corals suggests that under the right conditions deep-water corals can grow reasonably quickly.

58. Parrotfish have numerous teeth forming a parrot-like beak, and nibble at coral. It might appear that they are purposefully eating it. In fact, their only purpose is to get at the algae and coral polyps upon which they feed. Although they do ingest some hard coral incidentally, they do not digest it. Rather, they excrete it later as coral sand. In addition to calcium carbonate, opal (silica) is another mineral synthesized biologically to make shells and skeletons. Again, it is of no nutritional value and is not recycled. Like calcium carbonate, it simply piles up on the seafloor. The silica phytoliths in grasses are likewise not absorbed by the digestive tracts of terrestrial herbivores, but rather pass through and are subsequently excreted (Jones and Handreck 1965). While it is often thought about nature that "nothing is wasted," this is not true. It is instead correct to say that "little of value to any organism is wasted."

59. And also from dead coralline algae such as *Halimeda*. In fact, coralline algae probably contribute more of the rubble than corals, but the story applies equally well to them (even more so, as it happens, because they are photosynthetic organisms, unlike corals which are animals harboring photosynthetic symbionts).

60. This process for coral atoll formation was first proposed by Charles Darwin (Darwin 1842). However, it was only after very many decades of intellectual debate and investigation, wonderfully described by Dobbs (2005), that Darwin's theory was vindicated triumphantly in the 1950s by deep drilling down into the Eniwetok Atoll in the South Pacific. This drilling, which brought up 1,405 meters of reef limestone before reaching the underlying volcanic rock, at last provided convincing evidence that Darwin's theory is correct.

61. In the few cases where this occurs, such as Henderson Island in the southeast Pacific, it is not because corals have learned to live out of water. Instead, these strange

elevated coral islands occur where tectonic uplift has followed subsidence and has raised an atoll many meters above sea level. A covering of soil and vegetation develops on top of the once-living corals.

62. (Jenkyns and Wilson 1999)

63. (Volk 2004)

64. To add yet another clarification to the long list for this chapter, it should be noted that these "islands of genetic homogeneity" could also, from a different perspective, be seen as rafts of genetic heterogeneity. Consider a cow, for instance. The DNA is identical in the nucleus of each cow cell, whether the cell comes from the cow's limbs, from its brain, or from any of its various organs. And that is why the cow acts as a coordinated individual. But while recognizing this genetic uniformity, we also need to be aware that there are many additional genes present in a cow—firstly because some machinery in each cow cell is inherited from distant microbial ancestors and hence has genes of its own (the endosymbiosis mentioned in chapter 1), and secondly because all sizeable organisms represent potential resources and opportunities for other organisms. Any mature cow, certainly in the natural wild state, is infested with a large collection of various parasites, each with its own genes. The same is also to some extent true of termite, ant, or wasp colonies. When we add the genes of useful (enslaved?) bacteria or fungi hosted within cows and termite colonies, for instance to help them digest the cellulose in grass (which they can't do by themselves), it is apparent that even these islands of genetic homogeneity contain many "foreign" genes. All of this does not, however, detract from the overriding point, that there is no correspondence between this state of affairs and that for the planet as a whole. There is of course no coordinated network of cells stretching around the planet, each with near-identical DNA within their nuclei, each communicating and cooperating with one another. There is no superorganism in that sense either.

65. The only exceptions to this are mutually beneficial symbioses resistant to freeloading, and forced symbioses where one species "enslaves" the other. Neither of these offers an explanation for cooperation at the spatial scale required for Gaia.

CHAPTER 3. LIFE AT THE EDGE

1. It is at best arguable as to whether these constraints are properly described as "narrow." For instance, a pH range of 3 to 9 corresponds to a six-orders-of-magnitude range in hydrogen ion concentration ($pH = \log_{10}(1/[H^+])$).

2. The terms *environment* and *habitat* are used interchangeably here to refer to the complete collection of physical, chemical, and *biological* factors that affect an organism during its life. Both terms are assumed to include the physical landscape and the chemical composition of the medium inhabited by the organisms. They are also deemed to include, for a particular organism (for example, an antelope), competing organisms at the same trophic level (zebra for example, for antelopes), as well as predators (for example, lions in this case), parasites, and also whatever they eat (grass, in our current example). Both terms are used to refer not only to the place where the organisms live but also to the assemblage of other organisms with which they interact (the "biological landscape").

3. (Holland 1984)

4. Care has to be taken to distinguish between the abilities of organisms to *tolerate* extreme environmental conditions for restricted lengths of time, and the abilities of or-

ganisms to *grow and reproduce* while living constantly in the extreme conditions. Many microbes and animals possess adaptations allowing them to "shut down" during inhospitable periods, after which they can reactivate once conditions improve. Hibernation in mammals is one such adaptation (partial deactivation in this case), as is the formation of reproductive cysts, seeds, spores, and eggs, all of which can have the capacity to remain dormant until suitable growth conditions are sensed. Likewise, the ability of nematode worms and others to resist temporary dessication (Wharton 2002) and tardigrades to survive a trip to absolute zero (−273°C) and back. But the ability to tolerate stressful conditions temporarily is not the same as the ability to complete multiple life cycles (to reproduce) in those environments. It is for this reason that Emperor penguins are not included in table 3.2; while they can survive months of exposure to the Antarctic winter (temperatures as low as −50°C), they cannot survive and reproduce if permanently in such conditions. To complete their life cycle they need periods of much milder conditions (including an ice-free ocean), which they find during the Antarctic spring and summer when they travel north from the Antarctic interior. Likewise, although larch and spruce trees can survive Siberian winters in which temperatures sometimes dip below −50°C, they could not survive an environment that was permanently so cold.

5. A second important distinction is between those organisms that are *optimally adapted* to extreme environmental conditions and those that are merely able to *survive* them. Some organisms can just about maintain population growth in extreme conditions, but grow much better in nonextreme conditions. However, "it is a tenet of extremophilic microbiology that true extremophiles are those that actually *require* the extreme for growth and survival" (section 1.2 of Madigan 2000; see also Madigan and Marrs 1997). To put it another way, some organisms are facultative extremophiles (they can cope with extreme conditions if need be), whereas the true specialists are obligate extremophiles (they can survive only in their extreme environments, however brutal). True hyperthermophiles are *stimulated* to grow faster by high temperatures. The halophile *Halobacterium salinarum* requires at least 150‰ salinity in order to be able to carry out its normal metabolic activities and reproduce.

6. The tree of life has undergone many revisions over the years. Originally there were considered to be just two Kingdoms (the previous highest level of classification): plants and animals. This was subsequently revised to five Kingdoms: plants, animals, fungi, protista, and monera (bacteria). Then, a few years ago, genetic evidence showed that the prokaryotes consisted of two very distinct lineages, bacteria and archaea, which must have separated far back in evolutionary time. This led to the currently accepted proposal, by Carl Woese, of three Domains (with the new term *Domain* referring to an even more fundamental level of classification than a Kingdom): Bacteria, Archaea, and Eukarya (the latter containing all plants, animals, fungi, and protists).

7. See Zettler et al. (2002) for an unexpectedly high diversity of eukaryotic organisms in the highly acidic (pH = 2) Rio Tinto. This river has a deep red color and much higher concentrations of heavy metals (for example, 3–20 g of iron per kg of water) than are usually found in freshwaters. This ecosystem in rather hostile chemical conditions is ancient and is not a result of human mining; the unusual eukaryotic diversity may therefore be explained by the longer duration for organisms to arrive at the ecosystem and evolve there.

8. (Koschorreck and Tittel 2002)

9. (Smith and Baco 2003)

10. As an aside, this fact has practical implications for those trying to reconstruct the history of Earth from ocean sediment records. In oxygen-rich sediments, burrowing worms and other higher animals are able to survive. This is a disadvantage (to paleoclimatologists) because the animals disturb and mix the sediments. This phenomenon of churning of the sediments is known as *bioturbation*. Bioturbation is a bane to paleoceanographers, because it smudges together different annual bands of deposited sediment and hence makes dating much more difficult. Where the rate of sediment build-up is slow, bioturbation makes it impossible to resolve fine-scale events, because material deposited many thousands of years apart becomes churned together. In those places where a piece of seafloor has for a long time been anaerobic, on the other hand, the material falling to the seafloor is preserved in exquisite detail. Usually there is some seasonality to the rate at which particles arrive at the seafloor. If the location is near to the coast then high rainfall on land in spring may lead to high rates of terrestrial sediments being washed out to the ocean every spring, creating a gray layer once per year in coastal marine sediments. Alternatively, in an open-ocean environment there may be higher rates of phytoplankton growth in the spring and hence an organic-rich layer once per year. Anaerobic sediments are frequently characterized by distinct annual banding. The bands can be extremely helpful in terms of being able to count years down through the sediments.

11. (Peck, Webb, and Bailey 2004; Peck 2002)

12. These case histories of dodos, whales, and blind cavefish all borrow heavily from the relevant chapters of Dawkins and Wong (2004); see the chapter on hippos for the story of how whales came to be. David Quammen also covers the dodo story (Quammen 1996). The genetics of blindness in cavefish is covered by Yamamoto, Stock, and Jeffery (2004).

13. (Hunt, Bell, and Travis 2008)

14. (Lahti 2005)

15. (Shimamura et al. 1997; Nikaido, Rooney, and Okada 1999)

16. Genetic studies have thrown light on the different ways in which dolphins and whales lost their legs over time (Thewissen et al. 2006).

17. Darwin was wrong on one point, however. He was unaware, along with everyone else at the time, of the mechanism underlying evolution. As a result he was tempted into attributing regressive evolution in part to the now long-discredited Lamarckian mechanism of disuse. According to Lamarck, animals that do not use a particular organ during their lifetime (such as a rat that had got lost inside a cave, and could not use its eyes) would give birth to progeny possessing less developed versions of that organ. Nowadays of course we know that there is no such direct link between an organism's experiences and its genetic make-up, but this was not known in Darwin's time. In recent times interest has been focused instead on whether regressive evolution occurs because of natural selection for economy of resources (e.g., a cave-dwelling animal that saves resources by not building eyes will then have more resources to allocate elsewhere) or because of evolutionary drift (that is to say, a lack of selection pressure for eyes in caves would mean that any cavefish possessing genetic mutations leading to nonfunctioning eyes would be equally successful, and so mutant genes giving rise to mutant eyes would no longer be deleted from the population).

18. From Arthur C. Clarke's book *2010: Odyssey Two*.

19. In fact this concept is embodied in the very name "extremophiles," from "extreme" and "lovers" ("philos" in Greek).

20. The chances of life on Europa are thought to be finite but small. Europa may host life, but most probably doesn't. For in-depth technical discussions see Gaidos, Nealson, and Kirschvink (1999); Chyba and Hand (2001). Ganymede and Callisto, two other large moons of Jupiter, are also thought to possess salty liquid oceans beneath surface ice (Kivelson et al. 2000).

21. Douglas Adams's response, in a radio interview, is reconstructed from memory by Kirchner (Kirchner 2002).

22. All life shares certain similarities in, for instance, the genetic code (RNA and DNA), that make it unlikely that life has evolved from scratch several times. We also see fossilized stromatolites in rocks more than three billion years old that resemble living stromatolites on Earth today. The thread of life may at times have been stretched extremely thin (for instance during "Snowball Earth" events, chapter 8), but it seems that it was never completely broken.

23. Lovelock's quote is open to another interpretation, depending on his intended meaning of the phrase "narrow constraints for cell survival." It is not clear whether he was referring to the limits to survival for those organisms currently prevailing on Earth, or to the habitable envelope for life as a whole, after allowing evolution time to adapt to the changing conditions. If he was referring to the latter, the ultimate limits to life, then habitat commitment is not relevant. But even in this case I would argue that there is still not strong evidence in favor of Gaia. The ultimate constraints (tables 3.1. and 3.2) are, I would say, wide rather than narrow, encompassing as they do a range of more than 100°C of temperature, from salt-free solutions right up to fully saturated saline solutions, and a pH range from < 0 to 12 (i.e., greater than a 1,000,000,000,000-fold variation in hydrogen ion concentration). This is more than just a "glass half full" versus "glass half empty" question, but to some extent it is still a subjective choice. Of course it is fortunate that the Earth was originally (when life first evolved) a suitable crucible for life, and it is either fortunate or else down to a process like Gaia that it stayed that way over the ensuing billions of years. This last question falls square within the remit of chapter 9.

CHAPTER 4. TEMPERATURE PACES LIFE

1. As quoted on page 148 of Lovelock (1991), from conversations between Lovelock and Schneider at the US National Center for Atmospheric Research (NCAR) in 1986. Lovelock's response to the criticism can be read in full in the ensuing pages of his 1991 book, but is summarized as "Gaia likes it cold." He has also compared interglacials to "fevers" (of the planetary superorganism).

2. Which is the most informative indicator of whether an ecosystem or biota is thriving or is just scraping along? I don't think we really know, and statements of the Gaia hypothesis have not been very specific. In terms of this chapter's theme, fortunately, the answer doesn't really matter. As I will attempt to show, all of the above quantities are enhanced (to a greater or lesser degree) by warmth and depressed by cold. When assessed over the whole globe, not even a single one of them is enhanced by cold.

3. The general argument in this chapter is that warmer climates are generally more favorable to life. An important caveat needs to be stated at the outset, however, which is that this should not be taken to imply that the current global warming is "a good idea." In this chapter I am comparing the suitability of climates that stay warm or cold, without considering the suitability of transitions from one to the other. The current ongoing climate modification through injecting carbon dioxide rapidly into the atmosphere, if continued at the current (incredible, geologically speaking) rate, will continue to force an extremely rapid global warming with little time for an "orderly transition" of vegetation and animals. Even though tree distributions shifted poleward in response to climate improving upon emergence from the last ice age (Gates 1993), and even though vegetation appears in some places to already be on the move due to global warming (Sturm, Racine, and Tape 2001), such natural adjustments of vegetation will be hampered in the future by the precipitate rate of climate change and by "habitat fragmentation" (agricultural and urban land acting as barriers to dispersal). Each farm, town, and airfield represents a hurdle across which plant species will need to send seeds in order for their descendants to live closer to the poles than did their forebears. It is likely that many species, especially trees, will not be able to migrate sufficiently quickly, and consequently some species will find themselves stranded in unsuitable climatic zones and will die out as a result. Many species restricted to mountainous regions are being squeezed out of habitat space as they migrate to higher altitudes in response to global warming (Lenoir et al. 2008); eventually there will be nowhere higher for them to go. Crop production may improve in some areas but there are also likely to be many associated problems, not least with many crop plants having narrow temperature optima, and hence the areas they are currently grown in will become unsuitable. Some places will become much drier in the future, others wetter. A full discussion is far beyond the scope of this book, but taken together the evidence is clear that the current alarming rate of global warming is on the whole much more harmful than beneficial, in terms of its likely impacts on plant and animal life over the next few centuries.

4. (Arrhenius 1889). The exponential relationship applies for temperature expressed on the Kelvin scale (K), where $T(K) = 273.15 + T(°C)$. This leads to the definition of "absolute zero" ($0K = -273.15°C$) at which all chemical reactions take place infinitely slowly (i.e., not at all). Chemical reactions are very much possible at the freezing point of water (273.15K, 0°C) as long as they don't require liquid water itself, of course.

5. As so often in nature, there are some exceptions. It is not always completely either/or. Some species are neither a straightforward ectotherm nor a straightforward endotherm. Some large fish and sharks swim *continuously* to keep their bodies warm, which allows them to remain active even in cold waters.

6. Another example comes from the manufacture of salt by precipitating it out of seawater in a salt pan. The rate of precipitation is controlled, primarily, not by the present temperature of the water but rather by the amount of evaporation that has previously taken place (which itself is of course a function of the previous temperature history of the salt pan, but this is by the by). It is the degree to which the initial seawater has already been concentrated into a strong brine that mainly determines how readily salt crystals form in it. If two brines of the same strength are kept at different temperatures, we would expect crystals to form more rapidly in the warmer brine, but if two different strength

brines are kept at different temperatures, then we cannot be sure without additional knowledge which will precipitate out salt more rapidly.

7. See for instance West, Brown, and Enquist (1997).

8. The 0.75 relationship also has important consequences for the organisms concerned. A result of the more rapid metabolism of smaller animals (when calculated as rate of metabolism per gram of body tissue) is that smaller animals need to eat more, relative to their size. A shrew must eat more than its own body weight of food each day in order to stay alive, whereas an elephant need only eat one-fiftieth. This topic of how body size impacts on metabolic rates is described interestingly and at greater length by Whitfield (2006).

9. They argue that it derives from limitations as to how much "exchange surface," whether area of lung tissue for exchanging carbon dioxide and oxygen, or area of intestinal wall for exchanging nutriment, can be packed into bodies of different sizes (West, Brown, and Enquist 1997).

10. See for instance Gillooly et al. (2001) and Gillooly et al. (2006). The combined effects of body size and temperature on metabolism, together with resource stoichiometry, are now being used to generate predictions about a wide range of biological rates and ecological processes (Brown et al. 2004).

11. (Gillooly et al. 2002)

12. "Kleiber's Law," the finding that metabolic rate scales as the 0.75 power of body mass, does not apply in absolutely all instances. Up until recently it was thought that this law applied equally across all species of plants and animals. Recent measurements of the respiration rates of about four hundred tree seedlings and saplings have shown, on the contrary, to the surprise of many, an exception: the respiration rates of the seedlings and saplings are proportional instead to the 1.0 power of body mass, i.e., directly proportional to body mass (Reich et al. 2006). This may be because seedlings and saplings do not possess a "space-filling" architecture (see comment and reply about Reich paper: Enquist et al. [2007] and replies by Reich et al. [2007] and Hedin [2007]). Analysis of a large dataset of mammalian metabolic rates, containing five orders of magnitude range in body mass, suggests a slightly curved fit on a logarithmic graph, or in other words neither a strict 0.75 nor a strict 2/3 (as some others have suggested) nor a strict 1.0 power relationship. In fact the curvature suggests it may not be a straightforward power relationship at all (Kolokotrones et al. 2010).

13. (López-Urrutia 2003). The discontinuity at ~0°C also manifests itself in other, less direct, ways. For instance, permanently frozen soil (permafrost) prevents trees from sending down taproots and hence excludes large trees. Ground-lying herbs and mosses in polar environments are also seasonally covered by snow and hence restricted to short growing seasons.

14. (Johnston et al. 1998; Clarke 2003; Peck 2002). Mitochondria, the powerhouses of cells, produce ATP, the energy form muscles need. In the same way that oil coming out of the ground is refined into petroleum before being fed into the cylinder of an internal combustion engine, so mitochondria perform a comparable process of refining pyruvate into ATP before muscle fibers can use it.

15. There is a suggestion in the data that when normalized to the same temperature, higher-latitude ecosystems respire more rapidly than do lower-latitude ecosystems (within the 30 degrees of latitude range of the study sites used) (Enquist et al. 2003). So, when comparing a warmer location with an annual temperature range of 10°–30°C to a colder

location that experiences 0–15°C, the respiration rates at 15°C are greater in the colder than the warmer location. Some of this pattern may result from evolutionary adaptations to low temperatures (section 4.1.4). Some of this pattern may also derive from the study set containing some relatively barren (desert and grassland) sites at the lower latitudes but not at the higher latitudes, such that like was not compared with like (different vegetation densities at different latitudes). It is only recently that global datasets such as this have started to become available. It feels like a tremendous privilege to stand on high and look out across such an immensity of data, collected thanks to the unstinting efforts of so many, and to get a first glimpse of how important processes such as ecosystem respiration vary across the globe. Nevertheless, it is still difficult to extrapolate reliably from this dataset to determine if high-latitude forests really do have higher temperature–normalized respiration rates. Setting aside this question for the time being, the data nevertheless argue for ecosystems respiring more rapidly on warmer nights (e.g., 20°C) than on cooler nights (e.g., 10°C), within their seasonal temperature range.

16. Wet and moist tropical forests generate on average circa 800 g C m^{-2} of new biomass every year, compared to 430–650 g C m^{-2} y^{-1} for other forest types, 1,300 g C m^{-2} y^{-1} for wetlands, 320 g C m^{-2} y^{-1} for temperate steppes, 130 g C m^{-2} y^{-1} for tundra, on average 80 g C m^{-2} y^{-1} for deserts, and near zero for bare rock and ice. From table 5.2: Primary production and biomass estimates for the world, in Schlesinger (1997). Whereas tropical and high-latitude forests can have similar biomass densities (note 21 below), primary production rates are typically higher in tropical forests.

17. This does not, despite appearances, contradict the Energy Equivalence Rule (EER), that suggests that the population metabolism of a species, measured as a rate per unit area, is invariant and does not vary with size of the organism (Damuth 1981; 1991). Thus populations of larger organisms, although spread more sparsely across the landscape, consist of individuals each of which has a greater metabolic rate per individual. The idea is that the two effects cancel out so that the overall per area metabolic rate of each species is invariant with organism size. For example, a large dataset of measurements of the rate of water pumping by plants shows that the per area per species pumping rate (total xylem flux of all plants of the same species per hectare) is largely constant, even when compared across eleven orders of magnitude difference in organism size (Enquist, Brown, and West 1998). If the EER also holds true with respect to changes in temperature, then this would seem to clash with the idea of a decline in primary production in the cold. However, any ecosystem comprises multiple species, and species richness (biodiversity) declines strongly with temperature (section 4.3). Once this is taken into account, the EER can be reconciled with a temperature effect on the whole ecosystem rate of primary production.

18. This suggestion derives from a comparison of soil nutrient contents and plant primary production rates on different islands in the Hawaiian chain. Because the island chain forms as plate tectonics pushes a portion of ocean crust across a stationary volcanic "hotspot," the islands line up in sequence of age. The study found that the concentration of phosphorus in soils increased up to 120,000 years and declined thereafter (Chadwick et al. 1999). Eventually, in the oldest soils, phosphorus from weathering of the underlying rocks was replaced as the dominant source by inputs in dust, blown in from Asian deserts more than 6,000 km away. By inference, rain forests in the Amazon basin, which are also situated on old, highly weathered rocks, must also obtain most of

their phosphorus from wind-blown dust. The meager inputs of phosphorus do not seem to hold back biomass and primary production to any great degree, either in the luxuriant Amazon or in the Hawaiian forests: primary production was found to be phosphorus-limited on the older islands, but even so the rate of primary production on the 4.1-million-year-old island of Kauai was only 20% less than the maximum rate, which occurred on a 150,000-year-old site (Vitousek et al. 1997).

19. (Houghton and Skole 1990)

20. A study of phytoplankton and the rates at which they are growing and being consumed (by voracious microzooplankton) in Puget Sound, near to Vancouver, found that even though 50% of the population is often being eaten per day, growth is even more rapid, to the extent that the phytoplankton population would more than double in size every day if it weren't being eaten so rapidly (Strom et al. 2001). Phytoplankton populations typically turn over very rapidly in other locations too, even if not usually quite so rapidly as this.

21. The packing density of trees in forests is set in large part by competition for light, with survival and growth of new seedlings (required in order to increase the density) usually prevented by scarcity of light under the canopy of a full-grown forest. Perhaps as a result of this, neither numbers of trees nor standing biomass per hectare appear to vary greatly with latitude between ~40°S and ~60°N (Enquist and Niklas 2001). This pattern has been revealed through analysis of the Gentry dataset (Gentry 1988a), which was obtained by the cataloguing of all individual plants and their diameters, within 2 m × 50 m strips. Several repeat strips were measured within each of 227 closed-canopy forests spread over six different continents. As part of this apparent biomass constancy between forests, the stocking densities of some Alaskan forests are similar to those of Amazonian forests. This analysis reveals a surprising degree of constancy between different latitudes and temperatures in terms of biomass per hectare. It needs to be kept in mind, however, that this analysis based on Gentry plots is restricted to considering only well-stocked dense (closed-canopy) forests, and by design therefore excludes both savannahs and also large tracts of sparser high-latitude forests.

22. (Begon, Harper, and Townsend 1996)

23. (Colinvaux 1993). As Colinvaux also points out, this strong pattern can be readily appreciated when comparing the diversity of species in just a single hectare of Amazon rain forest (~300 separate species of largish trees) with that of the whole of temperate Europe and North America (again, startlingly, ~300 species) (Gentry 1988b).

24. There is a well-established principle of ecology, first established in a famous study by Robert MacArthur and E. O. Wilson of biodiversity on islands of differing size. This principle is that, other things being equal, species number on an island increases with surface area of that island. For islands this has more to do with reduced chances of extinction of larger populations. Similar principles (but this time because the frequency of speciation events is greater in a larger area) probably also apply to continents or the globe as a whole. The larger the area, the larger the number of species typically supported. Because of the spherical shape of the Earth, and the definition of latitude in terms of angles, it is clear that the amount of surface area per degree latitude decreases toward the poles, with for instance more than ten times as much planetary surface area between 0° and 10°N as between 80° and 90°N. This potentially confounding effect needs to be taken into account when analyzing latitude-diversity relationships (Rosenzweig 1995), as does diminishing area towards the top of cone-shaped mountains, when considering altitude-diversity relationships. Area-corrected species counts must be used.

25. (Peck and Brey 1996; Peck, Webb, and Bailey 2004)

26. Evidence for such a faster tempo to evolution in the tropics has been suggested to reside within the DNA itself. A study by Shane Wright and colleagues analyzed DNA of various types of trees in New Zealand, where the authors are based, as well as in several other tropical and temperate locations (Wright, Keeling, and Gillman 2006). By counting the rate at which genetic mutations accumulate in the DNA, and comparing between locations, they found that, on average, tropical trees are evolving at slightly more than twice the rate of temperate trees. There are (at least) two potential explanations for this (the authors compared species of similar abundance at the different latitudes, so they could discount the phenomenon of faster evolution in smaller populations as an explanation): firstly, as discussed in the main text, if the generations tend to pass more quickly in hotter climates, then this could explain the observation of faster evolution. Secondly, if the rate of mutations is controlled by the respiration rate (respiration leads to the production of "free radicals," which can induce mutations in DNA) then, because respiration rate tends to be stimulated by temperature, so too will mutation rate. Regardless of which of the underlying explanations turns out to be most important, the evidence of Wright and colleagues suggests the possibility of a role for temperature in producing the extraordinary diversity seen in the tropics. The rate of evolution has been linked to the metabolic theory of ecology in several papers in the last few years (for an example see Allen et al. 2006). Fossil evidence also agrees with this contention; the first appearances of new species in the fossil record occur predominantly at low latitudes (Jablonski 1993), presumably because of greater rates of speciation at higher temperatures. According to this view, it might not be a coincidence that the famously rapid rates of evolution of cichlids in the East African rift lakes (~1,000 species in Lake Malawi have originated from a single common ancestor within at most one million years [Kornfield and Smith 2000]), are in lakes situated within a few degrees of the equator.

27. (Roy et al. 1998; Buzas and Culver 1999)

CHAPTER 5: ICEHOUSE EARTH

1. (Hedges and Kiel 1995; Muller-Karger et al. 2005)

2. Shelf seas are defined in table 5.1 as areas where the seafloor is between 0 and 200 m deep. Needless to say, when sea level dropped by 120 m then any seafloor between 0 and 120 m deep became land. But, conversely, any seafloor between 200 and 320 m deep became new shelf sea during the ice age. The two amounts were not equal; otherwise the area of shelf seas would have stayed the same. The ocean's seafloor is not evenly distributed between different depths. There is a disproportionate amount of what is currently very shallow seafloor, greatly exceeding that currently between 200 and 320 m depth.

3. The area of tundra was two to three times greater at the height of the last ice age than at present (23 down to 8 million square kilometers according to Adams et al. [1990]; an approximate halving according to Hassol et al. [2004]).

4. (Adams et al. 1990). Lakes and bogs are pollen traps, and the accumulating layers of mud at their bottoms contain a fascinating record of how the local vegetation has changed over the time since the lakes first formed. Small pollen grains from birch, hazel, pine, oak, and grass all got blown into the standing water, settled under gravity to the bottom muds, and remained trapped there ever since. Extracting long, cylindrical cores down through these muds, and then analyzing the material in the cores under a micro-

scope, allows a pollen record to be generated. Piecing together numerous local records of this sort formed the basis for the vegetation reconstruction of Adams et al. (1990). Some later analyzes suggest rather more modest increases following the last ice age of between 600 and 1,100 Gt C (Peng, Guiot, and Van Campo 1998 and references therein; Beerling 1999a). One alternative estimate suggests an even larger increase, of 1,500 Gt C (Adams and Faure 1998).

5. (Kohfeld et al. 2005). We do not yet have a fully consistent picture, however, and some recent studies propose that the glacial oceans were instead more stagnant than at present (i.e., a reduced supply of nutrients from depth), leading to an accumulation of nutrients and carbon dioxide in deep waters that was then released when the world warmed again (see for instance Galbraith et al. 2007).

6. (Gersonde et al. 2003). Sea ice extended an extra 5° of latitude toward the equator (as far north as South Georgia in winter) in the Atlantic and Indian sectors of the Southern Ocean.

7. The Cretaceous has been chosen in this book as the exemplar warm climate period. The fecundity of life (in particular the luxuriance of the vegetation) during a second warm climate period, the Early Eocene Climatic Optimum, is the subject of chapter 7 ("Paradise Lost") of Beerling (2007).

8. (Spicer 2003; Francis and Poole 2002)

9. (Rich, Vickers-Rich, and Gangloff 2002)

10. (Tarduno et al. 1998; Huber 1998)

11. Forests do not lead inexorably to coals. Far from it. Some coals are being formed today (such as in the peat swamps of Borneo) but not under most forests on Earth today (McCabe and Parrish 1992). In fact, a number of different conditions all have to be satisfied in order for coals to form from forests. Firstly, the forests have to be standing in swamps, and more particularly in peat swamps in which the swamp water is low in oxygen so as to inhibit bacterial breakdown of the organic matter. In a forest situated on well-drained land, fallen tree trunks will rot away to nothing in fairly short order due to the efficiency with which fungi and other organisms break down the wood and bark (although perhaps not in the Carboniferous and Permian before the capacity to decompose lignin evolved—to be discussed further in chapter 7). A second requirement is that the water levels must be high year-round. Large parts of the Amazon are seasonally flooded, but this is not sufficient to protect organic remains from decomposition, because decomposers are still able to do their work in the dry season. A further requirement is that the land must be gradually sinking (subsiding) over geological time. If the land is gradually rising, the peat swamps will eventually rise above sea level, at which point the peat will be decomposed under aerobic conditions. Only if the land is slowly sinking can successive layers of plant remains accumulate over millennia while the surface remains simultaneously waterlogged. There are other requirements, but the main point here is that it is far from automatic that forests lead to coals.

12. (Miller et al. 2005)

13. (Jenkyns et al. 2004; Herman and Spicer 1996)

14. While Arctic Ocean surface temperatures averaged about 15°C in summertime, tropical oceans experienced surface temperatures $\geq$ 30°C (Schouten et al. 2003).

15. The surface waters of the ocean are always oxygen rich, because of a ready supply from the abundant oxygen in the atmosphere. If the ocean waters cycle rapidly between surface and deep then it is hard for biological utilization of oxygen to strip out all of the

oxygen from deep water before that water is replaced by new, oxygen-rich water from the surface. When the ocean circulation is much slower then oxygen depletion has more time to act and anoxia is likely to become more widespread. On the other hand, if there is little productivity in the surface waters then only a little organic material will fall into deeper waters and there will be a diminished demand for the limited oxygen. Oxygen exhaustion in today's ocean occurs mostly below specific areas of very high phytoplankton production. When prodigious quantities of organic detritus fall into the waters below and are decomposed by bacteria, then this increases the chance that demand for oxygen will exceed supply. Finally, the temperature of seawater has a pronounced effect on its "solubility," or in other words its propensity for absorbing gases from the atmosphere. Cold waters tend to hold lots of oxygen and other gases; warm waters expel most of their gases to the air above and hold on to relatively little. This temperature effect controls how much oxygen is present in sinking surface seawater as it first enters the deep ocean. So the widespread anoxia of the Ocean Anoxic Events could be evidence of reduced circulation (implying reduced nutrient supply to the surface and, hence, lower productivity). Conversely, it could be evidence for greater productivity. Finally, it could instead be down to higher surface temperatures.

16. (Beerling 1999b). Application of the same technique to more recent times suggests that total biological productivity was about 80% of present-day values during the peak of the last ice age (Blunier et al. 2002).

17. Lamnid sharks (white sharks, porbeagles, mako sharks, and salmon sharks) and tuna concentrate their red muscle mass centrally in their torsos. By swimming *continuously* this muscle mass can keep their bodies warm and allow them to remain active even in cold waters. The red muscles of these fish seem to work best at temperatures much higher than those of the seawater they inhabit, again indicating (partial) endothermy (Bernal et al. 2005).

18. (Peck and Brey 1996)

19. Some special adaptations help these organisms to cope with the cold, but often at a cost. "Ice fish" (members of the Antarctic notothenioid family of fish) are the only vertebrates on Earth that do not synthesize hemoglobin with which to transport oxygen in the blood. Although the lack of hemoglobin lowers their ability to supply oxygen to their tissues, they can tolerate this because of the low metabolic rates (low oxygen demand) in the cold conditions. To this extent the cold is advantageous and allows them to dispense with a costly adaptation that all other vertebrates need. In other regards, however, the extreme cold is potentially hazardous. Typical Southern Ocean temperatures are −1°C or lower (seawater does not freeze until it gets a bit colder: −1.9°C). The blood of normal fish would freeze at such temperatures. To overcome this hazard, they and other Antarctic fish synthesize special antifreeze molecules (glycoproteins) in their blood that inhibit ice crystal formation, so saving themselves from freezing solid (Fothergill 1993). And, interestingly, this doesn't just occur off Antarctica. It has been discovered that Arctic cod also synthesize similar but independently evolved antifreeze glycoproteins (Chen, DeVries, and Cheng 1997).

20. It is not clear whether this argument should be expected to hold over longer timescales. Whenever there is upwelling there must also, by conservation of mass, be an equivalent amount of "downwelling," i.e., sinking of surface waters to form new deep waters, although this often takes place in a geographically distant area of the ocean. Surface waters are the source of all deep waters, given sufficient time (the "overturning

time" of the oceans is presently about 1,000 years). Therefore, once enough time has elapsed following the transition to a colder or warmer climate, we would expect the deep ocean temperatures to also fall or rise in response, as deep waters are replaced with sinking surface waters bearing a new temperature signature. And once the deep waters have also cooled or warmed then the effect of climate on stratification (and hence productivity) should be annulled. It may be that Lovelock's argument about marine primary production being favored by a colder climate only holds true over rather short timescales, in a geological sense, i.e., timescales shorter than the overturning circulation time of the ocean. What we do know for sure is that marine primary production was most definitely not insubstantial during warm climates on Earth such as the Cretaceous (when vast layers of chalk rocks were formed from planktonic remains), as discussed in section 5.3.2. We also know that in the Cretaceous deep ocean temperatures were between about 10° and 20°C higher than they are now (Huber, Norris, and MacLeod 2002). Deep-sea temperatures were 10–15°C warmer than today during the Eocene (55–34 million years ago), according to the ratio of ^{16}O to ^{18}O isotopes in the calcium carbonate shells of foraminifera (Zachos et al. 2001). Finally, deep-sea temperatures were probably ~3°C cooler during the last glacial maximum than they are today, according to the ratio of ^{16}O to ^{18}O isotopes in pore waters of marine sediment cores (Adkins, McIntyre, and Schrag 2002) and the Mg/Ca ratios of the calcium carbonate shells of seafloor-dwelling foraminifera (Martin et al. 2002). In summary, what this all tells us is that when the surface of the Earth warms or cools, the deep ocean eventually follows suit, leading to elimination of the effect on stratification and nutrient supply.

21. (Shine 1985; Blackburn 2000)

22. In fact the difference in food requirements is perhaps even larger; a review of metabolic rates measured in the field suggests that mammals and birds need more than ten times as much food per day as do reptiles of equivalent size (Nagy, Girard, and Brown 1999). Looked at in another way, this observation leads to the prediction that for the same food supply, range of body sizes, and temperature, a planet populated solely by ectotherms might hold ten times as much biomass as a planet populated solely by endotherms. For this reason the Cretaceous may have been home to much larger populations of dinosaurs (if cold blooded) than would be expected from comparison to the size of mammal and bird populations supported on today's Earth.

23. (Fothergill 1993)

24. Many attempts have been made over time to carry out nitrogen fixation artificially, as part of an industrial process to produce nitrogen-rich fertilizers for agriculture. It was not until the early twentieth century that Fritz Haber first succeeded in converting air (dinitrogen gas) into ammonia. Even today the best N_2-fixation method that human ingenuity has been able to come up with is still crude compared to the natural process; industrial N_2 fixation requires very high temperatures and pressures. The difficulty of the reaction may explain why nitrogenase has only ever evolved once.

25. Until recently it was thought, or rather it was assumed, that the main process creating dinitrogen, and in the process destroying the valuable stores of fixed nitrogen on which most life depends, was a process called denitrification. In the last few years several studies have provided more and more evidence that in fact it is a totally different process, known as anammox (ANaerobic AMMonium OXidation, first discovered in waste-treatment reactors) that is the main destroyer of fixed nitrogen (e.g., Dalsgaard, Thamdrup, and Canfield 2005 and references therein; Kuypers et al. 2005). Although

this work is revolutionizing our understanding of some aspects of the nitrogen cycle, it makes little difference to the argument being put forward here. Both denitrification and annamox are carried out by different guilds of bacteria, and so both are under biological control. In combination they nevertheless destroy more fixed nitrogen than is desirable from the perspective of the collective good.

26. Vast quantities of fixed nitrogen are created (e.g., through nitrogen fixation) and destroyed (e.g., through denitrification and anammox) every year. The "residence time" (total amount of fixed nitrogen in the oceans divided by annual rate of addition/removal) is only a few thousand years. If one flux were to change while the other stayed constant, fixed nitrogen stocks would rapidly accumulate or become depleted.

27. (Vitousek et al. 2002)

28. (White 1993)

29. Several groups of researchers have found this (see for instance Mills et al. 2004; Graziano et al. 1996; Moisander et al. 2003). Because of the great expense of sending research ships to sea, and the laborious nature of the nutrient enrichment experiments, only a disappointingly small number have so far been carried out.

30. The tendency for fires to start, for instance in forest undergrowth, is a steep function of the oxygen concentration of air. Without the dilution of every 1 part of oxygen by 4 parts of nitrogen, the increased dominance of oxygen in the atmosphere would lead to massive conflagrations and prevent vegetation growing on the land surface. For this reason it is possible to picture denitrifying and anammox bacteria as Earth's fire-prevention officers (they convert fixed nitrogen into dinitrogen), rather than as destroyers of the Earth biota's precious resource, fixed nitrogen. The fire-prevention view is compromised, however, by the realization (compare numbers in the table below) that it would take only approximately one-hundredth of one percent of the vast atmospheric reservoir of dinitrogen to double the ocean's fixed nitrogen content. Nitrogen starvation could be everywhere alleviated (both in the oceans and on land) without significantly increasing the fire risk. There is no reason why fire prevention could not easily coexist with provision of fixed nitrogen to the masses.

TABLE 5.3.
Split of Earth's nitrogen between different reservoirs

Reservoirs	Amount (Tg N)	Percentage of Total
Atmosphere, N_2	3,950,000,000	79.5
Sedimentary rocks	999,600,000	20.1
Ocean		
N_2	20,000,000	0.4
NO_3^-	570,000	< 0.1
Biomass		
Soil organics	190,000	< 0.1
Land biota	10,000	< 0.1
Marine biota	500	< 0.1

Note: Amounts of nitrogen in rocks and in different parts of the biosphere (taken from Galloway 2005).

31. The reason for this close correspondence between the ratio of need and the ratio of supply will be discussed in chapter 6. As it happens, there is usually a small deficiency of fixed nitrogen in the supply, compared to phosphate, and so the more frequently observed situation in the oceans is that fixed nitrogen runs outs just slightly before phosphate, and that nitrogen starvation is slightly more severe than phosphorus starvation. To call this optimal is akin to saying that the survivors of a plane crash in the desert, if slowly dying of thirst, would be better off if simultaneously also dying of hunger. It is true that doubling the stocks of fixed nitrogen would not greatly increase planetary marine fertility, because phosphorus is also scarce and there is no large alternative reservoir for phosphorus as there is for nitrogen. But it would alleviate one pervasive shortage that need not exist.

32. A similar conclusion has been drawn about another warm epoch: "There is no reason I know of to say that the Eocene hothouse was an inhospitable world for living things on Earth" (Archer 2009).

33. Note that while a hotter planet would (eventually) be better for life as a whole, it would not necessarily be better for humankind. We are one species out of many, and what is good for us and our civilization need bear no relation to what is good for life on Earth as a whole. Unlike most life we are endotherms; moreover, the genus *Homo* has evolved during the last two million years or so, a period dominated by ice ages. We should therefore not be surprised about any differences between our temperature preferences and those of the majority of life.

Chapter 6. Given Enough Time . . .

1. The spectrum of electromagnetic radiation emitted by a planet contains information about the composition of its atmosphere. Different gases absorb electromagnetic radiation at different wavebands (different locations in the spectrum). The full spectrum of radiation emitted from a planet (or a star) can therefore be analyzed and examined to look for conspicuous gaps (known as Fraunhofer lines). The wavelengths at which radiation is missing reveal which gases are present in the atmosphere of the remote planet or star. Sunlight, for instance, includes copious quantities of all hues of yellow light in the spectrum, with the striking exception of two missing bands at circa 589 and 589.6 nm. These missing wavelengths are a total giveaway for the presence of large quantities of sodium in the upper layers of the sun.

2. With some caveats. Remote detection by atmospheric analysis is not perfect. Firstly, in order to perturb the atmosphere there must of course be enough life to produce an appreciable impact on the atmospheric composition. If life is present in just a few scattered oases on an otherwise lifeless planet then its chemical impact will most likely be masked by larger contributions from inorganic processes. Secondly, the alien life must also produce large quantities of oxygen (rather than some other gas), and it does not seem inevitable that an alien biota would also be O_2-generating.

3. Although the atmosphere of Mars is dominated by carbon dioxide, there is in fact not all that much carbon dioxide in its atmosphere. This is because the Martian atmosphere is rather thin, or in other words there is little gas of any kind. There would be more gas in Mars's atmosphere (as well as an even greater percent of CO_2) if Mars were not so cold. But, because of the extreme cold, much of the CO_2 on Mars now exists in solid rather than gaseous form, as solid CO_2 in its icecaps. The reason for the much thin-

ner atmosphere on Mars than on Earth and Venus is still debated, but may relate to the weak magnetic field of Mars having allowed the solar wind to slowly strip away the atmosphere.

4. This is a simplified account. There are in reality additional sources and sinks of oxygen, such as the liberation of oxygen associated with formation and burial of pyrite (FeS_2, with the S having come from SO_4). But in any case the main point still stands: the immense amount of carbon taken up annually via photosynthesis, a small fraction of which is not subsequently respired, acts to prop up the high concentrations of oxygen and resist the decline in concentrations that would otherwise ensue.

5. Given the preceding discussion, it might seem odd to suggest that CH_4 and O_2 can coexist for millennia in ice bubbles, without mutually annihilating. However, the tendency of CH_4 and O_2 to react together in the atmosphere depends critically on high-energy sunlight generating reactive intermediaries. Because sunlight does not penetrate far into ice, CH_4 and O_2 coexist indefinitely in ice core bubbles.

6. There are several lines of geological evidence for very low oxygen levels during the Archean, summarized by Holland (2006).

7. Actually, it may not be quite so simple. There is general agreement that the advent of O_2-evolving photosynthesis was responsible for the oxygenation of Earth, but it may not have all been down to photosynthesis. It is possible that processes other than photosynthesis also contributed to the initial rise in oxygen. One hypothesis (Catling, Zahnle, and McKay 2001) proposes that methane (CH_4) produced by early life, which could persist for longer in the initial low-oxygen atmosphere, released hydrogen when it was broken down by UV radiation and some of that hydrogen leaked to outer space. Because the hydrogen in the CH_4 originally came from water (H_2O), the loss of some of the light hydrogen molecules to space would have left free oxygen behind. A second, somewhat similar, hypothesis (Hoehler, Bebout, and Des Marais 2001) proposes that H_2 released by microbial mats would have led to a net gain in oxygen on Earth, assuming that some of that hydrogen was able to rise through the oxygen-poor atmosphere and escape to space.

8. Geologists estimate, from surveys of rocks and their make-up, that ca. 12,500,000 Gt of organic carbon are dispersed within Earth's sedimentary rocks, i.e., about 1×10^{21} moles of organic carbon (Li 2000; see also Kump, Kasting, and Crane 2004). The total amount would be equivalent to a staggering five million cubic kilometers of solid carbon, if the carbon had the density of graphite (~ 2.2 g cm^{-3}), and if all the disseminated carbon were to be assembled into one place. It is quite hard to grasp mentally the amount of material in a solid cube one km on a side, never mind the amount of material in five million such cubes. Suffice it to say that truly vast quantities of organic residues are interspersed within the ancient sedimentary rocks of the Earth's crust.

9. How much organic carbon must have been buried in order to leave behind 21% oxygen in the atmosphere? The atmosphere contains 1.8×10^{20} moles of dry air, 21% of which amounts to 3.8×10^{19} moles of O_2. But this is only the tip of the iceberg, $\sim 2\%$ of the total amount of excess O_2 production required: the first oxygen was released into an oxygen-scavenging environment containing reduced substances (e.g., dissolved iron in the oceans, pyrite on land) that readily combined with (and thereby removed) any oxygen that appeared. It was only once the land surface and the ocean were swept clear of such reduced substances that the photosynthetically produced oxygen was free to accumulate in the ocean and atmosphere. It is estimated that 1.6×10^{21} moles of O_2 were

produced in total, ~83% of which now resides as Fe_2O_3 in geological deposits such as Red Beds, ~15% of which now sits in SO_4^{2-} molecules in the ocean, leaving only ~2% in the atmosphere (Schidlowski 1983; see also chapter 2 of Schlesinger 1997). If we assume a molar ratio of 1.2:1 of O_2 production to particulate organic carbon (POC) burial then 1.3×10^{21} moles of POC needed to have been buried in order to have produced all that oxygen. This compares fairly well with estimates of the amount of POC spread throughout the Earth's crust (see previous note).

10. It would take about two million years in total. Initially the ~600 Gt of organic carbon in tree trunks, leaves, etc., and the ~1,600 Gt of organic carbon in soils would draw down large quantities of oxygen as they were broken down and not replaced. But the enormous amount of oxygen in the atmosphere means that there would still be plenty (more than 99% of the original amount) left, even after *all* of the organic matter on Earth had been returned to its inorganic constituents, and combined with oxygen in the process. Drawing down the large remainder of atmospheric oxygen would then depend on the (fairly slow) rate at which volcanic gases and new (unoxidized) rocks are brought to the surface.

11. Nitrate will be used in this chapter to refer to the sum of nitrate (NO_3) plus nitrite (NO_2) plus ammonium (NH_4), all of which forms are readily taken up by phytoplankton. This sum is conventionally known as DIN (dissolved inorganic nitrogen), or fixed nitrogen, but in deep waters it is nearly all nitrate, with the other forms of DIN being of negligible importance. To keep things simple I contrast nitrate (by which I mean all the bioavailable nitrogen forms other than N_2) and N_2 as alternative sources of nitrogen.

12. Measured N:P ratios vary somewhat from place to place (Geider and La Roche 2002), between different types of phytoplankton (Quigg et al. 2003), and according to whether the phytoplankton are growing rapidly or slowly (Klausmeier et al. 2004). However, despite many reasons to expect more variability, measurements of the relative accumulations of N and P in deep water as it moves slowly through the ocean suggest a considerable degree of constancy in the ratio, centered on a value of about 16 (Anderson and Sarmiento 1994).

13. A third possibility, given the propensity of evolution to fit organisms to environments (chapter 3), is that plankton have evolved to build themselves with the very same nitrogen to phosphorus ratio as is present in seawater. However, such an explanation seems rather unlikely. If true, then it might be expected to hold true for all elemental ratios in plankton and seawater, which it most certainly doesn't.

14. (Tyrrell 1999)

15. All computer models are only as good as the assumptions they contain. If an assumption, whether implicit or explicit, is incorrect, then the model behavior is unlikely to resemble that of the real world. In this case the model contains two main explicit assumptions. Firstly, it is assumed that nitrogen fixers can continue to grow well even in the complete absence of all forms of fixed nitrogen, i.e., even when there is no nitrate or ammonium in the water, whereas nonfixing phytoplankton become limited. This assumption is safe and not controversial. The second assumption is that if nitrate is abundant in the water, nitrogen fixers will tend to be out-competed by nonfixers. It is generally accepted that nitrogen fixers face additional difficulties (for instance, the large energy input required to break apart the strong triple bond between the two N atoms in an N_2 molecule, and also the inconvenient need to keep oxygen away from the nitrogen-fixing apparatus because oxygen deactivates it), and that nitrogen-fixing organisms are

generally rather slow growing in nature. However, it is unlikely that in all circumstances the competition between fixers and nonfixers is determined only by the abundances of nitrate and phosphate. Other factors are also important. Sensitivity of the competition to iron availability is a case in point (see note 18 below).

16. During growth, the nitrogen within biomass of new N_2 fixers is obtained from the plentiful stock of N_2 dissolved within seawater. Upon death of a N_2-fixing algal cell, the N within either becomes incorporated into N within the organic matter of zooplankton, or else is released into the surrounding seawater and thereafter broken down by bacterial action. In either case it will eventually (for instance upon death of the zooplankton, or upon death of whatever ate them, and so forth) be remineralized to dissolved ammonium ions (NH_4) in seawater. Other guilds of bacteria make a living from catalyzing the chemical reactions (called nitrification) that convert ammonium ions to nitrite (NO_2) and then finally nitrite ions to nitrate (NO_3) ions, completing the process whereby N_2 fixation indirectly converts N_2 to nitrate.

17. Nitrogen, in common with several other elements, is a required ingredient for the construction of new organisms. It cannot be done without. The enzymes that cells synthesize are built up from amino acids, all of which contain nitrogen.

18. Additional factors (over and above the levels of nitrate and phosphate) affect the competition between N_2 fixers and nonfixers. In particular, levels of dissolved iron also affect which group is most competitive. The most conspicuous of oceanic N_2 fixers, the colonial cyanobacterium *Trichodesmium*, occurs preferentially in areas receiving a lot of dust from the atmosphere (e.g., downwind of the Sahara), which are also areas where there tends to be a lot of dissolved iron in the surface waters. This can be seen especially clearly along a north–south line through the eastern Atlantic, where *Trichodesmium* co-occurs with high levels of dissolved iron (Moore et al. 2009).

19. The anammox reaction actually converts nitrite (NO_2) and ammonium (NH_4) ions back to N_2, but this is not important because I am using nitrate as a shorthand for ($NO_3 + NO_2 + NH_4$) in this chapter.

20. The evidence for nitrogen fixers being abundant at these times ranges from the anomalous isotopic composition of all (bulk) nitrogen in the sediments (Rau, Arthur, and Dean 1987) to the anomalous isotopic composition of phytoplankton-specific nitrogen (Ohkouchi et al. 2006) to the abundant presence in the sediments of distinctive organic molecules ("biomarkers") that are produced only by cyanobacteria, the group of phytoplankton containing the N_2 fixers (Kuypers et al. 2004).

21. If the mixing up of deep water brings up much more phosphate than nitrate, such that removal into phytoplankton blooms exhausts nitrate long before exhausting phosphate, then we would expect blooms of N_2 fixers to thrive in the ensuing phosphate-rich, nitrate-deficient waters. Such a situation can in fact be seen every year in the Baltic Sea, where deep waters are anoxic and stirring up of deep water in winter drives the prebloom surface NO_3:PO_4 ratio to a value of about 8. Diatom blooms in spring then remove NO_3 and PO_4 in a 16:1 ratio until they exhaust nitrate but not phosphate. Nitrogen-fixer blooms follow predictably later in the summer (Larsson et al. 2001).

22. Much of this debate has centered on the behavior of the nitrogen cycle over glacial-interglacial cycles. A corollary of the argument pursued so far is that we should probably expect losses of nitrate (primarily via denitrification and anammox) to be balanced by inputs (primarily via N_2 fixation) during the relatively stable last few thousand years since the end of the last ice age. But the residence time of nitrate in the oceans is

TABLE 6.2.

Major fluxes in the present-day marine nitrogen cycle

Process	Flux magnitude $(Tg\ N\ y^{-1})$
N SOURCES	
Water-column N_2 fixation	120 ± 50
Seafloor N_2 fixation	15 ± 10
River input as DON	35 ± 10
River input as PON	45 ± 10
Atmospheric deposition	50 ± 20
Total Sources	265 ± 55
N SINKS	
Water-column denitrification	180 ± 50
Denitrification in sediments	65 ± 20
Burial of N in sediments	25 ± 10
N_2O loss to atmosphere	4 ± 2
Total Sinks	275 ± 55

Note: An estimate of the magnitude of each flux is given (adapted from Gruber 2004). Italicized fluxes are produced by living organisms.

very short, about 3,000 years, and so large changes since the time of the last ice age are possible. An alternative proposal (Codispoti 1995; Altabet, Higginson, and Murray 2002; Codispoti 2007) has it that the ocean has been steadily losing nitrate during the current interglacial, but in contrast was steadily gaining it during the last ice age. Oceanographers are currently struggling to ascertain, from direct rate measurements, whether sinks and sources of N are balanced or not, but so far the accuracy and coverage of the measurements is not adequate for the task ("our ability to quantify the marine nitrogen cycle balance remains limited" [Gruber 2004; see also Karl et al. 2002; Mahaffey, Michaels, and Capone 2005]). An alternative approach has been to examine whether the nitrogen isotopic composition of organic particles falling to the seafloor has changed over time, as would be expected if the N cycle is not in steady state. The $\delta^{15}N$ measurements are more suggestive of a balanced cycle (Altabet 2007), but the question is far from being resolved.

23. (Yool and Tyrrell 2003)

24. Many studies have sampled the water flowing out to sea down the Amazon and other rivers and studied the relative abundance of different dissolved substances in the water. The most abundant elements dissolved inorganically in river water are, in order (most abundant first): C, Ca, Na, Cl, and Si. If the ocean were to act simply as a trap for these materials supplied down rivers, we might expect to see the same ranking of elements when seawater composition is analyzed. However, the major elements dissolved inorganically in seawater (the major contributors to sea salt), ranked in order of molar

concentration, are: Cl, Na, Mg, SO_4, and Ca (Sarmiento and Gruber 2006). Ions that are heavily used by phytoplankton, for instance silicic acid, which diatoms use to make opal shells, are strongly depleted in ocean water compared to the expectation from river water.

25. This beautiful example of coevolution of life and environment is apparent in geological evidence (Siever 1991; Raven and Giordano 2009). The concentration of dissolved silicic acid in seawater fell from perhaps 2,000 μmol kg^{-1} in the Precambrian down to ~1 μmol kg^{-1} in most surface waters today, as abiotic precipitation gave way to biotic precipitation by first sponges and radiolaria and then finally by the much more numerous diatoms. The appearance and abundance of siliceous sponges in the fossil record changed significantly at the same time as diatoms became numerous. Sponges became scarce and spindly when diatoms became abundant. Bizarrely, when modern siliceous sponges are grown in artificially high silicic acid seawater (reminiscent of Precambrian oceans) they form more robust skeletons resembling fossil sponges from long ago (Maldonado et al. 1999). Habitat commitment appears not to have been so important in this particular case.

26. The proposal that land plants are a "geological force," that they have been prime movers in shaping the Earth's environment over geological time, is set out in detail by David Beerling (2007).

CHAPTER 7. EVOLUTIONARY INNOVATIONS AND ENVIRONMENTAL CHANGE

1. Milankovitch cycles are periodic variations in the way the Sun heats the Earth. They are produced by subtle changes over time in Earth's tilt and orbit. One of these cycles is ultimately responsible for the oscillation in Earth's climate between frigid ice ages and warmer interglacials (to be described in section 8.4). We know this because the actual timings of past ice ages match up closely with the calculations made by the Serbian mathematician Milutin Milanković. Earth's orbit is not quite circular; instead, it is slightly elliptical in shape. What's more, the degree of deviation from an exact circle varies on an approximately 100,000-year timescale. This has consequences for the degree of seasonality of Earth's climate; as the orbit gets more elliptical, there is a greater difference between the Earth-Sun distance at the point of closest approach and at the point of greatest separation. This is known as the eccentricity cycle. Milanković also worked out two other periodic variations (known as obliquity and precession), which both affect the degree of tilt of the Earth's spin axis relative to the plane of the Earth's orbit around the Sun. Again, there are consequences for the degree of seasonality on Earth (the severity of polar winters and the mildness of polar summers).

2. (Catling, Zahnle, and McKay 2001). See note 7 of chapter 6 for some other hypotheses.

3. For instance, converting it from Fe to Fe_2O_3.

4. Chemolithotrophs carry out specific chemical reactions in order to harness the energy they yield. The "litho" part of the name (from the Greek word "lithos" for rock or stone) refers to the fact that the molecules they break down needn't come from a biological source. And so, for instance, some of the very first organisms on Earth may have exploited FeS and H_2S that was freely available in the anoxic environment to carry out the reaction $FeS + H_2S \rightarrow FeS_2 + H_2$, which releases energy. Most organisms get their energy either from capturing sunlight (plants), from eating plants (herbivores), from eating animals (carnivores), from parasitizing the above (parasites), or from consuming

their remains (scavengers and detritivores). Only chemolithotrophs can pursue an existence that is not dependent on either direct sunlight or other organisms.

5. Profitable in the sense of containing potential energy, i.e., able to yield a net energy gain to the organism if it is able to induce a particular reaction involving that chemical.

6. (Kulp et al. 2008)

7. It is fundamental, but rarely, in nature, quite that simple. Equation 7.1 is a simplified representation of the much more complex biochemical reactions that take place in reality, but it encapsulates the common aspects and overall net effect.

8. This is a straightforward enough idea, but it has been argued that the situation may have been more complicated than this, with both sinks as well as sources of O_2 needing to be taken into account. It has been argued that oxygen couldn't build up in the early atmosphere until, as well as a strong source, there was also a decline in oxygen consumption through a shift toward fewer submarine volcanoes, because these submarine volcanoes produced large quantities of oxygen-grabbing ("reduced") chemicals (Kump and Barley 2007).

9. Note that atmospheric oxygen (and hence, presumably, the density of the ozone layer) did not rise to full modern levels until ~600 million years ago. After the Great Oxidation Event, atmospheric oxygen probably rose to between 1 and 40% of the current value, but it took nearly another 2 billion years before it rose to modern levels (Kump 2008).

10. One of the most important of these lines of evidence is variation in the ratios of different isotopes of sulfur in sediments. Prior to ~2.4 billion years ago the sulfur isotopic ratios (the abundances of ^{33}S and ^{34}S relative to ^{32}S) show considerable variation over time in their "mass-independent fractionation," whereas ever since that time they do not. An understanding of how these isotopes behave suggests that the considerable variability before ~2.4 Bya would only have been possible in an atmosphere containing less than 0.001% as much oxygen as at present (Farquhar, Bao, and Thiemens 2000; Bekker et al. 2004). Other evidence, such as changes in molybdenum cycling, also points to a similar date for the GOE.

11. There are a few bacterial groups that are multicellular, at certain points in their life cycles at least, such as the slime molds, but these are the exception to the rule.

12. (Rasmussen et al. 2008)

13. Multicellular animals (metazoans) are also usually completely absent in present-day anoxic environments. It was previously thought that absolutely all oxygen-free environments are too extreme to support any metazoans, but in the last few years a few specialists have been found to be able to tolerate an absence of oxygen (e.g., Danovaro et al. 2010). They have evolved away from use of mitochondria toward other means of generating energy in the cell.

14. Ozone molecules consist of three oxygen atoms (O_3), hence the alternative name trioxygen. They are produced mostly in the upper atmosphere (the stratosphere) from sunlight-induced breakdown of the much more abundant diatomic oxygen molecules (O_2). The overwhelming majority of oxygen in the atmosphere is present as O_2 (~21% of air), rather than O_3 (at most several parts per million). As will be discussed later in chapter 10, the presence of ozone in the atmosphere ("the ozone layer") is important in preventing harmful ultraviolet (UV) light from reaching the surface of the Earth. It is almost as if the present-day Earth is surrounded by a layer of UV-sensitive sunglasses. Before the rise in atmospheric oxygen, there would have been little O_2 from which to

create O_3, and therefore no ozone layer barrier preventing penetration of UV to the Earth's surface. We experience the consequences of exposure to excess UV in terms of sunburn and melanomas. Part of the reason that UV causes cancers is that it disrupts DNA, through causing adjacent base pairs to combine. The beneficial effects of more abundant oxygen in having promoted eukaryote success are discussed in greater detail in chapter 17 of Kump, Kasting, and Crane (2004).

15. (Kasting 2004; although see also Haqq-Misra et al. 2008)

16. (Kopp et al. 2005)

17. The exact date of the first appearance of life on Earth is, however, the subject of vigorous debate. Probably the earliest unequivocal evidence of life (in the form of stromatolites that are clearly biologically produced) dates from about 2.7 billion years ago. There are also credible indications of stromatolites at 3.4 billion years ago (Allwood et al. 2006).

18. (Wellman, Osterloff, and Mohiuddin 2003). It has been claimed, based on isotopic evidence, that the first greening of Earth may have taken place much earlier, as early as 850 million years ago (Knauth and Kennedy 2009). However, the evidence for this is indirect, and the claim is not yet substantiated.

19. The evolution and spread of the first woody land plants has been considered by Kenrick and Crane (1997) and Meyer-Berthaud, Scheckler, and Wendt (1999).

20. The discovery of *Wattieza* trunks and crowns is described by Stein et al. (2007).

21. The evidence for global cooling comes in part from signs of glaciation during the Late Devonian, Carboniferous, and Early Permian (summarized in Stanley 1999). The evidence for a fall in CO_2 comes from a variety of sources (see review by Royer, Berner, and Beerling 2001). Stomatal densities of modern-day plants grown at different CO_2 concentrations show a correlation between the density of stomata (number of "breathing holes" per unit leaf area) and the CO_2 concentration of the air they are grown in. The stomatal densities of the fossil leaves from the Permo-Carboniferous are most similar to those of equivalent modern plants that have been grown at fairly low CO_2 concentration of 300–350 ppmv (Beerling 2002). This is a quite direct method of estimating past CO_2, although suffering from the fact that no plant species from that long ago have survived through to the present day. The method can therefore only be calibrated against the most similar plants living today, but not against plants of the very same species.

22. The rates of organic carbon burial into rocks can be calculated as follows. Firstly, the prevalence of different rock types (e.g., shales, sandstones, coal) formed during that time is estimated from geological mapping of the continents. Secondly, the average organic carbon content (%) at that time of each rock type is calculated according to chemical analyses. These two sets of numbers are combined and then multiplied by an estimated rate of sedimentation (rate of formation of new rocks) to calculate the global rate of organic carbon burial (Berner and Canfield 1989). Another way of estimating the amount of organic carbon burial is from inspection of the carbon isotope record (organic matter has a low $\delta^{13}C$, and hence its removal to sediments in large quantities raises the $\delta^{13}C$ of the carbon left behind in the Earth surface environment) (Berner 2003).

23. By comparison, the total amount of carbon in today's atmosphere (390 ppmv of CO_2 at time of writing) is ~840 Gt C, and the total amount in the atmosphere 380 Mya, before the burial event started, was between 2,000 and 10,000 Gt C. If the ocean at that time held fifty times more inorganic carbon than the atmosphere, as it does today, then the ocean + atmosphere total would still have been completely drained between eight and

forty times over by the excess burial. How was this possible? How could more carbon be buried in additional organic matter than was originally present in the atmosphere and ocean? Part of the reason for the discrepancy may be due to inaccuracies in some of the carbon estimates. But a tenfold inaccuracy seems rather unlikely. The numerical discrepancy can in fact be accounted for without questioning the numbers. Sedimentary rocks, although they survive for enormous lengths of time, do not survive forever. Eventually they are brought back up to the surface and exposed to wind and rain and microbes, and organic carbon in them is returned to atmospheric CO_2. Some are brought back to the surface and eroded within just a few million years, others survive for many hundreds of millions of years, and just a few have survived for several billions of years, as just described in section 7.1. The average time between laying down and erosion of sedimentary rocks is something like 100 million years. Therefore, while enormous amounts of carbon were buried between 380 and 250 Mya, a large fraction of that buried carbon must have been released again before the end of the 130 My period.

24. (Robinson 1990; Retallack 1997; Robinson 1991)

25. A fascinating recent study lends some support to this idea. Thirty-one fungal genomes were compared in order to understand how and when the capacity to degrade lignin evolved among white rot fungi, the only fungi capable of breaking down lignin. As well as identifying key enzymes required, a combination of molecular clocks and fossils was used to try and put a date on when the lignin decomposition capabilities first appeared. Although the analyses are only able to constrain the date very weakly (to within ± 100 million years), the best guess is that the ability to degrade lignin first appeared somewhere around 300 Mya, which would put it towards the end of the period of high organic carbon burial (note 23) and earlier than the first definitive white rot fossils at ~260 and ~230 Mya (Floudas et al. 2012).

26. Another peculiar feature of wood is that, unlike most other biologically produced materials, it contains little or no nitrogen or phosphorus or other nutrients. It is almost all carbon, hydrogen, and oxygen (because cellulose, hemicellulose, and lignin are made from these three elements). This means that abundant quantities of wood can be formed without requiring additional supplies of the relatively scarce quantities of critical nutrients such as N and P.

27. (Scott and Glasspool 2006)

28. (Berner 1999; updated in Berner 2006 and Berner 2009). See also chapter 5 of Lane (2002).

29. (Dudley 1998)

30. (Schwartzman and Volk 1989). Some grasses may however have benefited from the enhanced rate of mobilization of silicon, due to their use of silicon to stiffen their stems and/or leaves.

31. (Berner 2001). This reaction lumps together the dissolving of silicate rocks, transport of the released calcium and silicate ions down rivers to the sea, and their stimulation of silicate and calcium carbonate mineral formation in the sea.

32. The origin of rooted plants had numerous effects. The meshwork of roots threading through soil would have stabilized it (as it does today) and allowed soil to build up where previously it would have been rapidly removed through erosion. The organic debris trapped in the new soils would have been a carbon sink, but only a small one. The carpeting of the land surface with soils would not have been in all ways favorable to increased weathering, given that it would in some sense have protected the underlying rock

from exposure to rain and some of the extremes of temperature. However, the increased acidity of the water coming into contact with the rock surfaces would most likely have been the dominant impact on weathering rates (as discussed in Schwartzman and Volk 1989). Other effects related to the rise of plants would have included longer retention times of rock particles within stabilized soils (and hence extended exposure to acids) and higher rates of average rainfall due to increased evapotranspiration (Berner, Berner, and Moulton 2003).

33. These were two prominent pathways by which the spread of the first forests impacted on atmospheric CO_2. But there were other pathways, in addition to those mentioned in the last note. At the same time that trees roots facilitated an increased rate of weathering of silicate rocks, other constituents of rocks were also released at a faster rate. One consequence would have been a more rapid release of carbon from kerogens (fossil organic carbon within rocks), tending to elevate atmospheric CO_2. A second additional consequence would have been a more rapid extraction of phosphorus from rocks. As this was washed into the sea it would have brought about algal blooms, sequestering additional amounts of CO_2 (Algeo and Sheckler 1998).

34. The two carbon fluxes (due to silicate weathering and due to burial of land-derived carbon in marine sediments) are of fairly similar magnitude on Earth today (both in the range 0.01–0.10 Gt C y^{-1}: Kump, Brantley, and Arthur 2000; Schlünz and Schneider 2000), but may have proceeded at different rates in the aftermath of the first arrival of forests on land. The relative importance of the two processes in removing atmospheric CO_2 was discussed by Berner (1998).

35. (Came et al. 2007)

36. (Algeo, Scheckler, and Maynard 2001, and references therein)

37. Chapters 14 and 15 of Stanley (1999) review the geological evidence of Devonian and Carboniferous glaciations.

38. Despite the overall environmental degradation, there was also, it appears, a silver lining. Beerling, Osborne, and Chaloner (2001) argue that low CO_2 paved the way for the evolution of the large leaves that we see today.

39. Although this is not to say that new links will not be established in the future. For instance, it is suggested that an even earlier colonization of the land surface, by lichens, had transformational consequences in terms of causing a fall in CO_2 leading to the Neoproterozoic Snowball Earth events (Lenton and Watson 2011).

CHAPTER 8. A STABLE OR AN UNSTABLE WORLD?

1. Paleoscientists use a variety of techniques to try and narrow down the exact age of ancient rocks or ice, or of items within them. These techniques include radiocarbon decay, decay of other radioactive substances, oscillations that are tied to Milankovitch cycles (note 1 of chapter 7), counting of annual layers (if the deposits contain them), identification of characteristic fossils that are found only within restricted time windows, and looking for evidence of magnetic reversals (the Earth's magnetic field reverses polarity, that is, magnetic north and south swap positions, usually every few hundred thousand years, and because the dates of these reversals have been more accurately dated they give "tie-points" through time). Often a combination of approaches is employed. However, many of these methods only work for fairly recent objects or sediments (for instance, radiocarbon dating does not work for material older than about

50,000 years), and some are of no use at certain times (for instance, there are long periods when the Earth's magnetic field remained stable in orientation). The problem of trying to accurately date adjacent layers in older sediments in order to calculate how rapidly they were laid down is therefore a difficult one.

2. For instance, we know that the waxing and waning of the ice ages has had little effect on the type of sediment laid down over large parts of the seafloor. Seafloor muds in most parts of the world do not change color between ice ages and interglacials, as can be seen in the appearance of cylindrical cores recovered from drilling vertically down through ocean sediments. Evidently considerable changes in environmental conditions can take place without greatly altering the community of planktonic creatures that sends a rain of shells and corpses down to the seafloor.

3. See http://earthobservatory.nasa.gov/Features/Paleoclimatology_Understanding/ for a glimpse into the diverse array of techniques.

4. Part of the attraction is the extreme longevity of the record. Whereas the oldest ice on Earth is about 1 million years old, fluid inclusions survive from hundreds of millions of years back in Earth history, although obviously without anything near the fine temporal resolution (annual cycles) of the ice core data.

5. (Chapter 2 of Warren 2006). For instance, as an initial modern seawater solution becomes more and more concentrated, the first salts to precipitate out are always calcium carbonate ($CaCO_3$) and then gypsum ($CaSO_4 \cdot 2H_2O$), followed by halite (NaCl). The sequential formation of these salts first exhausts dissolved bicarbonate (HCO_3^-, of which there is initially about 2 mmol kg^{-1} in seawater), putting an end to $CaCO_3$ formation, and then depletes the solution of Ca^{2+} and SO_4^{2-} until one or the other runs out (always Ca^{2+}, in today's ocean, because modern seawater contains 29 mmol kg^{-1} of SO_4^{2-} but only 10.6 mmol kg^{-1} of Ca^{2+}).

6. Analyses of fluid inclusions formed in modern times show that they faithfully record the composition of evaporated modern seawater, with some understood exceptions (Horita, Zimmermann, and Holland 2002). This is seen both in fluid inclusions in salt formed naturally in a supratidal sabkha (salt flat) in Baja California (Timofeeff et al. 2001), and also in fluid inclusions in salt formed artificially in a salt works in the Bahamas (von Borstel, Zimmermann, and Ruppert 2000).

7. Seawater (Mg/Ca) ratios have been reconstructed (primarily from fluid inclusion data) by Horita, Zimmermann, and Holland (2002) and by Lowenstein et al. (2001). The former article also contains reconstructions of other major ion concentrations through the last 540 million years.

8. (See for instance the discussions in Goldstein 2001 and chapter 2 of Warren 2006)

9. (Timofeeff et al. 2006; Lowenstein et al. 2001)

10. Some Mg atoms inadvertently slip in and occasionally substitute for the Ca atoms in the $CaCO_3$, and this is possible because Mg and Ca atoms both have the same +2 electrical charge and are also rather similar in other ways.

11. (Dickson 2002); see also the associated overview article by Kerr (2002).

12. (Coggon et al. 2010)

13. This note is a digression from the main topic. The Mg/Ca data (and also other records in the chapter) lead to an interesting conclusion, separate from those associated with evaluating Gaia. It is easy, especially given that all human civilizations rose and fell within a stable interglacial period, to assume that the Earth environment is now stable. Perhaps, to continue the unsafe assumption, Earth's environment underwent a lot of

change in the past but then settled down and would have remained constant hereafter. But, if we look closely at figure 8.3, and in particular at the right-hand end of it, we can see that there is no evidence at all of (Mg/Ca) stabilizing as we get closer to the present time. If anything the opposite appears to be the case: it is accelerating away. Despite the Mg/Ca ratio being higher now than at any previous time during the Phanerozoic, it still appears to be rising at a rate that is as rapid (in geological terms) as at any time during the Phanerozoic (the last 540 million years). This is not anything that need concern us on human timescales. The point is not that (Mg/Ca) will change detectably in our life-times (it will not), but rather that there is nothing special about this present instant in geological history. We are not at any endpoint in environmental evolution; *Homo sapiens* has appeared at one particular point in Earth history, and apart from the fact that the last 10,000 years or so have seen an unusually warm and stable climate, this point is not otherwise particularly special. We also have no strong reason to believe that the 100-million-year trend toward ever-cooler climates (section 8.3) had reached its culmination. One computer model suggests that the Earth may have been heading into a permanent and more severe ice age before we started altering the climate. The natural progression toward ever-cooler climates over the last 50 to 100 million years had not ended, the study suggests, but rather the current phase of oscillations between glacials and interglacials is just one waypoint in the long transition from the more or less permanently ice-free greenhouse world of the Cretaceous and Paleocene toward the permanently glaciated planet that the world was heading toward before humans started disinterring fossil fuels. If this view is correct then the oscillations between ice ages and interglacials seen over the last few million years were just a precursor to the stable and permanent installation of northern hemisphere ice sheets dwarfing those seen at the height of the last ice age (Crowley and Hyde 2008). As with Mg/Ca, there is no reason to suspect that the trend in Earth temperature had leveled out. It may have taken very many millions of years before climate and ocean chemistry would have settled down and reached an accommodation with whatever modifications are being forced upon them.

14. (Sandberg 1983)

15. Coral reefs today are mostly made of the aragonite form of calcium carbonate, as are the shells of a few swimming organisms such as pteropods. But most plankton (especially the exceptionally numerous coccolithophores and foraminifera) construct their shells out of the calcite form of $CaCO_3$.

16. The evidence for long-term variations in the type of calcium carbonate organisms secrete is summarized by Stanley and Hardie (1998) and Porter (2007). For instance, although the majority of reef-forming organisms today synthesize the aragonite form of $CaCO_3$, in past "calcite seas" it is argued that the dominant reef formers mostly synthesized the calcite form. Changeovers may have been achieved mostly by evolutionary replacement of calcitic organisms by aragonitic ones. However, during an experiment in which aragonitic corals were placed in "calcitic seawater" (low Mg/Ca ratio) they were found to switch over to producing calcite coral (Ries, Stanley, and Hardie 2006). The laboratory results have been supported by fossil corals from the Late Cretaceous (when the Mg/Ca ratio was very low compared to today), which are calcitic in contrast to the aragonitic corals observed today (Stolarski et al. 2007). Similar laboratory results have also been obtained for coralline algae (Stanley, Ries, and Hardie 2002).

17. Large numbers of calcifying organisms have come and gone since Cambrian time (fig. 8.3). These include aragonite-synthesizing groups such as the labechiid stromatopo-

roids, phylloid and dasycladacean algae, chaetetid and sphinctozoan sponges, and modern scleractinian corals and calcareous and coralline algae. Major calcite-synthesizing groups over geological time include trepostome and cryptoporate bryozoans, tabulate, heliolitid and rugose corals, demosponges, and rudist bivalves. Some of the swings in dominance from one group to another (in terms of being major contributors to reef building) may have been driven by mass extinction events, following which the new groups that evolved and rose to abundance were those better suited to the prevailing water chemistry (prevailing Mg/Ca ratio). In other cases certain groups have been extremely abundant during certain geologic intervals but have subsequently faded away to rarity (without immediately going extinct) as the seawater Mg/Ca has swung away from a favorable composition for them (Stanley 2006). See also chapter 10 of Stanley (1999) for an introductory-level discussion of the interactions between calcifiers and the chemistry of the seawater medium they inhabit.

18. It should be noted however that this concept of calcite synthesizers being more successful in low Mg/Ca seas, aragonite synthesizers more successful in high Mg/Ca seas, is not without its detractors. A statistical analysis has in fact pointed to a lack of *any* correlation over geological time between seawater Mg/Ca and the type of calcium carbonate secreted, except for the major reef-building organisms (Kiessling, Aberhan, and Villier 2008) although this conclusion is itself contested (Porter 2010).

19. There is a strong and well-understood feedback called "carbonate compensation" that has kept the $CaCO_3$ saturation of the oceans within a fairly narrow range during the last 100 million years while there have been abundant calcifying plankton on Earth. This negative feedback must have acted to oppose large changes in the degree of saturation with respect to $CaCO_3$. Even before 100 Mya, in what Zeebe and Westbroek (2003) refer to as the Neritan ocean, it is likely that a somewhat weaker feedback operated to stabilize $CaCO_3$ saturation. In either case it would seem impossible for the ocean to have remained chemically unfavorable for calcification over any extended period of time, because any cessation in burial would have swiftly led to increases in the $CaCO_3$ saturation of the ocean (or, in other words, a swift return toward more favorable conditions). Zeebe and Westbroek (2003) treat this subject at length.

20. There is considerable uncertainty about the long-term pattern of Earth temperature. Some methods of deriving past seawater temperatures come up with ancient (3.5 Bya) ocean temperatures as high as 70°C (Robert and Chaussidon 2006), whereas others suggest it was instead fairly cool (≤ 40°C) (Hren, Tice, and Chamberlain 2009; Blake, Chang, and Lepland 2010). Our general ignorance of conditions on the early Earth is mitigated in one regard: there is one aspect of long-term Earth climate that we think we know fairly well, which is how the rate of heating of the Earth by the Sun has changed over time. Widely accepted models of stellar dynamics (how stars change over time) suggest that four billion years ago the Sun was only ~70% as hot as it is now (Gough 1981). This has led to what is known as the Faint Young Sun Paradox: the difficulty in reconciling the geological evidence of a mainly ice-free early Earth (Huronian glaciation excepted) with the calculation of a cooler early Sun, as a consequence of which the early Earth should have been covered with ice. If the higher ancient temperatures above are correct then this problem is only exacerbated. How could the early Earth have been ~50°C warmer under a Sun that was ~30% cooler? This paradox could potentially be resolved by the presence of a strong greenhouse on the early Earth, which then declined over time.

21. (Zachos et al. 2001)

22. (Friedrich, Norris, and Erbacher 2012)

23. Although a little complex to understand, the $\delta^{18}O$ signal recorded in calcium carbonate shells is extremely useful. $\delta^{18}O$ is a measure of the frequency of "unusual" oxygen atoms in a sample, where "unusual" in this case refers to those oxygen atoms that have ten neutrons and eight protons in the nucleus, as opposed to the normal complement of eight neutrons and eight protons. ^{18}O atoms are therefore slightly (12.5%) heavier than ^{16}O atoms, and this makes a difference to their propensity to participate in different chemical reactions. For instance, because $H_2^{18}O$ water molecules are slightly heavier than $H_2^{16}O$ water molecules they evaporate less readily. On the other hand, $H_2^{18}O$ condenses more readily then $H_2^{16}O$. This reluctance of $H_2^{18}O$ to become airborne, or once there to remain airborne, leads to its scarcity in precipitation and explains why measurements of recently formed Antarctic ice show it to have 5% fewer ^{18}O atoms than ocean water. While fewer ^{18}O atoms make it into ice, more make it into biogenic $CaCO_3$ formed from the ^{18}O-enriched seawater left behind. These shells therefore have a higher $\delta^{18}O$ when there is a lot of ice on the planet. A second feature is that when marine organisms make $CaCO_3$ shells then the oxygen in those shells tends to contain a greater fraction of ^{18}O atoms than does the seawater in which the shell makers are living. What is more, the offset between the seawater $\delta^{18}O$ and the shell $\delta^{18}O$ varies with temperature. This property is most helpful, because the precise value of $\delta^{18}O$ in a fossil shell (and this can be measured with great accuracy) is related to the temperature of the water in which the shell maker grew. Variations of shell $\delta^{18}O$ through geological time therefore reflect variations in both Earth surface ocean temperature and ice sheet volume. But at least both effects work the same way, so that both cooling and ice sheet growth move $\delta^{18}O$ in the same direction. There are other secondary caveats to interpreting the climate record from shell $\delta^{18}O$, such as the possibility of shells becoming encrusted with extra $CaCO_3$ (of different $\delta^{18}O$) while they reside in the accumulating sediments on the seafloor. Nevertheless, shell $\delta^{18}O$ gives an excellent record of climate that agrees with other proxies (for instance shell $\delta^{18}O$ is low during periods of past glaciation).

24. A variety of different approaches to reconstructing atmospheric CO_2 over many millions of years in the past are reviewed by Royer, Berner, and Beerling (2001).

25. (Petit et al. 1999). Longer (older) cores have since been recovered.

26. Even though, thanks in large part to the ice cores, we know in great detail how temperature and greenhouse gas levels changed over time, there is as yet no scientific consensus about exactly why those changes occurred. There is a lack of agreement about the mechanisms underlying the glacial-interglacial cycles. While we know that their timing is paced by subtle periodic variations in the Earth's orbit and tilt (Milankovitch cycles), the associated variations in sunlight intensity are far too small to explain by themselves the amplitude of Earth's temperature variation (about 5°C). It appears that the Earth system, including the biota, amplifies the physical forcing (positive feedback). By causing there to be higher amounts of atmospheric CO_2 and methane when the Earth is warmer, and less CO_2 and methane during colder phases, the system as a whole exacerbates the changes due to the orbital forcing.

27. The lurches in climate are replicated in two cores drilled 30 km apart, strongly suggesting a real signal rather than any very local artifact. The climate spikes are however much more pronounced in temperature reconstructions from Greenland ice cores than from Antarctic ice cores (Stocker 2000), suggesting that much of the "lurchiness"

may have been regional in nature, for instance confined to the Northern Hemisphere. This is what we would expect if they were caused by temporary changes in the transport of heat northward in the North Atlantic Ocean, which would have had greater impacts on the Northern than the Southern Hemisphere.

28. As described in Brian Fagan's book *The Long Summer: How Climate Changed Civilization* (Fagan 2003).

29. It is clear that temperatures dropped on Greenland, but there is still ongoing debate about whether temperatures dropped everywhere during the Younger Dryas. It now seems more likely that they dropped only in the Northern Hemisphere. Some climate records from the Southern Hemisphere do show synchronous changes with the Greenland ice cores (for instance the timings of advances of some glaciers in New Zealand (Ivy-Ochs et al. 1999), but others do not show any change [e.g., sea surface temperatures near Australia (Calvo et al. 2007) or point to warming coincident with the Younger Dryas (retreat of another New Zealand glacier (Kaplan et al. 2010))]. Methane concentrations change at this time in both Greenland and Antarctica, suggesting an overall global impact on methane emissions. A second caveat is that whereas temperatures dropped by 10°C over the Greenland ice sheet, temperature changes elsewhere in the Northern Hemisphere may have been less extreme.

30. The evidence for rapid changes comes in part from simple visual inspection of the annual layers in the ice cores, in which very rapid changes in snow accumulation rate, for instance, can be seen across the space of just a few layers. Other information comes from chemical analyses of the ice and of the bubbles of air trapped within the ice (Taylor et al. 1997). The ability of this latter technique to evaluate the rapidity of very rapid changes is limited however by diffusion of gases within new snow. Diffusion continues until the snow compacts due to the weight of even newer snow falling on top. Compaction eventually plugs gaps and seals holes, preventing further diffusion. This issue of high initial porosity causes mixing between one year's bubbles and the next and sets a limit to how finely the timing of such events can be resolved. A fascinatingly ingenious method of overcoming some of these problems, in order to more finely resolve the rapidity of these occasional rapid jumps in temperature, makes use of the fact that sudden temperature changes bring about layers of bubbles with distinctive isotopic composition. This new method has allowed finer resolution of sudden climate jumps. It has revealed that the sudden warming at the end of the Younger Dryas took place within only a few decades, that is, a single human lifetime (Severinghaus et al. 1998).

31. The cause of the Younger Dryas is still disputed, although different mechanisms have been suggested. One suggestion is that sudden catastrophic drainage of a large glacial lake may have been responsible. A large body of melt water, positioned where the Great Lakes are today, and held back behind walls of ice, may have suddenly breached, leading to a rapid outflow of low-density freshwater into the North Atlantic. This would have temporarily prevented deep-water formation there and hence shut down the overturning circulation of the ocean. While several sets of data support this interpretation (see for instance McManus et al. 2004) its acceptance has been hindered by an inability to find scour channels through which such drainage took place. Such a large flood should have cut canyons and channels, but none have yet been found that date from the correct time (Broecker 2006).

32. Two books describing the dramatic human impacts of the rather modest climate changes of the last 1,000 years are *The Little Ice Age: How Climate Made History, 1300–1850* by Brian Fagan (2000) and *The Little Ice Age* by Jean Grove (1988).

33. Another example is the emergence from the last ice age. On the whole the warming was fairly gradual, or at least such is the appearance according to Antarctic ice cores. On Greenland, however, a portion of the warming took place exceedingly rapidly at about 14.5 thousand years ago (kya), during what is known as the Bølling Transition. Examination of the isotopic fractionations of argon and nitrogen gases in Greenland ice cores suggests that the climate warmed by about 9°C within only a few decades (Severinghaus and Brook 1999). If we go further back to the height of the last glacial period, more than twenty Dansgaard-Oeschger events have been identified within Greenland ice cores (Dansgaard et al. 1993; Stocker 2000). These abrupt changes typically consisted of rapid warmings (5°C or more over just a few decades) followed by slower relaxations back to full glacial cold.

34. (Hammer et al. 1997)

35. The pH scale is a logarithmic one (pH = $\log_{10}(1.0/[H^+])$) such that a pH change of 1.0 corresponds to a tenfold difference in the hydrogen ion concentration.

36. Dissolved boron is relatively abundant (concentration of about 400 µmol kg^{-1}) in seawater throughout the oceans. Most of that boron is held in $B(OH)_3$ and $B(OH)_4^-$ molecules. The relative abundance of the two molecules is not constant, however, but rather varies dynamically as a function of the pH of seawater. If the pH changes then so too does the $[B(OH)_3]$: $[B(OH)_4^-]$ ratio. There is a difference of about 2‰ in the isotopic compositions of the two molecules, with larger fractions of the ^{11}B isotope in $B(OH)_3$. For this reason the $\delta^{11}B$ of precipitated calcium carbonate also varies with the pH of the seawater solution out of which it is precipitated. The potential for a paleorecorder of ocean pH arises because the $\delta^{11}B$ of ancient corals or foraminifera shells (assumed to indicate the ancient seawater pH) can be measured with great accuracy using a mass spectrometer.

37. (Lemarchand et al. 2000; Pagani et al. 2005)

38. In a paper with Richard Zeebe (Tyrrell and Zeebe 2004) we also calculated that surface ocean pH was lower 60 Mya than it is now, although in our case we estimate that the difference from the present was somewhat smaller, about 0.5 units less than today. Our estimate was arrived at from a completely independent approach based on the fluctuating major ion chemistry described in section 8.2.

39. Coincidentally, the pH of average global surface seawater is predicted to undergo almost the exact opposite change (i.e., decline from about 8.2 to 7.5) in the future, due to ocean acidification brought about by fossil fuel CO_2 emissions (Caldeira and Wickett 2003). The severity of future ocean acidification obviously depends crucially on how much CO_2 we eventually emit to the atmosphere, and how rapidly we emit it.

40. Although not the very first to consider the possibility of global-scale glaciation during the Neoproterozoic, Joseph Kirschvink was the first to coin the term *Snowball Earth* (Kirschvink 1992). Many more scientists became fascinated by the topic following publication of a paper in *Science* by Paul Hoffmann and colleagues (Hoffman et al. 1998).

41. (Nisbet 2003)

42. See for instance photographic evidence shown in Hoffman and Schrag (2002).

43. A diamictite is a rock made up of a heterogeneous mixture of assorted particles and rock fragments of various sizes, including many of larger size (> 2 mm).

44. Geologists have spent considerable time considering other possible causes of the horizontal magnetic alignment of these sedimentary rocks containing dropstones, such as resetting of the magnetic field by subsequent alteration of the rocks a long time after they were originally laid down. But various tests suggest that the horizontal magnetism is indeed pristine and original. Supporting evidence for a long-lasting and complete Snowball comes from a post-Snowball spike in iridium levels in seafloor sediments. This element is principally delivered to Earth via a constant rain of cosmic dust falling down from space (although also from comets and asteroids). In order to generate such a spike in iridium, it is argued that the iridium must have piled up for millions of years on top of the ice, before being suddenly released to marine sediments in a large dump as the ice melted. The spike in iridium supports the argument for a complete Snowball because ice flow toward any areas of open sea would be expected on a partial Snowball, preventing such a large build-up of iridium (Bodiselitsch et al. 2005).

45. (Williams 1993)

46. Evidence from the orientation of magnetic fields in rocks dating from that time does not support the idea that Earth's inclination with respect to the sun was much different from today (Evans 2006).

47. A microbial "deep biosphere" (see summary of recent research by Mascarelli [2009]) ekes out a living in locations many hundreds of meters down within the Earth's crustal rocks, underneath either seafloor (Roussel et al. 2008) or land surface (e.g., Lin et al. 2006). Even some multicellular life survives cocooned far down within rocks; roundworms have recently been discovered at more than a kilometer below the surface in South African gold mines (Borgonie et al. 2011). It is hard to see how this deep biosphere could be anything other than extremely resilient, given that it is geothermally heated from below and presumably almost unaffected by any climatic fluctuations occurring a long way above at the surface. It must be very hard to destroy absolutely all life on Earth, when all life includes this "hidden-away" and protected deep biosphere.

48. (Olcott et al. 2005; Corsetti, Olcott, and Bakermans 2006).

49. (Hoffman and Schrag 2002)

50. Interestingly, from the point of view of evaluating Gaia, the saving feedback that probably brought Earth back from the brink (that prevented it remaining a Snowball forever) was most likely completely abiological. Life probably had no hand in the rescue. Because hardly any carbon can be buried in a frozen ocean, and all weathering fluxes (except perhaps those on the seafloor: Le Hir et al. [2008]) would presumably also have been suppressed by the ice cover, there was nothing to prevent CO_2 from building up in the atmosphere as it continued to be emitted from volcanoes (this scenario for recovery is perfectly reasonable, but there is as yet little hard evidence (see Kasemann et al. [2010] for an initial study). The presumed inexorable accumulation of volcanic emissions would eventually have led to a supergreenhouse sufficient to overcome even the supreme shininess of a Snowball reflecting the majority of sunlight directly back to space. Despite the bright white, pearly Earth rejecting nearly all solar heating, it did, eventually, warm up again. The Snowball eventually melted, and Earth came to accept solar heat once more. But if this interpretation is correct then it was no thanks to Gaia: the process that saved the day was an inorganic, geological one (Watson 2004).

51. This is predicted if the "runaway icehouse" effect ran in reverse. In other words, if a completely ice-covered Earth (very high albedo, reflecting almost all incoming sun-

light back to space) coexisted with an increasingly strong CO_2 greenhouse (tending to elevate temperatures), then at some point the increasing greenhouse effect should have raised temperatures sufficiently to start melting a little of the ice. Once ice started to melt, the darker land or sea surface exposed underneath would absorb a greater fraction of the sun's heat, causing even greater warming and melting in a positive feedback. The positive feedback would lead to a sudden transition from ice covered to ice free, and the associated sudden change in albedo would lead to a step change in temperature. This may well be how Snowball Earth episodes ended, although proxy data on ocean acidity cast doubt on this interpretation (Kasemann et al. 2010).

52. (Maloof et al. 2010)

53. We can be completely confident, because the common DNA-based genetic code is shared by all known life on Earth, that all of it has evolved from just one common ancestor. To be clear, this is not the same as saying that life only ever evolved once on Earth; we do not know whether or not this is true. It is also totally different from saying that the first life form ever to appear on the planet (or indeed any of the progenitors, if there were more than one) contained DNA. That such a complex coding system would have spontaneously sprung fully formed into existence is vanishingly unlikely. Primitive forms must have had much simpler means of coding the assembly instructions for descendants, with more sophisticated coding systems eventually evolving from the more primitive ones. Nevertheless, the universality of DNA among all extant organisms tells us that all organisms living today share a common ancestor. From this we can logically deduce that at least one part of the planet (at least one nook or cranny) has remained habitable throughout the totality of time ever since that last common ancestor lived. This deduction does not, however, preclude the possibility that all life on Earth was wiped out at some time before that last common ancestor and then subsequently restarted from scratch.

So how long ago did the last common ancestor live? The chronology of evolution is hinted at from the amount of observed variation in presumed nonfunctional (neutral) DNA, when the DNAs of organisms in very different lineages are compared. Times since divergence between lineages can be calculated, based on the following assumptions: (1) a constant background rate of mutation of neutral DNA in all organisms, (2) that the neutral DNA has no impact at all on the fitness of the organisms and hence is not subjected to any purifying evolutionary pressure to weed out mutations, and (3) that the neutral DNA of two lineages therefore keeps on diverging at a known rate after they split apart. Such reconstructions of evolutionary chronology can be calibrated in places by the fossil record. Even so, there are many potentially troublesome caveats, not the least of which is gene swapping by a process known as "horizontal gene transfer," whereby DNA is exchanged between otherwise unrelated species that are distant from each other on the evolutionary tree. Together these caveats render dating of the last universal ancestor by this technique problematic in the extreme. In terms of more direct and less contentious evidence, there is a continued presence of conspicuous stromatolites in the fossil record throughout the last nearly three billion years. This record includes 2.7-billion-year-old fossils with a clear microbial origin (Lepot et al. 2008). These fossils in some cases look similar to the structures of modern stromatolites still actively growing today in a few scattered supersaline bays and lagoons around the world. This continued record of fossil stromatolites resembling modern ones suggests an uninterrupted presence of life over at least the last 2.7 billion years.

CHAPTER 9. THE PUZZLE OF LIFE'S LONG PERSISTENCE

1. In this context, the term *reservoir* does not have its normal meaning: an artificial lake for storing water. Instead it is used here to refer to the total amount of a substance in a specific volume. So, for instance, we have the reservoir of oxygen in the atmosphere, the reservoir of carbon dissolved in the oceans, or the reservoir of water locked up in ice sheets.

2. The term *residence time* relates to how quickly a particular reservoir can be emptied or filled by a particular input or output. So, to give an example, a bathtub of volume 150 liters, filling from a tap that delivers 10 liters of hot water per minute, would have a residence time (with respect to the input through the tap) of 150/10 = 15 minutes. A residence time is always calculated as the total amount of substance in the volume divided by a rate of input or output. Residence times are useful because they give a preliminary indication of how quickly we can expect the amount or concentration of a substance in a volume to be completely drained or completely replenished. Or, in other words, they give an indication of how rapidly the given environmental variable can vary.

3. (Wigley 2005)

4. A "mole of something" refers to a specific number of items of that thing (in this case a very large number, Avogadro's number: $\sim 6.022 \times 10^{23}$), in the same way that a dozen refers to twelve or the prefix kilo- to 1,000. Therefore 37×10^{18} moles of O_2 corresponds to 2.2×10^{43} molecules of O_2.

5. In actual fact the true timescale would be much, much longer than this, because there is only a finite amount of organic material on Earth that is available to be decomposed. All of the dead organic matter on the planet would decompose in a matter of decades, after which only fluxes (2) and (4) in the main text would operate, and the rate of oxygen removal would slow.

6. For a start, the residence time of atmospheric CO_2 is extended when we take into account the fact that it is in close contact with the oceanic reservoir of carbon that is ~50 times larger. The surface ocean and the atmosphere exchange CO_2 relatively rapidly, and any large imbalances between the two are leveled out within about a year or so (if new processes don't introduce additional imbalances in the meantime). The ocean circulation brings all the deep water to the surface (and therefore into contact with the atmosphere) over a period of several centuries or a thousand years. The terrestrial biosphere and soils also contain carbon that exchanges with the atmosphere on relatively short timescales. Taking all of these additional considerations into account we end up with a new residence time for carbon dioxide (with respect to photosynthesis and respiration) of about 400 years.

But even this more advanced picture is still too simplified. There is less than 10% as much carbon in biomass and soils together (~2,300 Gt C) as there is in the atmosphere and ocean together (~39,000 Gt C). Assuming that we are talking about times after which decomposers of lignin had evolved, there are no conceivable circumstances under which ten times as much organic matter as at present would pile up (e.g., in tree trunks) on the surface of the Earth but fungi and bacteria would not increase in number to cause them to rot. Conversely, as just considered in note 5 for oxygen, even if absolutely all of the 2,300 Gt C in plants and soils were to be converted into CO_2, the total amount of carbon in the atmosphere+ocean would only rise by less than 10%. For these reasons, it is actually necessary to look beyond photosynthesis and respiration for scenarios in which atmospheric CO_2 will be drastically affected. The vast amounts of carbon in rocks, which dwarf the amount in the ocean and atmosphere together, have the capacity. Geologists therefore consider residence times with regard to geological sources and sinks of CO_2, such as volcanic inputs or weathering removal. The residence time of CO_2 with respect to these slower (but ultimately potentially much more powerful) fluxes is much longer (Berner and Caldeira calculate about 400,000 years). Even this much longer timespan, however, although long with respect to human lifespans, is rather short compared to the more than 500 million years of elapsed time since there has last been a Snowball Earth. Over this whole immense interval of time the rate of creation of CO_2 must have always been balanced rather precisely by the rate of removal of CO_2. If the sinks had exceeded the sources by only 25% for about 1.6 million years then all the CO_2 in the ocean and atmosphere would have been exhausted, in the absence of any protective compensating feedbacks. Although atmospheric CO_2 has fluctuated considerably *within* ice age cycles (~50% more CO_2 in the atmosphere during interglacial highs than during ice age lows), the value averaged *across* ice age cycles (comparing successive interglacial highs) has stayed more constant, declining by less than 10% over the last 600,000 years (Zeebe and Caldeira 2008).

7. The argument that our planet must possess protective negative feedbacks to regulate long-term temperature is set out in Berner and Caldeira (1997). Wally Broecker and Abhijit Sanyal reached similar conclusions (Broecker and Sanyal 1998).

8. If the evidence for Snowball/Slushball episodes is correct, then the (ocean+ atmosphere) CO_2 was indeed almost exhausted at the beginnings of those events. If Earth possesses a CO_2 controller then it must have been asleep at the wheel back in the Neoproterozoic. It has in fact been argued that atmospheric CO_2 and climate may have been more tightly controlled in recent eons (the last few hundred million years) than they were back in the Neoproterozoic (Ridgwell, Kennedy, and Caldeira 2003). This is because the Earth now possesses an extra feedback on carbon (the so-called carbonate compensation feedback) that it didn't possess back then, because planktonic calcifiers hadn't yet evolved. The Earth does indeed appear to have been less prone to extreme CO_2 depletion and dramatic climate swings over the last few hundred million years, although serious glaciations have still occurred, including those following the first spread of forests across the land (section 7.2) and the "current" ice ages of the Pleistocene.

9. See for instance the clearly written recent overview by Stephen Hawking and Leonard Mlodinow (2010).

10. (Watson 2004; see also Watson 1999)

11. Cosmologists also debate a similar question. The emergence of intelligent life in the universe is thought to have been contingent on certain physical constants having values in rather narrow ranges (Rees 1999). The anthropic principle comes into play if it is assumed that there are multiple universes, all with different values of the fundamental physical constants, but that life only exists in those (few?) universes like our own with suitable values of the constants. If there are numerous universes and if it is only in universes with suitable constants that life can emerge, anthropic reasoning leads to the conclusion that it is no surprise to find ourselves in such a universe.

12. Not in the same way that we would have been if, for instance, we had visited this planet in a spacecraft after having chosen it at random from many millions in outer space.

13. (Davies 2006)

14. On many occasions it must have been a close call. Of my two grandfathers, for instance, one was buried alive in a trench in the First World War but survived to become a parent later in life, whereas the other was killed in the Second World War only months after fathering my mother. In either case only very slight differences in fortune would have broken the line of descent to myself. Innumerable such chance events must all have turned out the "right" way (from my perspective) during the immensity of evolutionary time.

15. Made worse, in general, by the great overproduction of progeny in all species, with most young never reaching adulthood (section 2.1). We can begin to calculate the probabilities over the more recent geological past, made easier if we limit ourselves to the last 200,000 years since the evolution of *Homo sapiens*. Natural mortality rates prior to reaching reproductive age can be estimated from a number of techniques, including anthropological studies of infant and child mortality in hunter-gatherer societies, and fascinating detective work that estimates the age at death of numerous fossil skeletons. Synthesizing all of this information suggests that something like 40% were infertile or else died before reaching sexual maturity. So what then are the odds of perfect, unbroken bloodlines? Taking an average parental age at birth of twenty-five years leads to: $P = 0.6^{(200000/25)}$ where P is the probability of an unbroken chain of descent, with absolutely every ancestor in the chain surviving to reproduce. The resulting probability (value of P) is less than 10^{-1000}, that is, so small as to have more than a thousand zeros between the decimal point and the first nonzero digit. The a priori (looking forward) chances of any one posited line of descent being completely successful (every link in the chain being unbroken) are therefore minuscule. They are astronomically small. They do not even begin to approach the chances of purchasing a winning ticket in a national lottery. And even this astoundingly small number is a large overestimate, because it assumes that only one ancestor in each previous generation had to survive, as opposed to an increasing number further back in time due to branching of the family tree (2 parents, 4 grandparents, 8 great-grandparents, etc.).

16. (Gould 1985)

17. (Larson 2007)

18. For a recent review see Deming and Seager (2009) or the later chapters of Kasting (2010).

19. Because different atmospheric compounds absorb light at specific wavelengths, the light (or more generally, the radiation) emitted by both stars and planets has distinct gaps in its spectra, called Fraunhofer lines. Light is seen at all wavelengths except for these missing narrow bands. Careful analysis of the missing wavelengths in the light emitted by a particular body, or shining through a particular atmosphere, can show whether the Fraunhofer lines for oxygen, carbon dioxide, sodium, hydrogen, helium, and so on occur or not, and can therefore indicate whether or not these compounds are present in the alien atmosphere.

20. The simultaneous presence of, for instance, distinct Fraunhofer lines for both oxygen and methane could be taken as strongly suggestive evidence of the presence of life, although life need not have evolved in such as way as to produce oxygen.

21. In a solar system orbiting a single central star, the Goldilocks Zone refers to the range of planetary orbital radii from the star that are believed to be compatible with the presence of life. Planets orbiting very close to their star will tend to be so hot that all water evaporates. Planets orbiting at great distances from their parent star will necessar-

ily receive so little heat that any water on the surface must be frozen. In between there exists a range of orbital distances within which liquid water is a possibility. The Goldilocks Zone of red dwarf stars obviously lies much closer in than that of stars much brighter than our own Sun.

22. Although there are large amounts of water on present-day Mars, the temperature and atmospheric pressure are too low to allow bodies of standing water to remain liquid. The water is present instead as ice. The white polar caps seen at both the south and north poles of Mars are partly solid carbon dioxide, but are also in large part made up of solid water ice (Plaut et al. 2007). It has been estimated that if it were possible to melt these polar caps then the melting would release sufficient water to cover the whole planet with a layer of water greater than 10 meters thick (if the planet were completely flat, which it is not). What is less certain is whether there ever was liquid water on Mars at some time in the past. Some landscape features on Mars look reminiscent of river valleys, flood channels, deltas, or other features usually associated with flows of liquid water. Some minerals found on Mars form in the presence of water and may be indicative of its past presence. There is still much controversy, however, and none of the evidence found so far gives definite proof that rivers once flowed across the Martian surface or that waves once lapped against ancient Martian seashores.

23. The extinction of absolutely all life on Earth might in fact require a more extreme perturbation than at first thought. Life has penetrated into all sorts of obscure and remote environments on Earth, including microorganisms living deep within rocks hundreds of meters beneath the seafloor and land surface. The discovery of these abundant communities protected deep within the Earth's crust, where environmental conditions are to a large extent isolated from those of the outside world, raises questions about whether even a full Snowball Earth would be expected to completely wipe out all life. See also note 47 to chapter 8.

24. An average galaxy is made up of hundreds of billions of stars, and there are hundreds of billions of galaxies in the observable universe. The total number of stars in the observable universe has been estimated at about 10^{21}, or, more recently, greater than 10^{23}. It is also easy to get from the number of stars in the observable universe to the number of planets; if we assume the solar system is not all that unusual in having about ten planets then we just add another order of magnitude.

25. It may have occurred to you that, if the whole of the universe originated at the same point, in the Big Bang, and then expanded outward at less than or equal to the speed of light, we should be able to see something of the whole of the universe. However, cosmologists believe that space itself has also been expanding (see for instance Hawking and Mlodinow 2010), leading to the possibility of much of the universe now being out of sight.

26. Some of the key evidence that supports this hypothesis of homogeneity and isotropy, known as the cosmological principle, is as follows: (1) the distribution of "radio galaxies" (those that emit a lot of radiation at radio wavelengths) was found to be randomly distributed across the sky; (2) the distribution of all galaxies was likewise seen to be uniformly distributed in space; and (3) the intensity of the cosmic microwave background radiation, relic thermal radiation from just after the Big Bang, is almost completely identical in all directions. None of this evidence was easily reconcileable with the view that we can see the whole of the universe. All of it sits comfortably with the idea that the universe is much larger than just that part we are able to observe.

27. Carbonate compensation to name but one (Zeebe and Westbroek 2003).

28. Such as the darkening of the planet's surface as seen from space (leading to greater heat absorption), as dark green forests replace bright ice on a warming planet (Betts 2000). A similar feedback is currently accelerating the warming of the Arctic. As sea ice melts due to warming, a highly reflective surface (ice) is replaced by a dark and highly heat-absorbing surface (seawater). The increased heat absorption exacerbates the initial warming. Another example of a destabilizing feedback involves life: as Paul Crutzen (2004) pointed out, when there is a high concentration of oxygen in the atmosphere leading to frequent forest fires, the nature of wood means that increasing quantities of charcoal get formed. Whereas complete burning or rotting of wood consumes all of the oxygen that was produced during its formation, the growth of wood followed by conversion to charcoal leads to a net release of oxygen. Charcoal is resistant to degradation and so its formation during high oxygen periods of Earth's history acted to further raise atmospheric oxygen, tending to make fires even more frequent.

29. The process of silicate weathering has long attracted the interest of those intrigued by how Earth's climate has been controlled, because it has the potential to provide a negative feedback control on climate. It can parallel the operation of a thermostat in the following way: when calcium silicate rocks are weathered (chemically dissolved at the Earth's surface) then this takes place according to the approximate average reaction:

$$CaSiO_3 + 2.CO_2 + H_2O \rightarrow Ca + 2.HCO_3 + SiO_2$$

with the two CO_2 molecules on the left-hand side of the equation coming ultimately from the atmosphere. One of these two molecules is eventually returned to the atmosphere as the extra dissolved calcium is washed into the sea, and once there instigates precipitation of a molecule of calcium carbonate ($CaCO_3$). But the removal of the other molecule of CO_2 from the atmosphere persists for probably many millions of years, until such time as metamorphic processes deep within the Earth's crust convert calcium carbonate rocks back to calcium silicate rocks. The important feature arising out of all these chemical reactions is that silicate weathering removes CO_2 from the atmosphere for extraordinarily long durations. The second critical property is that the rate of silicate weathering appears to be responsive to temperature (i.e., it appears to be accelerated in warmer conditions), which in turn is driven by atmospheric CO_2. The negative feedback is therefore suggested to operate as follows: high atmospheric CO_2 → warm Earth → intense silicate weathering → CO_2 removal. Silicate weathering therefore has the potential to constitute an automatic thermostat with the potential to keep Earth temperatures in check (Walker, Hays, and Kasting 1981). Although silicate weathering is the most widely discussed, it is by no means the only candidate for a thermostat; there is also interest in whether feedbacks connected with organic carbon burial have stabilized Earth temperature (e.g., Raymo 1994; Bowen and Zachos 2010).

30. (Broecker and Sanyal 1998)

31. There is plenty of evidence challenging the proposition that silicate weathering tightly regulates planetary temperature. Nevertheless, one source of doubt comes from measurements of the chemical compositions of rivers draining different landscapes. The relevance of river composition is that the abundance of different elements in river water is in large part determined by the rates of erosion and weathering of the landscapes draining into the rivers. More weathering of granites leads to more dissolved silicon in

rivers, for example. Because the tested landscapes varied in terms of climate, relief (average degree of flatness versus steepness of slopes) and rock type, the importance of temperature in dictating weathering rate could be examined. Some of these studies do support a strong positive effect of temperature on weathering rate (e.g., Velbel 1993), as do some laboratory experiments (e.g., Chen and Brantley 1998; White et al. 1999). But a source of doubt arises because other sets of measurements, for instance of rates of erosion in South America, find that the highest weathering rates come from the most mountainous regions, and that this is more important than whether the regions have a hot climate or not (Edmond et al. 1995). Weathering rate is clearly a function of much more than just temperature (Riebe, Kirchner, and Finkel 2004). The rate of weathering of fresh (in a geological sense) rock fragments and particles, ones only recently exposed to wind and rain, does appear to be sensitive to temperature, as well as wetness, acidity, and other factors. In other parts of the world, where the underlying rocks have not recently been brought to the surface, weathering is much slower, regardless of ambient conditions (West, Galy, and Bickle 2005). In the geologically ancient and quiescent Amazon basin, for instance, even though it is hot and wet and organic acids abound, weathering rates are nevertheless slow because the soils are already highly leached and the remaining particles of rock (those that haven't previously been dissolved) are resistant to further weathering. This should perhaps come as no surprise in the light of the earlier discussion in chapter 4, which pointed out that the Arrhenius rate law is overridden when chemical reactions are more strongly controlled by availability of substrate than by temperature. It appears that the global average weathering rate is affected by both (i) climatic factors such as global temperatures, and (ii) geological factors such as the global rate of mountain building (Edmond and Huh 1997).

This second factor is also behind a famous hypothesis. In a much-cited paper, Maureen Raymo and William Ruddiman (1992) pointed out that silicate weathering appears to have accelerated over the past 40 million years or so, during a time in which the climate has been cooling (part of the 100-million-year climate cooling trend documented in chapter 8). In order to explain this anomalous behavior (if it protects climate, then when climate cools silicate weathering should slow, in order to remove less CO_2), they invoked the collision of the Indian subcontinent into the Asian continental plate, which led to the "crumpling up" of the collision zone, that is, the uplift of the Himalayas and the Tibetan plateau. Their line of reasoning is that this exceptional event of mountain building produced a global-scale increase in silicate weathering and a consequent cooling of climate. The pushing-up of the Himalayas continually exposes large amounts of new, fresh, highly weatherable rock to direct contact with wind and rain. Thus they not only proposed that silicate weathering was not effective in opposing the climate change, they also went further to propose that silicate weathering was in fact causing it. The collision of India with Asia started somewhere between 40 and 50 million years ago, and so cannot account for the whole 100 million years of cooling; nevertheless, the climate-stabilizing property of silicate weathering is brought into question. There are further troubling details that are hard to reconcile with silicate weathering's proposed role as a climate enforcer, or at any rate suggest that it is a rather imperfect one. Contained within the long and gradual 100-million-year cooling, a one-off sharp cooling took place at the boundary between the Eocene and Oligocene epochs at about 34 million years ago. According to the hypothesized link between silicate weathering and planetary temperature, we would have expected silicate weathering to leap to Earth's defense at this time, figuratively speaking. We would have expected silicate weathering, if it is a protective feedback for climate, to suddenly start removing less CO_2 to help warm the climate back up. But careful work measuring two different proxies for weathering across the EO boundary detects instead an increase in silicate weathering following the sudden cooling (Ravizza and Peucker-Ehrenbrink 2003; Zachos et al. 1999). Other proxy studies have also failed to find the expected silicate weathering response to climate change. Different proxies suggest that weathering rates have either increased (Misra and Froelich 2012) or else been rather constant (Willenbring and von Blanckenburg 2010) over the last 12 million years, despite cooling; they have also been rather constant over the last half-million years (Foster and Vance 2009), despite oscillations between ice ages and interglacials. Taking all of these reservations into consideration, detailed inspection of the evidence therefore does not give anything like unequivocal support for silicate weathering as a strong safeguard of climate. There is still much room for debate. It is not possible to expand further upon this interesting topic here, but the important conclusion in this context is that the jury is still out. The evidence does not line up solidly behind silicate weathering as a highly effective climate stabilizer.

32. (Kump, Brantley, and Arthur 2000)

33. As exemplified in a thought-provoking paper (Edmond and Huh 2003), which took as its starting point that inputs and outputs of CO_2 are decoupled. They argued that volcanic input of CO_2 is presently abnormally low because relatively little calcium carbonate is being subducted. Furthermore, they argued that removal of CO_2 due to silicate weathering is presently abnormally high because of the rise of the Himalayas due to the collision of India into Asia.

34. (Ward and Brownlee 2000)

35. This list of coincidences can be extended (Kasting 2010), although it pays to be cautious in distinguishing conditions that truly make life impossible from those that would just have made Earth a very different planet. In the same way that there is a habitable zone in the solar system (a restricted range of distances from the Sun), there is also likely to be a habitable zone in our galaxy, the Milky Way. Too close to the center and the solar system would have frequently been wiped clean of life by supernovae or gamma-ray bursts among tightly packed adjacent stars; too far from the center and it would not have contained enough carbon and other heavy elements. It is also important that the Earth orbits the right kind of star. If Earth had orbited a bright blue star instead, one that burns brighter but much more quickly, then there would have been insufficient time for intelligent life to develop before the star underwent the transition to a supernova. Earth orbits a single, solitary star. Many stars occur as pairs (binary stars) orbiting each other, or in multiple star systems; if the Earth had been located in such a system, it is less likely that its orbit and therefore climate would have been stable over very many millions of years. Earth has an almost exactly circular orbit around the Sun; the orbit is slightly elliptical, but the difference between minimum and maximum distances is only about 2%. More highly elliptical orbits, like those of Mercury (50% difference), lead to great swings in planetary temperature between the times of closest approach to the Sun and farthest separation. The Earth is unusual, among the solar system planets at least, in possessing such a large moon. It is generally accepted that (a) the Moon stabilizes Earth's tilt with respect to the Sun, thereby helping to stabilize climate (e.g., Waltham 2011), and (b) Earth's large moon was created following a statistically improbable glancing impact between the early Earth and another planet about the size of Mars.

36. Taken from the title of another book (Taylor 1998) that looks at the nature of the solar system, planet Earth, the origin of life and its continued maintenance through Earth history, and why all are suitable for life. Taylor, following Stephen Jay Gould (section 9.3.4), dismisses the anthropic principle due to its frequent association with Intelligent Design. But then, rather confusingly, he proceeds to conclude that the long evolution of intelligent life has been mostly a matter of chance: "chance events have dominated both the development of the planets and the evolution of life," and "conscious intelligence has arisen accidentally."

Chapter 10. Conclusions

1. James Lovelock has won a large number of scientific prizes and awards. There are too many to mention here, but they include the Norbert Gerbier Prize of the World Meteorological Organization, the Wollaston Medal of the Geological Society of London (its highest honor), the Dr. A. H. Heineken Prize for Environmental Sciences, and the Japanese Blue Planet Prize. He has also won many prizes separately for his invention of the electron capture detector. He was elected a Fellow of the Royal Society in 1974, awarded a CBE in 1990, and made a Companion of Honour in 2003. Gaia has featured often in the pages of major journals and has been championed by many scientists. It has also been disputed or rejected, sometimes in strong terms, by many others, some of whom are quoted earlier in this book. The mixed scientific reception of Gaia to that point was described in Jon Turney's 2003 book *Lovelock and Gaia: Signs of Life*, published by Icon Books. Scientists who have criticized Gaia include James Kirchner, Richard Dawkins, W. Ford Doolittle, Heinrich Holland, Stephen Schneider, David Beerling, Ken Caldeira, and Stephen Jay Gould. Most recently, Peter Ward has proposed the opposite hypothesis, essentially that life operates against its own best interests, in his own book (2009). Yet in the same year a glowing biography of Lovelock appeared (Gribbin and Gribbin 2009).

2. This has also been concluded elsewhere: "The Gaia hypothesis is supported neither by evolutionary theory nor by the empirical evidence of the geological record" (Cockell et al. 2008).

3. (Kirchner 1991)

4. Peter Ward goes further (Ward 2009); he contends that the effect of life on the environment is usually harmful, often leading to a deterioration in global habitability. He blames life for several of the global mass extinctions. In my view, however, the evidence for this is not compelling either. Overall I see no more reason to believe in a general tendency of life to annihilate itself than of life to tend to look after itself. As far as I can see, neither evolution nor Earth's history compels us to either the former or the latter view.

5. Perhaps the closest parallel in Earth history to our current rate of CO_2 addition occurred about 55 Mya, during the Paleocene-Eocene Thermal Maximum (PETM), when it appears that there was also a sudden influx of carbon to the biosphere. However, even this event is not an exact parallel. The world was very different at that time (e.g., no polar ice sheets, a smaller Atlantic Ocean) and the rate of carbon input was almost certainly much less rapid than it is now.

6. The title of a paper by P. Friedlingstein and colleagues, "How Positive Is the Feedback between Climate Change and the Carbon Cycle?" is emblematic of just the sort of question we should be ready to ask (Friedlingstein et al. 2003).

7. For instance, higher atmospheric CO_2 is stimulating extra growth of the land vegetation at the present time. As trees increase their girths under high CO_2, this is sequestering some of the fossil fuel carbon emissions and opposing the rise in atmospheric CO_2 in a negative feedback. The potential of nitrogen fixation to stabilize the nitrate:phosphate ratio in the ocean (section 6.3) is another negative feedback tending to impart stability.

8. Indeed, an even more Panglossian view used to be common, that the world was so large and robust that our activities would not be able to perturb it. There was a common belief in the "Balance of Nature," whereby natural forces summed over the vastness of the whole planet were assumed to be so enormous that human influences were as nothing in comparison. For instance, even the eminent ecologist G. N. Hutchinson wrote in 1948 that "the self-regulating mechanisms of the carbon cycle can cope with the present influx of carbon of fossil origin" (Hutchinson 1948). Such a complacent view seems naive now, and has become increasingly impossible to sustain after Hutchinson's time, as more and more evidence has accumulated of our ability to greatly alter our environment. First we became aware of our potency at making local or regional alterations, for instance air pollution (smogs), acid rain leading to acid lakes, and loss of 75% of the surface area of the Aral Sea since the 1960s as river water was diverted for irrigation. The extensive reach of some of our actions also became increasingly obvious, as radioactive fallout from individual nuclear explosions was found to spread rapidly around the world, and the pesticide DDT was seen to permeate through food chains to accumulate in higher predators. By now we are all too aware that human activities can change the planet even at the global scale, by changing for instance the concentration of ozone in the stratosphere, the atmospheric CO_2 concentration, the ocean's pH, and the global temperature. (Some of this discussion is adapted from Weart 2003.)

9. A higher level of the atmosphere between 10 and 30 km above the Earth's surface.

10. The CFC-producing industry was by this time a rather large one, with a significant investment in facilities related to putting CFCs into aerosol cans, into fridges, into air-conditioning equipment, etc. Du Pont, in particular, had just spent $100 million on a new fluorocarbon plant in Texas, completed in late 1974. The industry strongly resisted any attempts at CFC regulation, partly by trying to discredit the science linking CFCs to ozone depletion and thereby to levels of cancer-producing UV radiation (Dotto and Schiff 1978; Roan 1989).

11. (Crutzen 1995)

12. The beginnings of the ozone controversy took place sufficiently long ago to make a recap worthwhile. The usefulness of chlorofluorocarbons (CFCs) as refrigerants was first demonstrated by Thomas Midgley in 1928; in a presentation to the American Chemical Society he graphically showed both their safety and their inertness by inhaling a mouthful of CFCs and then exhaling to blow out a candle. After this time CFCs became increasingly widely used in industry. By the early 1970s they were being used mainly as spray-can propellants and refrigerants, but also in air-conditioning systems, to make foam, and within fire extinguishers. Their usefulness derives in part from being inert (nontoxic, nonflammable, and noncorrosive) while at the same time having low boiling points, below, but not too far below, the freezing point of water (e.g., boiling point of $-40°C$ at atmospheric pressure). In 1971–72 James Lovelock measured CFC concentrations during a long ocean voyage and from those measurements deduced that the CFCs produced by humankind up to that time had not been destroyed to any great extent in

the lower atmosphere. The same general inertness that makes them so safe to use also protects them from being destroyed. This gives them time to percolate up into the stratosphere without being destroyed in transit. It was not until 1974 that Molina and Rowland published their key paper (Molina and Rowland 1974) pointing out the disturbing potential of CFCs to destroy the ozone layer. This paper initiated a whirlwind of controversy, resistance by industry to legislation, and intense further scientific research. Important support for Molina and Rowland's hypothesis came from a report by the US National Academy of Sciences, published in 1976, at which point the United States and a few other countries initiated legislation to ban CFCs in aerosol spray cans. Concerted international action did not immediately follow, however, until the issue became suddenly more urgent following the startling discovery of the ozone hole over Antarctica, published in 1985 (Farman, Gardiner, and Shanklin 1985). This took scientists by surprise, because it had not previously been realized that ozone depletion would be much more serious near the poles than near the equator. A large multidisciplinary scientific campaign was quickly organized to Antarctica, with the evidence obtained pointing clearly to CFCs as the culprits. In 1987 forty-three countries signed up to the Montreal Protocol on Substances that Deplete the Ozone Layer. This protocol has been a great success, and since that time emissions have declined to much lower levels. Because of the long lifetime of CFCs in the atmosphere, it will however take many decades before the ozone hole disappears and the ozone layer completely recovers. Satellites can now give us a good picture of ozone levels around the globe. As shown in the figure below (fig. 10.1), the ozone hole still remains nearly twenty years after the protocol was agreed, despite a great reduction in CFC production.

13. Using the Electron Capture Detector that he invented, Lovelock made some of the key measurements leading to discovery of the problem of ozone destruction by CFCs, but ironically was himself overly complacent about the likely dangers: "The presence of these [CFC] compounds constitutes no conceivable hazard" (Lovelock, Maggs, and Wade 1973). "Wherever we sailed we found it easy to detect and measure the fluorochlorocarbon gases, and this discovery led directly to the present possibly exaggerated concern over their capacity to deplete the ozone layer" (page 105 of Lovelock 1979). Lovelock was continuing to express doubt until only a few months prior to the announcement of the discovery of the ozone hole over Antarctica: "Had we known in 1975 as much as we know now about atmospheric chemistry, it is doubtful if politicians could have been persuaded to legislate against the emission of CFCs" (Lovelock 1984). This example demonstrates the tendency toward unwarranted complacency that a belief in Gaia can potentially produce. In his defense, after being insufficiently alarmist on this first occasion, Lovelock has more than made up for it since. A later book (Lovelock 2006) is apocalyptic in its predictions, in some respects unreasonably so. For example, his view that algal production will cease altogether above 500 ppm of CO_2 does not square with the facts (see section 5.3.2). Although there are plenty of other reasons to be very worried about such high levels of CO_2, it is not clear why global warming, even under the more pessimistic scenarios, would bring about the massive reduction in human population that he postulated.

14. For instance, another possible Achilles' heel might be the Amazon rain forest, with its suggested susceptibility to drying out and burning as a result of global warming (Cox et al. 2004).

15. There are other examples of how our developing understanding is difficult to reconcile with Gaia as an overarching paradigm. A second example of unexpected Earth

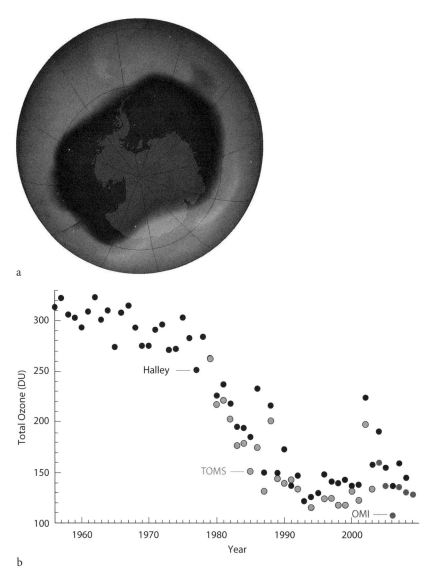

Figure 10.1. Ozone hole over Antarctica: (*a*) the largest ozone hole ever observed—image from September 21–30, 2006. Darker shades indicate least ozone; lighter shades correspond to higher ozone levels. (*b*) British Antarctic Survey scientists have made ground-based measurements (black circles) of atmospheric ozone since the late 1950s. Minimum annual values are shown here (note the nonzero origin on the *y*-axis). By the mid-1980s it had become apparent that the seasonal troughs in ozone were getting much more severe. For the last twenty-five years or so the ground-based measurements have been supplemented by satellite-based measurements (TOMS and OMI, gray circles). (NASA website.)

system behavior concerns the response to atmospheric CO_2. It is often assumed (implicitly) that if the Earth system has a perturbation imposed upon it, but is afterward left to its own devices, that it will eventually return back to the state it was in prior to the perturbation. We now know that this is not a safe assumption. We may have been misled by our familiarity with the way that pendulums pulled away from the vertical will always fall back toward it. Eventually, over time, as air friction takes its inexorable toll, a disturbed freestanding pendulum will always settle back toward a motionless state with the weight hanging down vertically. And this will happen time after time. Likewise, a metal spring continually recovers its original length after a stretching or compressing force is removed, as long as that force is not so great as to overstretch the spring. It is easy to subconsciously suspect that the Earth system will behave similarly (the assumption is even made explicitly in places, for instance page 269 of Hazen 2012). However, it is becoming apparent that it does not work that way. We have strong reasons to believe that the Earth system is not going to "return back to normal" after the end of fossil fuel burning, once all CO_2 emissions have ceased. The ocean chemistry feedbacks are too complicated to cover in detail here, but multiple modeling studies now agree on the fact that there will be a recovery back toward a new equilibrium state, one that is different from the equilibrium state before fossil fuels were taken out of the ground and burned. If the Earth were left to its own devices, atmospheric CO_2 would come down from its peak concentration, but then settle back down at a level that is still significantly higher than the preindustrial concentration, and then stay there for at least many tens or hundreds of thousands of years following the end of CO_2 emissions (Archer, Kheshgi, and Maier-Reimer 1997; Tyrrell, Shepherd, and Castle 2007; Archer 2009). The Earth does not, apparently, possess just one equilibrium state. In this way it is unlike a pendulum or spring. The overturning circulation of the ocean is likewise thought to have more than one equilibrium state, and possibly to exhibit hysteresis such that, following removal of a temporary perturbation, the system will not necessarily return back to the original state before the perturbation was applied. From these examples we can see that the Earth is not in all ways inherently self-rectifying and self-healing. Following removal of a disturbing force it does not in all cases recover back toward the original state. This understanding is not easily arrived at from a Gaian perspective, which engenders a confidence that the Earth will always repair itself (and return to normal) once human intervention ceases. This is another conceptual wrong turning to which a Gaian mindset can unconsciously predispose us.

16. And for this reason I believe that the first bullet point of the Amsterdam Declaration (see chapter 1), issued jointly by four overarching international scientific organizations in 2001, was misguided. In language that could have been taken from a publication supporting Gaia, it stated that "the Earth System behaves as a single, self-regulating system comprised of physical, chemical, biological and human components." The concept of the Earth system as a "single, self-regulating system" is not supported by hard science, and it is surprising that it was inserted. There are indeed self-regulating feedbacks, but there are also destabilizing feedbacks. The history of the Earth system does not tally with what would be expected of a single, self-regulating system.

FURTHER READING

CHAPTER 1

For an independent history of the Gaia idea and its reception by the scientific community, Jon Turney's book (Turney 2003) is recommended. For the same history from the perspective of two of the protagonists, see Lovelock's autobiography (Lovelock 2001) and Stephen Schneider's review of it (Schneider 2001). If you want to find out about Gaia by reading only one of Lovelock's books, the third is beautifully designed and illustrated (Lovelock 1991).

CHAPTER 2

Excerpts from some of the most pertinent classic papers about evolution and regulation of the Earth are contained in Connie Barlow's edited volume (Barlow 1991). Scott Turner's book (Turner 2000) provides an exposition of the ways in which many animals co-opt the external environment to their own benefit, by producing physical structures or by other modifications.

CHAPTER 3

Two useful reviews of extremophiles and limits to life are provided by Madigan and Marrs (1997), which is at more of an introductory level, and Rothschild (2009), which is more specialized although still accessible. For a greater depth of treatment and a thorough coverage see Wharton (2002). Some of the byways and alleyways (as well as the major thoroughfares) of evolution are explained by Dawkins and Wong (2004).

CHAPTER 4

No modern consideration of the effects of temperature on life could possibly be complete without reference to the exciting new developments in the field of metabolic ecology. This theoretical framework offers a fresh and stimulating perspective on many aspects of biology and ecology; it provides a new lens through which to view the world. The spate of developments has been admirably reviewed in John Whitfield's book (2006).

CHAPTER 5

A concise overview of ice ages and their effects, in an accessible style, is Doug Macdougall's *Frozen Earth: The Once and Future Story of Ice Ages* (Macdougall 2004). Alastair Fothergill's book on the scarce life on Antarctica (Fothergill 1993) includes fascinating details about the curious adaptations of Antarctic animals to their hostile environment. A comprehensive overview of the Cretaceous world can be found in the university-level textbook of the same name (Skelton et al. 2003). A personal account of the search for polar dinosaur remains on a remote coast of Australia is given in Rich and Vickers-Rich (2000).

CHAPTER 6

The strangeness of oxygen and methane co-occurring in Earth's atmosphere, from the perspective of inorganic chemistry, is well described in Lovelock's books. Two recent reviews describe our current knowledge of why the Earth, uniquely in the solar system, developed an oxygen-rich atmosphere: Canfield (2005) and Catling and Claire (2005). The latter also considers the processes, or balance of processes, maintaining abundant atmospheric oxygen. The slim volume by Postgate (1998) is recommended for its concise introduction to nitrogen fixation. The topic of whether the nitrogen cycle is currently (since the last ice age) in balance is one that is actively debated. Niki Gruber provided a clear articulation of one viewpoint (Gruber 2004) and Louis Codispoti the other (Codispoti 1995). More broadly, Tyler Volk (1998) elegantly describes many different ways in which the Earth's biota alter the Earth's environment.

CHAPTER 7

A brief but useful history of atmospheric oxygen and the "Great Oxidation Event" is compressed into just two pages by Lee Kump (2008). The GOE and the rise in oxygen are included in an article on life's ability to exert control over and effect change in the Earth's environment (Knoll 2003). The story of the evolution of land plants, the spread of forests across the land, and the to-and-fro interactions between the evolution of trees and the modification of their environment (atmospheric carbon dioxide and oxygen in particular) is set out in chapters 2 and 3 of David Beerling's book *The Emerald Planet: How Plants Changed Earth's History* (Beerling 2007). Evidence for the global environmental consequences of the first trees is described in chapter 16 of Lenton and Watson (2011). For an alternative approach to the subject of this chapter, see Peter Ward's provocative recent book (Ward 2009). Ward proposes the direct opposite to Gaia—namely, that life engineers its own demise through detrimental impacts on planetary habitability.

CHAPTER 8

The story of the discovery of episodes of abrupt climate change in Earth's history, and the increasing awareness that environmental change is not only slow and gradual but can also be sharp and sudden, is set out in Cox (2002). The story of how ice cores are recovered, and the lessons we can learn from these amazing climate records, is told in Richard Alley's book *The Two-Mile Time Machine: Ice Cores, Abrupt Climate Change, and Our Future* (Alley 2000). A more general introduction to some of the many different methods of deducing past climates is available at http://earthobservatory.nasa.gov/Features/Paleoclimatology/paleoclimatology_intro.php. For an accessible and entertaining portrayal of the Snowball Earth hypothesis and the scientific personalities involved, see Gabrielle Walker's book *Snowball Earth* (Walker 2003). A more technical review article, with wonderful photographs showing some of the direct geological evidence driving the interest in the hypothesis, is (Hoffman and Schrag 2002). An advanced-level review of our understanding of how ocean chemistry has changed over time can be found in Montanez (2002).

CHAPTER 9

An interesting discussion of whether our planet's continued habitability was coincidental or inevitable is contained in Andy Watson's article on "Gaia and Observer Self-Selection" (Watson 2004). The application of the anthropic principle to Gaia and continued Earth habitability is also discussed in chapter 5 of Lenton and Watson (2011) and by Simon Levin (1999). For discussion of the anthropic principle from a cosmological perspective, written in an accessible style, see (Rees 1999).

CHAPTER 10

The history of the discovery that human activities were destroying the ozone layer, and the subsequent controversy over banning chlorofluorocarbons, is described from a scientific perspective in Rowland (2006), and from a cultural and political perspective in Dotto and Schiff (1978). The story is retold in Walker (2007) together with the story of the discovery of the ozone hole. Spencer Weart gives an account of the slow shift toward accepting the reality of global warming (in part because of the need to overcome excessive faith in Earth's ability to automatically counteract any perturbations) in Weart (2003); see also an expanded and updated version at http://aip.org/history/climate/index.html.

REFERENCES

Adams, J. M., and H. Faure. 1998. A new estimate of changing carbon storage on land since the last glacial maximum, based on global land ecosystem reconstruction. *Global and Planetary Change* 17: 3–24.

Adams, J. M., H. Faure, L. Fauredenard, J. M. Mcglade, and F. I. Woodward. 1990. Increases in terrestrial carbon storage from the last glacial maximum to the present. *Nature* 348 (6303): 711–14.

Adkins, J. F., K. McIntyre, and D. P. Schrag. 2002. The salinity, temperature, and delta O-18 of the glacial deep ocean. *Science* 298 (5599): 1769–73.

Aletsee, L., and J. Jahnke. 1992. Growth and productivity of the psychrophilic marine diatoms *Thalassiosira antarctica* Comber and *Nitzschia frigida* Grunow in batch cultures at temperatures below the freezing point of seawater. *Polar Biology* 11 (8): 643–47.

Algeo, T. J., and S. E. Sheckler. 1998. Terrestrial-marine teleconnections in the Devonian: Links between the evolution of land plants, weathering processes, and marine anoxic events. *Philosophical Transactions of the Royal Society B—Biological Sciences* 353: 113–30.

Algeo, T. J., S. E. Scheckler, and J. B. Maynard. 2001. Effects of the Middle to Late Devonian spread of vascular land plants on weathering regimes, marine biotas, and global climate. In *Plants Invade the Land*, edited by P. G. Gensel and D. Edwards. New York: Columbia University Press.

Allen, A. P., J. F. Gillooly, V. M. Savage, and J. H. Brown. 2006. Kinetic effects of temperature on rates of genetic divergence and speciation. *Proceedings of the National Academy of Sciences of the United States of America* 103 (24): 9130–35.

Alley, Richard B. 2000. *The Two-Mile Time Machine: Ice Cores, Abrupt Climate Change, and Our Future*. Princeton, NJ: Princeton University Press.

Alley, R. B., D. A. Meese, C. A. Shuman, A. J. Gow, K. C. Taylor, P. M. Grootes, J.W.C. White, M. Ram, E. D. Waddington, P. A. Mayewski, and G. A. Zielinski. 1993. Abrupt increase in Greenland snow accumulation at the end of the Younger Dryas event. *Nature* 362 (6420): 527–29.

Allwood, A. C., M. R. Walter, B. S. Kamber, C. P. Marshall, and I. W. Burch. 2006. Stromatolite reef from the Early Archaean era of Australia. *Nature* 441 (7094): 714–18.

Altabet, M. A. 2007. Constraints on oceanic N balance/imbalance from sedimentary N-15 records. *Biogeosciences* 4 (1): 75–86.

Altabet, M. A., M. J. Higginson, and D. W. Murray. 2002. The effect of millennial-scale changes in Arabian Sea denitrification on atmospheric CO_2. *Nature* 415 (6868): 159–62.

Altshuler, D. L., and C. J. Clark. 2003. Darwin's hummingbirds. *Science* 300 (5619): 588–89.

Anderson, L. A., and J. L. Sarmiento. 1994. Redfield ratios of remineralization determined by nutrient data analysis. *Global Biogeochemical Cycles* 8 (1): 65–80.

Archer, David. 2009. *The Long Thaw: How Humans Are Changing the Next 100,000 Years of Earth's Climate*. Science Essentials. Princeton, NJ: Princeton University Press.

Archer, D., H. Kheshgi, and E. Maier-Reimer. 1997. Multiple timescales for neutralization of fossil fuel CO_2. *Geophysical Research Letters* 24 (4): 405–8.

Arrhenius, S. 1889. Über die Reaktionsgeschwindigkeit bei der Inversion von Rohzucker durch Säuren (On the reaction velocity of the inversion of cane sugar by acids). *Zeitschrift für Physikalische Chemie* 4: 226–48.

Astin, T. R., J.E.A. Marshall, H. Blom, and C. M. Berry. 2010. The sedimentary environment of the Late Devonian East Greenland tetrapods. *Geological Society Special Publications* 339 (1): 93–109.

Barlow, Connie C. 1991. *From Gaia to Selfish Genes: Selected Writings in the Life Sciences.* Cambridge, MA: MIT Press.

Bayly, Ian A. E., and W. D. Williams. 1973. *Inland Waters and Their Ecology.* Camberwell, Australia: Longman.

Beerling, D. J. 1999a. New estimates of carbon transfer to terrestrial ecosystems between the last glacial maximum and the Holocene. *Terra Nova* 11 (4): 162–67.

———. 1999b. Quantitative estimates of changes in marine and terrestrial primary productivity over the past 300 million years. *Proceedings of the Royal Society B—Biological Sciences* 266 (1431): 1821–27.

———. 2002. Low atmospheric CO_2 levels during the Permo-Carboniferous glaciation inferred from fossil lycopsids. *Proceedings of the National Academy of Sciences of the United States of America* 99 (20): 12567–71.

———. 2007. *The Emerald Planet: How Plants Changed Earth's History.* Oxford; New York: Oxford University Press.

Beerling, D. J., C. P. Osborne, and W. G. Chaloner. 2001. Evolution of leaf-form in land plants linked to atmospheric CO_2 decline in the Late Palaeozoic era. *Nature* 410 (6826): 352–54.

Begon, Michael, John L. Harper, and Colin R. Townsend. 1996. *Ecology: Individuals, Populations, and Communities.* 3rd ed. Oxford; Cambridge, MA: Blackwell Science.

Bekker, A., H. D. Holland, P. L. Wang, D. Rumble, H. J. Stein, J. L. Hannah, L. L. Coetzee, and N. J. Beukes. 2004. Dating the rise of atmospheric oxygen. *Nature* 427 (6970): 117–20.

Benton, M. J. 2011. The biosphere rebooted. *Nature* 471 (7338): 303–4.

Bernal, D., J. M. Donley, R. E. Shadwick, and D. A. Syme. 2005. Mammal-like muscles power swimming in a cold-water shark. *Nature* 437 (7063): 1349–52.

Berner, E. K., R. A. Berner, and K. L. Moulton. 2003. Plants and mineral weathering: Present and past. *Treatise in Geochemistry* 5: 169–88.

Berner, R. A. 1997. The rise of plants and their effect on weathering and atmospheric CO_2. *Science* 276 (5312): 544–46.

———. 1998. The carbon cycle and CO_2 over Phanerozoic time: The role of land plants. *Philosophical Transactions of the Royal Society of London Series B—Biological Sciences* 353 (1365): 75–81.

———. 1999. Atmospheric oxygen over Phanerozoic time. *Proceedings of the National Academy of Sciences of the United States of America* 96 (20): 10955–57.

———. The effect of the rise of land plants on atmospheric CO_2 during the Paleozoic. In *Plants Invade the Land*, edited by P. G. Gensel and D. Edwards. New York: Columbia University Press.

———. 2003. The long-term carbon cycle, fossil fuels, and atmospheric composition. *Nature* 426 (6964): 323–26.

———. 2006. GEOCARBSULF: A combined model for Phanerozoic atmospheric O_2 and CO_2. *Geochimica Et Cosmochimica Acta* 70 (23): 5653–64.

———. 2009. Phanerozoic atmospheric oxygen: New results using the GEOCARBSULF model. *American Journal of Science* 309 (7): 603–6.

Berner, R. A., and K. Caldeira. 1997. The need for mass balance and feedback in the geochemical carbon cycle. *Geology* 25 (10): 955–56.

Berner, R. A., and D. E. Canfield. 1989. A new model for atmospheric oxygen over Phanerozoic time. *American Journal of Science* 289 (4): 333–61.

Betts, R. A. 2000. Offset of the potential carbon sink from boreal forestation by decreases in surface albedo. *Nature* 408 (6809): 187–90.

Blackburn, D. G. 2000. Reptilian viviparity: Past research, future directions, and appropriate models. *Comparative Biochemistry and Physiology—A: Molecular and Integrative Physiology* 127 (4): 391–409.

Blake, R. E., S. J. Chang, and A. Lepland. 2010. Phosphate oxygen isotopic evidence for a temperate and biologically active Archaean ocean. *Nature* 464 (7291): 1029–32.

Blunier, T., B. Barnett, M. L. Bender, and M. B. Hendricks. 2002. Biological oxygen productivity during the last 60,000 years from triple oxygen isotope measurements. *Global Biogeochemical Cycles* 16 (3): 1029.

Bodiselitsch, B., C. Koeberl, S. Master, and W. U. Reimold. 2005. Estimating duration and intensity of Neoproterozoic Snowball glaciations from Ir anomalies. *Science* 308 (5719): 239–42.

Bolhuis, H., E.M.T. Poele, and F. Rodriguez-Valera. 2004. Isolation and cultivation of Walsby's square archaeon. *Environmental Microbiology* 6 (12): 1287–91.

Borgonie, G., A. Garcia-Moyano, D. Litthauer, W. Bert, A. Bester, E. van Heerden, C. Moller, M. Erasmus, and T. C. Onstott. 2011. Nematoda from the terrestrial deep subsurface of South Africa. *Nature* 474 (7349): 79–82.

Bostrom, Nick. 2002. *Anthropic Bias: Observation Selection Effects in Science and Philosophy.* Studies in Philosophy. New York: Routledge.

Bowen, G. J., and J. C. Zachos. 2010. Rapid carbon sequestration at the termination of the Palaeocene-Eocene Thermal Maximum. *Nature Geoscience* 3 (12): 866–69.

Boyd, P. W., T. Jickells, C. S. Law, S. Blain, E. A. Boyle, K. O. Buesseler, K. H. Coale, J. J. Cullen, H.J.W. de Baar, M. Follows, M. Harvey, C. Lancelot, M. Levasseur, N.P.J. Owens, R. Pollard, R. B. Rivkin, J. Sarmiento, V. Schoemann, V. Smetacek, S. Takeda, A. Tsuda, S. Turner, and A. J. Watson. 2007. Mesoscale iron enrichment experiments 1993–2005: Synthesis and future directions. *Science* 315 (5812): 612–17.

Brauer, A., G. H. Haug, P. Dulski, D. M. Sigman, and J.F.W. Negendank. 2008. An abrupt wind shift in western Europe at the onset of the Younger Dryas cold period. *Nature Geoscience* 1 (8): 520–23.

Broecker, W. S. 2006. Was the Younger Dryas triggered by a flood? *Science* 312 (5777): 1146–48.

Broecker, W. S., and A. Sanyal. 1998. Does atmospheric CO_2 police the rate of chemical weathering? *Global Biogeochemical Cycles* 12 (3): 403–8.

Brown, J. H., J. F. Gillooly, A. P. Allen, V. M. Savage, and G. B. West. 2004. Toward a metabolic theory of ecology. *Ecology* 85 (7): 1771–89.

Buckling, A., R. C. Maclean, M. A. Brockhurst, and N. Colegrave. 2009. The *Beagle* in a bottle. *Nature* 457 (7231): 824–829.

Buzas, M. A., and S. J. Culver. 1999. Understanding regional species diversity through the log series distribution of occurrences. *Diversity and Distributions* 8: 187–95.

Caldeira, K., and M. E. Wickett. 2003. Anthropogenic carbon and ocean pH. *Nature* 425 (6956): 365.

Calvo, E., C. Pelejero, P. De Deckker, and G. A. Logan. 2007. Antarctic deglacial pattern in a 30 kyr record of sea surface temperature offshore South Australia. *Geophysical Research Letters* 34 (13).

Came, R. E., J. M. Eiler, J. Veizer, K. Azmy, U. Brand, and C. R. Weidman. 2007. Coupling of surface temperatures and atmospheric CO_2 concentrations during the Palaeozoic era. *Nature* 449 (7159): 198–201.

Canfield, D. E. 2005. The early history of atmospheric oxygen: Homage to Robert A. Garrels. *Annual Review of Earth and Planetary Sciences* 33: 1–36.

Catling, D. C., and M. W. Claire. 2005. How Earth's atmosphere evolved to an oxic state: A status report. *Earth and Planetary Science Letters* 237 (1–2): 1–20.

Catling, D. C., K. J. Zahnle, and C. P. McKay. 2001. Biogenic methane, hydrogen escape, and the irreversible oxidation of early Earth. *Science* 293 (5531): 839–43.

Chadwick, O. A., L. A. Derry, P. M. Vitousek, B. J. Huebert, and L. O. Hedin. 1999. Changing sources of nutrients during four million years of ecosystem development. *Nature* 397 (6719): 491–97.

Chen, L. B., A. L. DeVries, and C.H.C. Cheng. 1997. Evolution of antifreeze glycoprotein gene from a trypsinogen gene in Antarctic notothenioid fish. *Proceedings of the National Academy of Sciences of the United States of America* 94 (8): 3811–16.

Chen, Y., and S. L. Brantley. 1998. Diopside and anthophyllite dissolution at 25 degrees and 90 degrees C and acid pH. *Chemical Geology* 147 (3–4): 233–48.

Choquenot, D. 1991. Density-dependent growth, body condition, and demography in feral donkeys—testing the food hypothesis. *Ecology* 72 (3): 805–13.

Chyba, C. F., and K. P. Hand. 2001. Life without photosynthesis. *Science* 292 (5524): 2026–27.

Clarke, A. 2003. Costs and consequences of evolutionary temperature adaptation. *Trends in Ecology and Evolution* 18 (11): 573–81.

Clutton-Brock, T. 2002. Breeding together: Kin selection and mutualism in cooperative vertebrates. *Science* 296 (5565): 69–72.

———. 2009. Cooperation between non-kin in animal societies. *Nature* 462 (7269): 51–57.

Clutton-Brock, T. H., P.N.M. Brotherton, R. Smith, G. M. McIlrath, R. Kansky, D. Gaynor, M. J. O'Riain, and J. D. Skinner. 1998. Infanticide and expulsion of females in a cooperative mammal. *Proceedings of the Royal Society of London Series B—Biological Sciences* 265 (1412): 2291–95.

Cockell, C., R. Corfield, N. Edwards, and N. Harris. 2008. *An Introduction to the Earth-Life System*. Cambridge: Cambridge University Press.

Codispoti, L. A. 1995. Is the ocean losing nitrate? *Nature* 376 (6543): 724.

———. 2007. An oceanic fixed nitrogen sink exceeding 400 Tg Na-1 vs the concept of homeostasis in the fixed-nitrogen inventory. *Biogeosciences* 4 (2): 233–53.

Coggon, R. M., D.A.H. Teagle, C. E. Smith-Duque, J. C. Alt, and M. J. Cooper. 2010. Reconstructing past seawater Mg/Ca and Sr/Ca from mid-ocean ridge flank calcium carbonate veins. *Science* 327 (5969): 1114–17.

Colinvaux, Paul A. 1993. *Ecology 2*. New York: J. Wiley.

Cornwallis, C. K., S. A. West, K. E. Davis, and A. S. Griffin. 2010. Promiscuity and the evolutionary transition to complex societies. *Nature* 466 (7309): 969–72.

Corsetti, F. A., A. N. Olcott, and C. Bakermans. 2006. The biotic response to Neoproterozoic Snowball Earth. *Palaeogeography, Palaeoclimatology, Palaeoecology* 232 (2–4): 114–30.

Cox, John D. 2002. *Climate Crash: Abrupt Climate Change and What It Means for Our Future*. Washington, DC: Joseph Henry Press.

Cox, P. M., R. A. Betts, M. Collins, P. P. Harris, C. Huntingford, and C. D. Jones. 2004. Amazonian forest dieback under climate-carbon cycle projections for the 21st century. *Theoretical and Applied Climatology* 78 (1–3): 137–56.

Crowley, T. J., and W. T. Hyde. 2008. Transient nature of Late Pleistocene climate variability. *Nature* 456 (7219): 226–30.

Crozier, R. H. 2008. Advanced eusociality, kin selection, and male haploidy. *Australian Journal of Entomology* 47: 2–8.

Crutzen, P. 1995. My life with O_3, NO_x and other YZO_xs. Nobel lecture, December 8, 1995. http://www.nobelprize.org/nobel_prizes/chemistry/laureates/1995/crutzen-lecture.pdf.

———. 2004. Anti-Gaia. In *Global Change and the Earth System*, edited by W. Steffen. New York: Springer.

Currie, D. J. 1991. Energy and large-scale patterns of animal- and plant-species richness. *American Naturalist* 137: 27–49.

Dalsgaard, T., B. Thamdrup, and D. E. Canfield. 2005. Anaerobic ammonium oxidation (anammox) in the marine environment. *Research in Microbiology* 156 (4): 457–64.

Damuth, J. 1981. Population density and body size in mammals. *Nature* 290 (5808): 699–700.

———. 1991. Ecology—of size and abundance. *Nature* 351 (6324): 268–69.

Danovaro, R., A. Dell'Anno, A. Pusceddu, C. Gambi, I. Heiner, and R. M. Kristensen. 2010. The first metazoa living in permanently anoxic conditions. *BMC Biology* 8: 30.

Dansgaard, W., S. J. Johnsen, H. B. Clausen, D. Dahljensen, N. S. Gundestrup, C. U. Hammer, C. S. Hvidberg, J. P. Steffensen, A. E. Sveinbjornsdottir, J. Jouzel, and G. Bond. 1993. Evidence for general instability of past climate from a 250-kyr ice-core record. *Nature* 364 (6434): 218–20.

Darwin, C. R. 1842. *The Structure and Distribution of Coral Reefs: Being the First Part of the Geology of the Voyage of the Beagle, under the Command of Capt. Fitzroy, R. N. during the Years 1832 to 1836*. London: Smith Elder.

Davies, N. B. 2000. *Cuckoos, Cowbirds and Other Cheats*. London: T. & A. D. Poyser.

Davies, P.C.W. 2006. *The Goldilocks Enigma: Why Is the Universe Just Right for Life?* London: Allen Lane.

Dawkins, Richard. 1982. *The Extended Phenotype: The Gene as the Unit of Selection*. San Francisco: Freeman.

———. 2004. Extended phenotype—but not too extended: A reply to Laland, Turner, and Jablonka. *Biology and Philosophy* 19 (3): 377–96.

Dawkins, Richard, and Yan Wong. 2004. *The Ancestor's Tale: A Pilgrimage to the Dawn of Evolution*. London: Weidenfeld and Nicolson.

Deming, D., and S. Seager. 2009. Light and shadow from distant worlds. *Nature* 462 (7271): 301–6.

Dickson, J.A.D. 2002. Fossil echinoderms as monitor of the Mg/Ca ratio of Phanerozoic oceans. *Science* 298 (5596): 1222–24.

Dobbs, David. 2005. *Reef Madness: Charles Darwin, Alexander Agassiz, and the Meaning of Coral*. 1st ed. New York: Pantheon.

Doolittle, W. F. 1981. Is Nature really motherly? *Coevolution Quarterly* (Spring), 29: 58–63.

Dotto, Lydia, and Harold Schiff. 1978. *The Ozone War*. 1st ed. Garden City, NY: Doubleday.

Dudley, R. 1998. Atmospheric oxygen, giant Paleozoic insects, and the evolution of aerial locomotor performance. *Journal of Experimental Biology* 201 (8): 1043–50.

Edmond, J. M., and Y. Huh. 1997. Chemical weathering yields from basement and orogenic terrains in hot and cold climates. In *Tectonics, Uplift, and Climate Change*, edited by W. Ruddiman. New York: Plenum Publishing.

Edmond, J. M., M. R. Palmer, C. I. Measures, B. Grant, and R. F. Stallard. 1995. The fluvial geochemistry and denudation rate of the Guayana Shield in Venezuela, Colombia, and Brazil. *Geochimica Et Cosmochimica Acta* 59 (16): 3301–25.

Edwards, K. J., P. L. Bond, T. M. Gihring, and J. F. Banfield. 2000. An archaeal iron-oxidizing extreme acidophile important in acid mine drainage. *Science* 287 (5459): 1796–99.

Ehrlich, P. 1991. Coevolution and its applicability to the Gaia hypothesis. In *Scientists on Gaia*, edited by S. H. Schneider and P. J. Boston. Cambridge, MA: MIT Press.

Enquist, B. J., A. P. Allen, J. H. Brown, J. F. Gillooly, A. J. Kerkhoff, K. J. Niklas, C. A. Price, and G. B. West. 2007. Does the exception prove the rule? *Nature* 445 (7127): E9–E10.

Enquist, B. J., J. H. Brown, and G. B. West. 1998. Allometric scaling of plant energetics and population density. *Nature* 395 (6698): 163–65.

Enquist, B. J., E. P. Economo, T. E. Huxman, A. P. Allen, D. D. Ignace, and J. F. Gillooly. 2003. Scaling metabolism from organisms to ecosystems. *Nature* 423 (6940): 639–42.

Enquist, B. J., and K. J. Niklas. 2001. Invariant scaling relations across tree-dominated communities. *Nature* 410 (6829): 655–60.

Eppley, R. W. 1972. Temperature and phytoplankton growth in the sea. *Fishery Bulletin* 70: 1063–85.

Evans, D.A.D. 2006. Proterozoic low orbital obliquity and axial-dipolar geomagnetic field from evaporite palaeolatitudes. *Nature* 444 (7115): 51–55.

Fagan, Brian M. 2000. *The Little Ice Age: How Climate Made History, 1300–1850*. New York: Basic Books.

———. 2003. *The Long Summer: How Climate Changed Civilization*. New York: Basic Books.

Farman, J. C., B. G. Gardiner, and J. D. Shanklin. 1985. Large losses of total ozone in Antarctica reveal seasonal ClOx/NOx interaction. *Nature* 315 (6016): 207–10.

Farquhar, J., H. M. Bao, and M. Thiemens. 2000. Atmospheric influence of Earth's earliest sulfur cycle. *Science* 289 (5480): 756–58.

Flannery, T. 2011. *Here on Earth: A New Beginning*. London: Allen Lane.

Floudas, D., et al. 2012. The Paleozoic origin of enzymatic lignin decomposition reconstructed from 31 fungal genomes. *Science* 336 (6089): 1715–19.

Foster, G. L., and D. Vance. 2006. Negligible glacial-interglacial variation in continental chemical weathering rates. *Nature* 444 (7121): 918–21.

Fothergill, Alastair. 1993. *A Natural History of the Antarctic: Life in the Freezer*. London: BCA.

Francis, J. E., and I. Poole. 2002. Cretaceous and Early Tertiary climates of Antarctica: Evidence from fossil wood. *Palaeogeography, Palaeoclimatology, Palaeoecology* 182 (1–2): 47–64.

Freeland, W. J., and D. Choquenot. 1990. Determinants of herbivore carrying capacity: Plants, nutrients, and *Equus asinus* in northern Australia. *Ecology* 71 (2): 589–97.

Friedlingstein, P., J. L. Dufresne, P. M. Cox, and P. Rayner. 2003. How positive is the feedback between climate change and the carbon cycle? *Tellus Series B—Chemical and Physical Meteorology* 55 (2): 692–700.

Friedrich, O., R. D. Norris, and J. Erbacher. 2012. Evolution of Middle to Late Cretaceous oceans: A 55 m.y. record of Earth's temperature and carbon cycle. *Geology* 40 (2): 107–10.

Futuyma, Douglas J., and Montgomery Slatkin. 1983. *Coevolution*. Sunderland, MA: Sinauer Associates.

Gaidos, E. J., K. H. Nealson, and J. L. Kirschvink. 1999. Life in ice-covered oceans. *Science* 284 (5420): 1631–33.

Galbraith, E. D., S. L. Jaccard, T. F. Pedersen, D. M. Sigman, G. H. Haug, M. Cook, J. R. Southon, and R. Francois. 2007. Carbon dioxide release from the North Pacific abyss during the last deglaciation. *Nature* 449 (7164): 890–93.

Galloway, J. N. 2005. The global nitrogen cycle. In *Biogeochemistry, Vol. 8 of Treatise on Geochemistry*, edited by W. H. Schlesinger: Amsterdam: Elsevier.

Gates, David Murray. 1993. *Climate Change and Its Biological Consequences*. Sunderland, MA: Sinauer Associates.

Geider, R. J., and J. La Roche. 2002. Redfield revisited: Variability of $C : N : P$ in marine microalgae and its biochemical basis. *European Journal of Phycology* 37 (1): 1–17.

Gensel, Patricia G., D. Edwards, and International Organization of Paleobotany. Conference. 2001. *Plants Invade the Land: Evolutionary and Environmental Perspectives*. Critical Moments and Perspectives in Earth History and Paleobiology. New York: Columbia University Press.

Gentry, A. H. 1988a. Changes in plant community diversity and floristic composition on environmental and geographical gradients. *Annals of the Missouri Botanical Garden* 75 (1): 1–34.

———. 1988b. Tree species richness of Upper Amazonian forests. *Proceedings of the National Academy of Sciences of the United States of America* 85 (1): 156–59.

Gersonde, R., A. Abelmann, U. Brathauer, S. Becquey, C. Bianchi, G. Cortese, H. Grobe, G. Kuhn, H. S. Niebler, M. Segl, R. Sieger, U. Zielinski, and D. K. Futterer. 2003. Last glacial sea surface temperatures and sea-ice extent in the Southern Ocean (Atlantic-Indian sector): A multiproxy approach. *Paleoceanography* 18 (3): 1061.

Gillon, J. 2000. Earth systems: Feedback on Gaia. *Nature* 406 (6797): 685–86.

Gillooly, J. F., A. P. Allen, V. M. Savage, E. L. Charnov, G. B. West, and J. H. Brown. 2006. Response to Clarke and Fraser: Effects of temperature on metabolic rate. *Functional Ecology* 20 (2): 400–404.

Gillooly, J. F., J. H. Brown, G. B. West, V. M. Savage, and E. L. Charnov. 2001. Effects of size and temperature on metabolic rate. *Science* 293 (5538): 2248–51.

Gillooly, J. F., E. L. Charnov, G. B. West, V. M. Savage, and J. H. Brown. 2002. Effects of size and temperature on developmental time. *Nature* 417 (6884): 70–73.

Goldstein, R. H. 2001. Clues from fluid inclusions. *Science* 294 (5544): 1009–11.

Goodall, Jane. 1986. *The Chimpanzees of Gombe: Patterns of Behavior*. Cambridge, MA: Belknap Press of Harvard University Press.

Gough, D. O. 1981. Solar interior structure and luminosity variations. *Solar Physics* 74 (1): 21–34.

Gould, Stephen Jay. 1985. *The Flamingo's Smile: Reflections in Natural History*. 1st ed. New York: Norton.

Graziano, L. M., R. J. Geider, W.K.W. Li, and M. Olaizola. 1996. Nitrogen limitation of North Atlantic phytoplankton: Analysis of physiological condition in nutrient enrichment experiments. *Aquatic Microbial Ecology* 11 (1): 53–64.

Gribbin, John R., and Mary Gribbin. 2009. *He Knew He Was Right: The Irrepressible Life of James Lovelock and Gaia*. London: Allen Lane.

Griffin, A. S., S. A. West, and A. Buckling. 2004. Cooperation and competition in pathogenic bacteria. *Nature* 430 (7003): 1024–27.

Grigg, G.C. 1973. Some consequences of the shape and orientation of "magnetic" termite mounds. *Australian Journal of Zoology* 21: 231–37.

Grove, Jean M. 1988. *The Little Ice Age*. London; New York: Methuen.

Gruber, N. 2004. The dynamics of the marine nitrogen cycle and its influence on atmospheric CO_2 variations. In *The Ocean Carbon Cycle and Climate*, edited by M. Follows and T. Oguz: Dordrecht; Boston: Kluwer.

Hamilton, W. D. 1964. The genetical evolution of social behaviour, I and II. *Journal of Theoretical Biology* 7: 1–16 and 17–52.

Hammer, C. U., P. A. Mayewski, D. Peel, and M. Stuiver. 1997. Preface to special issue on Greenland ice-core research. *Journal of Geophysical Research* 102 (C12): 26315–16.

Haqq-Misra, J. D., S. D. Domagal-Goldman, P. J. Kasting, and J. F. Kasting. 2008. A revised, hazy methane greenhouse for the Archean Earth. *Astrobiology* 8 (6): 1127–37.

Harding, S. P. 1999. Food web complexity enhances community stability and climate regulation in a geophysiological model. *Tellus Series B—Chemical and Physical Meteorology* 51 (4): 815–29.

Hassol, Susan Joy, Arctic Climate Impact Assessment, Arctic Monitoring and Assessment Programme, Program for the Conservation of Arctic Flora and Fauna, and International Arctic Science Committee. 2004. *Impacts of a Warming Arctic: Arctic Climate Impact Assessment*. Cambridge: Cambridge University Press.

Hawking, S. W., and Leonard Mlodinow. 2010. *The Grand Design*. New York: Bantam Books.

Hazen, R. M. 2012. *The Story of Earth: The First 4.5 Billion Years, from Stardust to Living Planet*. New York: Viking Books.

Hedges, J. I., and R. G. Keil. 1995. Sedimentary organic matter preservation: An assessment and speculative synthesis. *Marine Chemistry* 49 (2–3): 81–115.

Hedin, L. O. 2007. Does the exception prove the rule? Reply. *Nature* 445 (7127): E11.

Herman, A. B., and R. A. Spicer. 1996. Palaeobotanical evidence for a warm Cretaceous Arctic Ocean. *Nature* 380 (6572): 330–33.

Hirata, T., T. Kaneko, T. Ono, T. Nakazato, N. Furukawa, S. Hasegawa, S. Wakabayashi, M. Shigekawa, M. H. Chang, M. F. Romero, and S. Hirose. 2003. Mechanism of acid adaptation of a fish living in a pH 3.5 lake. *American Journal of Physiology-Regulatory Integrative and Comparative Physiology* 284 (5): R1199–R1212.

Hoehler, T. M., B. M. Bebout, and D. J. Des Marais. 2001. The role of microbial mats in the production of reduced gases on the early Earth. *Nature* 412 (6844): 324–27.

Hoffman, P. F., A. J. Kaufman, G. P. Halverson, and D. P. Schrag. 1998. A Neoproterozoic Snowball Earth. *Science* 281 (5381): 1342–46.

Hoffman, P. F., and D. P. Schrag. 2000. Snowball Earth. *Scientific American* 282: 68–75.

———. 2002. The Snowball Earth hypothesis: Testing the limits of global change. *Terra Nova* 14 (3): 129–55.

Holland, Heinrich D. 1984. *The Chemical Evolution of the Atmosphere and Oceans.* Princeton Series in Geochemistry. Princeton, NJ: Princeton University Press.

———. 2006. The oxygenation of the atmosphere and oceans. *Philosophical Transactions of the Royal Society B—Biological Sciences* 361 (1470): 903–15.

Horita, J., H. Zimmermann, and H. D. Holland. 2002. Chemical evolution of seawater during the Phanerozoic: Implications from the record of marine evaporites. *Geochimica Et Cosmochimica Acta* 66 (21): 3733–56.

Houghton, R. A., and D. L. Skole. 1990. Carbon. In *The Earth as Transformed by Human Actions*, edited by B. L. Turner, W. C. Clark, R. W. Kates, J. F. Richards, J. T. Mathews and W. B. Meyer. Cambridge: Cambridge University Press.

Hren, M. T., M. M. Tice, and C. P. Chamberlain. 2009. Oxygen and hydrogen isotope evidence for a temperate climate 3.42 billion years ago. *Nature* 462 (7270): 205–8.

Huber, B. T. 1998. Tropical paradise at the Cretaceous poles? *Science* 282 (5397): 2199–200.

Huber, B. T., R. D. Norris, and K. G. MacLeod. 2002. Deep-sea paleotemperature record of extreme warmth during the Cretaceous. *Geology* 30 (2): 123–26.

Hughes, W.O.H., B. P. Oldroyd, M. Beekman, and F.L.W. Ratnieks. 2008. Ancestral monogamy shows kin selection is key to the evolution of eusociality. *Science* 320 (5880): 1213–16.

Hunt, G., M. A. Bell, and M. P. Travis. 2008. Evolution toward a new adaptive optimum: Phenotypic evolution in a fossil stickleback lineage. *Evolution* 62 (3): 700–710.

Hunter, M. L., Jr., and P. Yonzon. 1993. Altitudinal distributions of birds, mammals, people, forests, and parks in Nepal. *Conservation Biology* 7: 420–23.

Hutchinson, G. N. 1948. Circular causal systems in ecology. *Annals of the New York Academy of Sciences* 50: 221–46.

Ivy-Ochs, S., C. Schluchter, P. W. Kubik, and G. H. Denton. 1999. Moraine exposure dates imply synchronous Younger Dryas glacier advances in the European Alps and in the Southern Alps of New Zealand. *Geografiska Annaler Series a—Physical Geography* 81A (2): 313–23.

Jablonski, D. 1993. The tropics as a source of evolutionary novelty through geological time. *Nature* 364 (6433): 142–44.

Jacklyn, P. M. 1992. Magnetic termite mound surfaces are oriented to suit wind and shade conditions. *Oecologia* 91 (3): 385–95.

Jenkyns, H. C., A. Forster, S. Schouten, and J.S.S. Damste. 2004. High temperatures in the Late Cretaceous Arctic Ocean. *Nature* 432 (7019): 888–92.

Jenkyns, H. C., and P. A. Wilson. 1999. Stratigraphy, paleoceanography, and evolution of Cretaceous Pacific guyots: Relics from a greenhouse earth. *American Journal of Science* 299 (5): 341–92.

Johnston, I. A., J. Calvo, H. Guderley, D. Fernandez, and L. Palmer. 1998. Latitudinal variation in the abundance and oxidative capacities of muscle mitochondria in perciform fishes. *Journal of Experimental Biology* 201 (1): 1–12.

Jones, L.H.P., and K. A. Handreck. 1965. The relation between the silica content of the diet and the excretion of silica by sheep. *Journal of Agricultural Science* 65 (1): 129–34.

Kaplan, M. R., J. M. Schaefer, G. H. Denton, D.J.A. Barrell, T.J.H. Chinn, A. E. Putnam, B. G. Andersen, R. C. Finkel, R. Schwartz, and A. M. Doughty. 2010. Glacier retreat in New Zealand during the Younger Dryas stadial. *Nature* 467 (7312): 194–97.

Karl, D., A. Michaels, B. Bergman, D. Capone, E. Carpenter, R. Letelier, F. Lipschultz, H. Paerl, D. Sigman, and L. Stal. 2002. Dinitrogen fixation in the world's oceans. *Biogeochemistry* 57/58 (1): 47–98.

Kasemann, S. A., A. R. Prave, A. E. Fallick, C. J. Hawkesworth, and K. H. Hoffmann. 2010. Neoproterozoic ice ages, boron isotopes, and ocean acidification: Implications for a Snowball Earth. *Geology* 38 (9): 775–78.

Kashefi, K., and D. R. Lovley. 2003. Extending the upper temperature limit for life. *Science* 301 (5635): 934.

Kasting, J. F. 2001. The rise of atmospheric oxygen. *Science* 293 (5531): 819–20.

———. 2004. When methane made climate. *Scientific American* 291 (1): 78–85.

———. 2010. *How to Find a Habitable Planet*. Science Essentials. Princeton, NJ: Princeton University Press.

Kasting, J. F., D. H. Eggler, and S. P. Raeburn. 1993. Mantle redox evolution and the oxidation-state of the Archean atmosphere. *Journal of Geology* 101 (2): 245–57.

Kenrick, P., and P. R. Crane. 1997. The origin and early evolution of plants on land. *Nature* 389 (6646): 33–39.

Kerr, R. A. 1988. No longer willful, Gaia becomes respectable. *Science* 240 (4851): 393–95.

———. 2002. Inconstant ancient seas and life's path. *Science* 298 (5596): 1165–66.

Kiessling, W., M. Aberhan, and L. Villier. 2008. Phanerozoic trends in skeletal mineralogy driven by mass extinctions. *Nature Geoscience* 1 (8): 527–30.

Kirchner, J. W. 1991. The Gaia hypotheses: Are they testable? Are they useful? In *Scientists on Gaia*, edited by S. H. Schneider and P. J. Boston. Cambridge, MA: MIT Press.

———. 2002. The Gaia hypothesis: Fact, theory, and wishful thinking. *Climatic Change* 52 (4): 391–408.

———. 2003. The Gaia hypothesis: Conjectures and refutations. *Climatic Change* 58 (1–2): 21–45.

Kirschvink, J. L. 1992. Late Proterozoic low-latitude global glaciation. In *The Proterozoic Biosphere: A Multidisciplinary Study*, edited by J. W. Schopf and C. Klein. Cambridge: Cambridge University Press.

Kivelson, M. G., K. K. Khurana, C. T. Russell, M. Volwerk, R. J. Walker, and C. Zimmer. 2000. Galileo magnetometer measurements: A stronger case for a subsurface ocean at Europa. *Science* 289 (5483): 1340–43.

Klausmeier, C. A., E. Litchman, T. Daufresne, and S. A. Levin. 2004. Optimal nitrogen-to-phosphorus stoichiometry of phytoplankton. *Nature* 429 (6988): 171–74.

Kleidon, A. 2002. Testing the effect of life on Earth's functioning: How Gaian is the Earth system? *Climatic Change* 52 (4): 383–89.

Knauth, L. P., and M. J. Kennedy. 2009. The Late Precambrian greening of the Earth. *Nature* 460 (7256): 728–32.

Knoll, A. H. 2003. The geological consequences of evolution. *Geobiology* 1 (1): 3–14.

Kohfeld, K. E., C. Le Quere, S. P. Harrison, and R. F. Anderson. 2005. Role of marine biology in glacial-interglacial CO_2 cycles. *Science* 308 (5718): 74–78.

Kolokotrones, T., V. Savage, E. J. Deeds, and W. Fontana. 2010. Curvature in metabolic scaling. *Nature* 464 (7289): 753–56.

Kopp, R. E., J. L. Kirschvink, I. A. Hilburn, and C. Z. Nash. 2005. The Paleoproterozoic Snowball Earth: A climate disaster triggered by the evolution of oxygenic photosynthesis. *Proceedings of the National Academy of Sciences of the United States of America* 102 (32): 11131–36.

Korb, J., and K. E. Linsenmair. 2000. Thermoregulation of termite mounds. *Insectes Sociaux* 47: 357–63.

Kornfield, I., and P. F. Smith. 2000. African cichlid fishes: Model systems for evolutionary biology. *Annual Review of Ecology and Systematics* 31: 163–96.

Koschorreck, M., and J. Tittel. 2002. Benthic photosynthesis in an acidic mining lake (pH 2.6). *Limnology and Oceanography* 47 (4): 1197–201.

Kovalevich, V. M., T. M. Peryt, and O. I. Petrichenko. 1998. Secular variation in seawater chemistry during the Phanerozoic as indicated by brine inclusions in halite. *Journal of Geology* 106: 695–712.

Krakauer, A. H. 2005. Kin selection and cooperative courtship in wild turkeys. *Nature* 434 (7029): 69–72.

Kulp, T. R., S. E. Hoeft, M. Asao, M. T. Madigan, J. T. Hollibaugh, J. C. Fisher, J. F. Stolz, C. W. Culbertson, L. G. Miller, and R. S. Oremland. 2008. Arsenic(III) fuels anoxygenic photosynthesis in hot spring biofilms from Mono Lake, California. *Science* 321 (5891): 967–70.

Kump, L. R. 2008. The rise of atmospheric oxygen. *Nature* 451 (7176): 277–78.

Kump, L. R., and M. E. Barley. 2007. Increased subaerial volcanism and the rise of atmospheric oxygen 2.5 billion years ago. *Nature* 448 (7157): 1033–36.

Kump, L. R., S. L. Brantley, and M. A. Arthur. 2000. Chemical weathering, atmospheric CO_2, and climate. *Annual Review of Earth and Planetary Sciences* 28: 611–67.

Kump, Lee R., James F. Kasting, and Robert G. Crane. 2004. *The Earth System.* 2nd ed. Upper Saddle River, NJ: Pearson Prentice Hall.

Kuypers, M.M.M., G. Lavik, D. Woebken, M. Schmid, B. M. Fuchs, R. Amann, B. B. Jorgensen, and M.S.M. Jetten. 2005. Massive nitrogen loss from the Benguela upwelling system through anaerobic ammonium oxidation. *Proceedings of the National Academy of Sciences of the United States of America* 102 (18): 6478–83.

Kuypers, M.M.M., Y. van Breugel, S. Schouten, E. Erba, and J.S.S. Damste. 2004. N_2-fixing cyanobacteria supplied nutrient N for Cretaceous oceanic anoxic events. *Geology* 32 (10): 853–56.

Lahti, D. C. 2005. Evolution of bird eggs in the absence of cuckoo parasitism. *Proceedings of the National Academy of Sciences of the United States of America* 102 (50): 18057–62.

Laland, K. N. 2004. Extending the extended phenotype. *Biology and Philosophy* 19: 313–25.

Lane, Nick. 2002. *Oxygen: The Molecule That Made the World.* Oxford; New York: Oxford University Press.

Langer, P., K. Hogendoorn, and L. Keller. 2004. Tug-of-war over reproduction in a social bee. *Nature* 428 (6985): 844–47.

Langmore, N. E., S. Hunt, and R. M. Kilner. 2003. Escalation of a coevolutionary arms race through host rejection of brood parasitic young. *Nature* 422 (6928): 157–60.

Larson, R. G. 2007. Perspectives on current issues: Is "anthropic selection" science? *Physics in Perspective* 9 (1): 58–69.

Larsson, U., S. Hajdu, J. Walve, and R. Elmgren. 2001. Baltic Sea nitrogen fixation estimated from the summer increase in upper mixed layer total nitrogen. *Limnology and Oceanography* 46 (4): 811–20.

Le Hir, G., G. Ramstein, Y. Donnadieu, and Y. Godderis. 2008. Scenario for the evolution of atmospheric pCO(2) during a Snowball Earth. *Geology* 36 (1): 47–50.

Lemarchand, D., J. Gaillardet, E. Lewin, and C. J. Allegre. 2000. The influence of rivers on marine boron isotopes and implications for reconstructing past ocean pH. *Nature* 408 (6815): 951–54.

Lenoir, J., J. C. Gegout, P. A. Marquet, P. de Ruffray, and H. Brisse. 2008. A significant upward shift in plant species optimum elevation during the 20th century. *Science* 320 (5884): 1768–71.

Lenton, Tim, and Andrew Watson. 2011. *Revolutions That Made the Earth.* Oxford; New York: Oxford University Press.

Lenton, T. M. 1998. Gaia and natural selection. *Nature* 394 (6692): 439–47.

———. 2002. Testing Gaia: The effect of life on Earth's habitability and regulation. *Climatic Change* 52 (4): 409–22.

Lenton, T. M., and D. M. Wilkinson. 2003. Developing the Gaia theory: A response to the criticisms of Kirchner and Volk. *Climatic Change* 58 (1–2): 1–12.

Lepot, K., K. Benzerara, G. E. Brown, and P. Philippot. 2008. Microbially influenced formation of 2,724-million-year-old stromatolites. *Nature Geoscience* 1 (2): 118–21.

Lerman, A., and F. T. Mackenzie. 2005. CO_2 air-sea exchange due to calcium carbonate and organic matter storage, and its implications for the global carbon cycle. *Aquatic Geochemistry* 11 (4): 345–90.

Levin, Simon A. 1999. *Fragile Dominion: Complexity and the Commons.* Reading, MA: Perseus Books.

Li, Y.-H. 2000. *A Compendium of Geochemistry: From Solar Nebula to the Human Brain.* Princeton, NJ: Princeton University Press.

Lin, L. H., P. L. Wang, D. Rumble, J. Lippmann-Pipke, E. Boice, L. M. Pratt, B. S. Lollar, E. L. Brodie, T. C. Hazen, G. L. Andersen, T. Z. DeSantis, D. P. Moser, D. Kershaw, and T. C. Onstott. 2006. Long-term sustainability of a high-energy, low-diversity crustal biome. *Science* 314 (5798): 479–82.

López-Urrutia, A. 2003. Allometry: How reliable is the biological time clock? *Nature* 424 (6946): 269–70.

Lovelock, James E. 1972. Gaia as seen through the atmosphere. *Atmospheric Environment* 6 (8): 579–80.

———. 1979. *Gaia: A New Look at Life on Earth.* Oxford; New York: Oxford University Press.

———. 1984. Causes and effects of changes in stratospheric ozone: Update 1983. *Environment* 26 (10): 25–26.

———. 1988. *The Ages of Gaia: A Biography of Our Living Earth.* 1st ed. The Commonwealth Fund Book Program. New York: Norton.

———. 1991. *Gaia, the Practical Science of Planetary Medicine.* London: Gaia Books.

———. 2001. *Homage to Gaia: The Life of an Independent Scientist.* 1st Oxford paperback ed. New York: Oxford University Press.

———. 2003a. Gaia and emergence: A response to Kirchner and Volk. *Climatic Change* 57 (1–2): 1–3.

———. 2003b. Gaia: The living Earth. *Nature* 426: 769–70.

———. 2004. "Reflections on Gaia." In *Scientists Debate Gaia: The Next Century*, edited by S. H. Schneider et al. Cambridge, MA: MIT Press.

———. 2006. *The Revenge of Gaia: Earth's Climate in Crisis and the Fate of Humanity*. New York: Basic Books.

Lovelock, J. E., R. J. Maggs, and R. J. Wade. 1973. Halogenated hydrocarbons in and over the Atlantic. *Nature* 241: 194–96.

Lovelock, J. E., and L. Margulis. 1974. Atmospheric homeostasis by and for the biosphere: The Gaia hypothesis. *Tellus* 26 (1/2): 2–10.

Lowenstein, T. K., M. N. Timofeeff, S. T. Brennan, L. A. Hardie, and R. V. Demicco. 2001. Oscillations in Phanerozoic seawater chemistry: Evidence from fluid inclusions. *Science* 294 (5544): 1086–88.

Lynas, Mark. 2007. *Six Degrees: Our Future on a Hotter Planet*. London: Fourth Estate.

Macdougall, J. D. 2004. *Frozen Earth: The Once and Future Story of Ice Ages*. Berkeley: University of California Press.

Madigan, M. T. 2000. Bacterial habitats in extreme environments. In *Journey to Diverse Microbial Worlds*, edited by J. Seckbach. Dordrecht; Boston: Kluwer Academic.

Madigan, M. T., and B. L. Marrs. 1997. Extremophiles. *Scientific American* 276 (4): 82–87.

Mahaffey, C., A. F. Michaels, and D. G. Capone. 2005. The conundrum of marine N-2 fixation. *American Journal of Science* 305 (6–8): 546–95.

Maldonado, M., M. G. Carmona, M. J. Uriz, and A. Cruzado. 1999. Decline in Mesozoic reef-building sponges explained by silicon limitation. *Nature* 401 (6755): 785–88.

Maloof, A. C., C. V. Rose, R. Beach, B. M. Samuels, C. C. Calmet, D. H. Erwin, G. R. Poirier, N. Yao, and F. J. Simons. 2010. Possible animal-body fossils in pre-Marinoan limestones from South Australia. *Nature Geoscience* 3 (9): 653–59.

Marañón, E., P. Cermeño, J. Rodriguez, M. V. Zubkov, and R. P. Harris. 2007. Scaling of phytoplankton photosynthesis and cell size in the ocean. *Limnology and Oceanography* 52: 2190–98.

Margulis, L., and J. E. Lovelock. 1974. Biological modulation of Earth's atmosphere. *Icarus* 21 (4): 471–89.

Martin, P. A., D. W. Lea, Y. Rosenthal, N. J. Shackleton, M. Sarnthein, and T. Papenfuss. 2002. Quaternary deep sea temperature histories derived from benthic foraminiferal Mg/Ca. *Earth and Planetary Science Letters* 198 (1–2): 193–209.

Mascarelli, A. L. 2009. Low life. *Nature* 459 (7248): 770–73.

Maynard Smith, John. 1964. Group selection and kin selection. *Nature* 201: 1145–47.

———. 1982. *Evolution and the Theory of Games*. Cambridge; New York: Cambridge University Press.

McCabe, P. J., and J. T. Parrish. 1992. Tectonic and climatic controls on the distribution and quality of Cretaceous coals. In *Controls on the Distribution and Quality of Cretaceous Coals*. Geological Society of America Special Paper 267, edited by P. J. McCabe and J. T. Parrish. Boulder, CO: Geological Society of America.

McKay, C. P., and C. R. Stoker. 1991. Gaia and life on Mars. In *Scientists on Gaia*, edited by S. H. Schneider and P. J. Boston. Cambridge, MA: MIT Press.

McManus, J. F., R. Francois, J. M. Gherardi, L. D. Keigwin, and S. Brown-Leger. 2004. Collapse and rapid resumption of Atlantic meridional circulation linked to deglacial climate changes. *Nature* 428 (6985): 834–37.

Mehdiabadi, N. J., C. N. Jack, T. T. Farnham, T. G. Platt, S. E. Kalla, G. Shaulsky, D. C. Queller, and J. E. Strassmann. 2006. Kin preference in a social microbe: Given the right circumstances, even an amoeba chooses to be altruistic towards its relatives. *Nature* 442 (7105): 881–82.

Meyer-Berthaud, B., S. E. Scheckler, and J. Wendt. 1999. Archaeopteris is the earliest known modern tree. *Nature* 398 (6729): 700–701.

Miller, K. G., M. A. Kominz, J. V. Browning, J. D. Wright, G. S. Mountain, M. E. Katz, P. J. Sugarman, B. S. Cramer, N. Christie-Blick, and S. F. Pekar. 2005. The Phanerozoic record of global sea-level change. *Science* 310 (5752): 1293–98.

Mills, M. M., C. Ridame, M. Davey, J. La Roche, and R. J. Geider. 2004. Iron and phosphorus co-limit nitrogen fixation in the eastern tropical North Atlantic. *Nature* 429 (6989): 292–94.

Misra, S., and P. N. Froelich. 2012. Lithium isotope history of Cenozoic seawater: Changes in silicate weathering and reverse weathering. *Science* 335 (6070): 818–23.

Mock, Douglas W. 2004. *More Than Kin and Less Than Kind: The Evolution of Family Conflict*. Cambridge, MA: Belknap Press of Harvard University Press.

Moisander, P. H., T. F. Steppe, N. S. Hall, J. Kuparinen, and H. W. Paerl. 2003. Variability in nitrogen and phosphorus limitation for Baltic Sea phytoplankton during nitrogen-fixing cyanobacterial blooms. *Marine Ecology—Progress Series* 262: 81–95.

Molina, M. J., and F. S. Rowland. 1974. Stratospheric sink for chlorofluoromethanes: Chlorine atomic-catalysed destruction of ozone. *Nature* 249 (5460): 810–12.

Montanez, I. P. 2002. Biological skeletal carbonate records changes in major-ion chemistry of paleo-oceans. *Proceedings of the National Academy of Sciences of the United States of America* 99 (25): 15852–54.

Montenegro, A., M. Eby, J. O. Kaplan, K. J. Meissner, and A. J. Weaver. 2006. Carbon storage on exposed continental shelves during the glacial-interglacial transition. *Geophysical Research Letters* 33 (8): L08703.

Moore, C. M., M. M. Mills, E. P. Achterberg, R. J. Geider, J. LaRoche, M. I. Lucas, E. L. McDonagh, X. Pan, A. J. Poulton, M.J.A. Rijkenberg, D. J. Suggett, S. J. Ussher, and E.M.S. Woodward. 2009. Large-scale distribution of Atlantic nitrogen fixation controlled by iron availability. *Nature Geoscience* 2 (12): 867–71.

Muchhala, N. 2006. Nectar bat stows huge tongue in its rib cage. *Nature* 444 (7120): 701–2.

Muller-Karger, F. E., R. Varela, R. Thunell, R. Luerssen, C. M. Hu, and J. J. Walsh. 2005. The importance of continental margins in the global carbon cycle. *Geophysical Research Letters* 32 (1): L01602.

Nagy, K. A., I. A. Girard, and T. K. Brown. 1999. Energetics of free-ranging mammals, reptiles, and birds. *Annual Review of Nutrition* 19: 247–77.

Nikaido, M., A. P. Rooney, and N. Okada. 1999. Phylogenetic relationships among cetartiodactyls based on insertions of short and long interspersed elements: Hippopotamuses are the closest extant relatives of whales. *Proceedings of the National Academy of Sciences of the United States of America* 96 (18): 10261–66.

Nisbet, E. 2003. Getting heated over glaciation. *Nature* 422 (6934): 812–13.

Nitecki, Matthew H. 1983. *Coevolution*. Chicago original paperbacks. Chicago: University of Chicago Press.

Nordstrom, K. M., V. K. Gupta, and T. N. Chase. 2004. Salvaging the Daisyworld parable under the dynamic area fraction framework. In *Scientists Debate Gaia: The Next Century*, edited by S. H. Schneider. Cambridge, MA: MIT Press.

Nowak, M. A., C. E. Tarnita, and E. O. Wilson. 2010. The evolution of eusociality. *Nature* 466 (7310): 1057–62.

Odling-Smee, F. John, Kevin N. Laland, and Marcus W. Feldman. 2003. *Niche Construction: the Neglected Process in Evolution.* Monographs in Population Biology. Princeton, NJ: Princeton University Press.

Ohkouchi, N., Y. Kashiyama, J. Kuroda, N. O. Ogawa, and H. Kitazato. 2006. The importance of diazotrophic cyanobacteria as primary producers during Cretaceous Oceanic Anoxic Event 2. *Biogeosciences* 3 (4): 467–78.

Olcott, A. N., A. L. Sessions, F. A. Corsetti, A. J. Kaufman, and T. F. de Oliviera. 2005. Biomarker evidence for photosynthesis during Neoproterozoic glaciation. *Science* 310 (5747): 471–74.

Paasche, E. 2001. A review of the coccolithophorid *Emiliania huxleyi* (Prymnesiophyceae), with particular reference to growth, coccolith formation, and calcification-photosynthesis interactions. *Phycologia* 40 (6): 503–29.

Pagani, M., K. Caldeira, R. Berner, and D. J. Beerling. 2009. The role of terrestrial plants in limiting atmospheric CO_2 decline over the past 24 million years. *Nature* 460 (7251): 85–88.

Pagani, M., D. Lemarchand, A. Spivack, and J. Gaillardet. 2005. A critical evaluation of the boron isotope-pH proxy: The accuracy of ancient ocean pH estimates. *Geochimica Et Cosmochimica Acta* 69 (4): 953–61.

Pearson, P. N., and M. R. Palmer. 2000. Atmospheric carbon dioxide concentrations over the past 60 million years. *Nature* 406: 695–99.

Peck, L. S. 2002. Ecophysiology of Antarctic marine ectotherms: Limits to life. *Polar Biology* 25 (1): 31–40.

Peck, L. S., A. D. Ansell, K. E. Webb, L. Hepburn, and M. Burrows. 2003. Movements and burrowing activity in the Antarctic bivalve molluscs *Laternula elliptica* and *Yoldia eightsi*. *Polar Biology* 27: 357–67.

Peck, L. S., and T. Brey. 1996. Bomb signals in old Antarctic brachiopods. *Nature* 380 (6571): 207–8.

Peck, L. S., K. E. Webb, and D. M. Bailey. 2004. Extreme sensitivity of biological function to temperature in Antarctic marine species. *Functional Ecology* 18 (5): 625–30.

Peltier, W. R. 2004. Global glacial isostasy and the surface of the ice-age earth: The ice-5G (VM2) model and GRACE. *Annual Review of Earth and Planetary Sciences* 32: 111–49.

Peng, C. H., J. Guiot, and E. Van Campo. 1998. Estimating changes in terrestrial vegetation and carbon storage: Using palaeoecological data and models. *Quaternary Science Reviews* 17 (8): 719–35.

Petit, J. R., J. Jouzel, D. Raynaud, N. I. Barkov, J. M. Barnola, I. Basile, M. Bender, J. Chappellaz, M. Davis, G. Delaygue, M. Delmotte, V. M. Kotlyakov, M. Legrand, V. Y. Lipenkov, C. Lorius, L. Pepin, C. Ritz, E. Saltzman, and M. Stievenard. 1999. Climate and atmospheric history of the past 420,000 years from the Vostok ice core, Antarctica. *Nature* 399 (6735): 429–36.

Plaut, I. 2000. Resting metabolic rate, critical swimming speed, and routine activity of the euryhaline cyprinodontid, *Aphanius dispar*, acclimated to a wide range of salinities. *Physiological and Biochemical Zoology* 73 (5): 590–96.

Plaut, J. J., G. Picardi, A. Safaeinili, A. B. Ivanov, S. M. Milkovich, A. Cicchetti, W. Kofman, J. Mouginot, W. M. Farrell, R. J. Phillips, S. M. Clifford, A. Frigeri, R. Orosei,

C. Federico, I. P. Williams, D. A. Gurnett, E. Nielsen, T. Hagfors, E. Heggy, E. R. Stofan, D. Plettemeier, T. R. Watters, C. J. Leuschen, and P. Edenhofer. 2007. Subsurface radar sounding of the south polar layered deposits of Mars. *Science* 316 (5821): 92–95.

Porter, S. M. 2007. Seawater chemistry and early carbonate biomineralization. *Science* 316 (5829): 1302.

———. 2010. Calcite and aragonite seas and the de novo acquisition of carbonate skeletons. *Geobiology* 8 (4): 256–77.

Postgate, J. R. 1998. *Nitrogen Fixation*. 3rd ed. Cambridge; New York: Cambridge University Press.

Quammen, David. 1996. *The Song of the Dodo: Island Biogeography in an Age of Extinctions*. New York: Scribner.

Queller, D. C., J. E. Strassmann, and C. R. Hughes. 1988. Genetic relatedness in colonies of tropical wasps with multiple queens. *Science* 242 (4882): 1155–57.

Quigg, A., Z. V. Finkel, A. J. Irwin, Y. Rosenthal, T. Y. Ho, J. R. Reinfelder, O. Schofield, F.M.M. Morel, and P. G. Falkowski. 2003. The evolutionary inheritance of elemental stoichiometry in marine phytoplankton. *Nature* 425 (6955): 291–94.

Randall, D. J., C. M. Wood, S. F. Perry, H. Bergman, G.M.O. Maloiy, T. P. Mommsen, and P. A. Wright. 1989. Urea excretion as a strategy for survival in a fish living in a very alkaline environment. *Nature* 337 (6203): 165–66.

Rasmussen, B., I. R. Fletcher, J. J. Brocks, and M. R. Kilburn. 2008. Reassessing the first appearance of eukaryotes and cyanobacteria. *Nature* 455 (7216): 1101–4.

Rau, G. H., M. A. Arthur, and W. E. Dean. 1987. N-15/N-14 variations in Cretaceous Atlantic sedimentary sequences: Implication for past changes in marine nitrogen biogeochemistry. *Earth and Planetary Science Letters* 82 (3–4): 269–79.

Raven, J. A., and M. Giordano. 2009. Biomineralization by photosynthetic organisms: Evidence of coevolution of the organisms and their environment? *Geobiology* 7 (2): 140–54.

Ravizza, G., and B. Peucker-Ehrenbrink. 2003. The marine Os-187/Os-188 record of the Eocene-Oligocene transition: The interplay of weathering and glaciation. *Earth and Planetary Science Letters* 210 (1–2): 151–65.

Raymo, M. E. 1994. The Himalayas, organic-carbon burial, and climate in the Miocene. *Paleoceanography* 9 (3): 399–404.

Raymo, M. E., and W. F. Ruddiman. 1992. Tectonic forcing of Late Cenozoic climate. *Nature* 359 (6391): 117–22.

Redfield, A. C. 1934. On the proportions of organic derivatives in sea water and their relation to the composition of plankton. In *James Johnstone Memorial Volume*, edited by R. J. Daniel, 176–92. Liverpool: University Press of Liverpool.

Rees, Martin J. 1999. *Just Six Numbers: The Deep Forces That Shape the Universe*. London: Weidenfeld and Nicolson.

Reeve, H. K., D. F. Westneat, W. A. Noon, P. W. Sherman, and C. F. Aquadro. 1990. DNA fingerprinting reveals high levels of inbreeding in colonies of the eusocial naked mole-rat. *Proceedings of the National Academy of Sciences of the United States of America* 87 (7): 2496–2500.

Reich, P. B., M. G. Tjoelker, J. L. Machado, and J. Oleksyn. 2006. Universal scaling of respiratory metabolism, size, and nitrogen in plants. *Nature* 439 (7075): 457–61.

———. 2007. Does the exception prove the rule? Reply. *Nature* 445 (7127): E10–E11.

Retallack, G. J. 1997. Early forest soils and their role in Devonian global change. *Science* 276 (5312): 583–85.

Rich, T. H., P. Vickers-Rich, and R. A. Gangloff. 2002. Polar dinosaurs. *Science* 295 (5557): 979–80.

Rich, Thomas H. V., and Patricia Vickers-Rich. 2000. *Dinosaurs of Darkness, Life of the Past.* Bloomington: Indiana University Press.

Ridgwell, A. J., M. J. Kennedy, and K. Caldeira. 2003. Carbonate deposition, climate stability, and Neoproterozoic ice ages. *Science* 302 (5646): 859–62.

Riebe, C. S., J. W. Kirchner, and R. C. Finkel. 2004. Erosional and climatic effects on long-term chemical weathering rates in granitic landscapes spanning diverse climate regimes. *Earth and Planetary Science Letters* 224 (3–4): 547–62.

Ries, J. B., S. M. Stanley, and L. A. Hardie. 2006. Scleractinian corals produce calcite, and grow more slowly, in artificial Cretaceous seawater. *Geology* 34 (7): 525–28.

Rippeth, T. P., J. D. Scourse, K. Uehara, and S. McKeown. 2008. Impact of sea-level rise over the last deglacial transition on the strength of the continental shelf CO_2 pump. *Geophysical Research Letters* 35 (24): L24604.

Roan, Sharon. 1989. *Ozone Crisis: The 15-Year Evolution of a Sudden Global Emergency.* Wiley Science Editions. New York: Wiley.

Robert, F., and M. Chaussidon. 2006. A palaeotemperature curve for the Precambrian oceans based on silicon isotopes in cherts. *Nature* 443 (7114): 969–72.

Robertson, D., and J. Robinson. 1998. Darwinian Daisyworld. *Journal of Theoretical Biology* 195 (1): 129–34.

Robinson, J. M. 1990. Lignin, land plants, and fungi: Biological evolution affecting Phanerozoic oxygen balance. *Geology* 18 (7): 607–10.

———. 1991. Phanerozoic atmospheric reconstructions: A terrestrial perspective. *Global and Planetary Change* 97 (1–2): 51–62.

Rosenzweig, Michael L. 1995. *Species Diversity in Space and Time.* Cambridge; New York: Cambridge University Press.

Roth-Nebelsick, A. 2005. Reconstructing atmospheric carbon dioxide with stomata: Possibilities and limitations of a botanical pCO_2-sensor. *Trees: Structure and Function* 19 (3): 251–65.

Rothschild, L. J. 2009. A biologist's guide to the solar system. In *Exploring the Origin, Extent, and Future of Life: Philosophical, Ethical, and Theological Perspectives*, edited by C. M. Bertka. Cambridge: Cambridge University Press.

Rothschild, L. J., and R. L. Mancinelli. 2001. Life in extreme environments. *Nature* 409 (6823): 1092–101.

Roussel, E. G., M. A. Cambon-Bonavita, J. Querellou, B. A. Cragg, G. Webster, D. Prieur, and R. J. Parkes. 2008. Extending the sub-sea-floor biosphere. *Science* 320 (5879): 1046.

Rowland, F. S. 2006. Stratospheric ozone depletion. *Philosophical Transactions of the Royal Society B—Biological Sciences* 361 (1469): 769–90.

Roy, K., D. Jablonski, J. W. Valentine, and G. Rosenberg. 1998. Marine latitudinal diversity gradients: Tests of causal hypotheses. *Proceedings of the National Academy of Sciences of the United States of America* 95 (7): 3699–702.

Royer, D. L., R. A. Berner, and D. J. Beerling. 2001. Phanerozoic atmospheric CO_2 change: Evaluating geochemical and paleobiological approaches. *Earth-Science Reviews* 54 (4): 349–92.

Royer, D. L., R. A. Berner, and J. Park. 2007. Climate sensitivity constrained by CO_2 concentrations over the past 420 million years. *Nature* 446 (7135): 530–32.

Sandberg, P. A. 1983. An oscillating trend in Phanerozoic non-skeletal carbonate mineralogy. *Nature* 305 (5929): 19–22.

Sarmiento, Jorge Louis, and Nicolas Gruber. 2006. *Ocean Biogeochemical Dynamics.* Princeton, NJ: Princeton University Press.

Schidlowski, M. 1983. Evolution of photoautotrophy and early atmospheric oxygen levels. *Precambrian Research* 20 (2–4): 319–35.

Schlesinger, William H. 1997. *Biogeochemistry: An Analysis of Global Change.* 2nd ed. San Diego: Academic Press.

Schlünz, B., and R. R. Schneider. 2000. Transport of terrestrial organic carbon to the oceans by rivers: Re-estimating flux and burial rates. *International Journal of Earth Sciences* 88 (4): 599–606.

Schneider, Stephen Henry. 1996. *Laboratory Earth.* London: Weidenfeld and Nicolson.

———. 2001. A goddess of earth or the imagination of a man? *Science* 291 (5510): 1906–7.

Schneider, Stephen Henry, et al. 2004. *Scientists Debate Gaia: The Next Century.* Cambridge, MA: MIT Press.

Schneider, Stephen Henry, Penelope J. Boston, and American Geophysical Union. 1991. *Scientists on Gaia.* Cambridge, MA: MIT Press.

Schneider, Stephen Henry, and Randi Londer. 1984. *The Coevolution of Climate and Life.* San Francisco: Sierra Club Books.

Schouten, S., E. C. Hopmans, A. Forster, Y. van Breugel, M.M.M. Kuypers, and J.S.S. Damste. 2003. Extremely high sea-surface temperatures at low latitudes during the Middle Cretaceous as revealed by archaeal membrane lipids. *Geology* 31 (12): 1069–72.

Schwartzman, D. W., and T. Volk. 1989. Biotic enhancement of weathering and the habitability of Earth. *Nature* 340 (6233): 457–60.

Scott, A. C., and I. J. Glasspool. 2006. The diversification of Paleozoic fire systems and fluctuations in atmospheric oxygen concentration. *Proceedings of the National Academy of Sciences of the United States of America* 103 (29): 10861–65.

Severinghaus, J. P., and E. J. Brook. 1999. Abrupt climate change at the end of the last glacial period inferred from trapped air in polar ice. *Science* 286 (5441): 930–34.

Severinghaus, J. P., T. Sowers, E. J. Brook, R. B. Alley, and M. L. Bender. 1998. Timing of abrupt climate change at the end of the Younger Dryas interval from thermally fractionated gases in polar ice. *Nature* 391 (6663): 141–46.

Shimamura, M., H. Yasue, K. Ohshima, H. Abe, H. Kato, T. Kishiro, M. Goto, I. Munechika, and N. Okada. 1997. Molecular evidence from retroposons that whales form a clade within even-toed ungulates. *Nature* 388 (6643): 666–70.

Shine, R. 1985. The evolution of viviparity in reptiles: An ecological analysis. In *Biology of the Reptilia*, vol. 15, edited by C. Gans and F. Billet. New York: Wiley.

———. 2005. Life-history evolution in reptiles. *Annual Review of Ecology, Evolution, and Systematics* 36: 23–46.

Siever, R. 1991. Silica in the oceans: Biological-geological interplay. In *Scientists on Gaia*, edited by S. H. Schneider and P. J. Boston. Cambridge, MA: MIT Press.

Skelton, P. W., et al. 2003. *The Cretaceous World.* Cambridge; New York: Open University; Cambridge University Press.

Smith, C. R., and A. R. Baco. 2003. Ecology of whale falls at the deep-sea floor. *Oceanography and Marine Biology* 41: 311–54.

Spicer, R. A. 2003. Changing climate and biota. In *The Cretaceous World*, edited by P. W. Skelton. Cambridge; New York: Open University; Cambridge University Press.

Stanley, Steven M. 1999. *Earth System History*. New York: W. H. Freeman.

———. 2006. Influence of seawater chemistry on biomineralization throughout Phanerozoic time: Paleontological and experimental evidence. *Palaeogeography, Palaeoclimatology, Palaeoecology* 232 (2–4): 214–36.

Stanley, S. M., and L. A. Hardie. 1998. Secular oscillations in the carbonate mineralogy of reef-building and sediment-producing organisms driven by tectonically forced shifts in seawater chemistry. *Palaeogeography, Palaeoclimatology, Palaeoecology* 144 (1–2): 3–19.

Stanley, S. M., J. B. Ries, and L. A. Hardie. 2002. Low-magnesium calcite produced by coralline algae in seawater of Late Cretaceous composition. *Proceedings of the National Academy of Sciences of the United States of America* 99 (24): 15323–26.

Steffen, W. L., et al. 2005. *Global Change and the Earth System: A Planet under Pressure*. Global Change: The IGBP Series. Berlin; New York: Springer.

Stein, W. E., F. Mannolini, L. V. Hernick, E. Landing, and C. M. Berry. 2007. Giant cladoxylopsid trees resolve the enigma of the Earth's earliest forest stumps at Gilboa. *Nature* 446 (7138): 904–7.

Stocker, T. F. 2000. Past and future reorganizations in the climate system. *Quaternary Science Reviews* 19 (1–5): 301–19.

Stolarski, J., A. Meibom, R. Przenioslo, and M. Mazur. 2007. A Cretaceous scleractinian coral with a calcitic skeleton. *Science* 318 (5847): 92–94.

Straka, R. P., and J. L Stokes. 1960. Psychrophilic bacteria from Antarctica. *Journal of Bacteriology* 80: 622–25.

Strom, S. L., M. A. Brainard, J. L. Holmes, and M. B. Olson. 2001. Phytoplankton blooms are strongly impacted by microzooplankton grazing in coastal North Pacific waters. *Marine Biology* 138 (2): 355–68.

Sturm, M., C. Racine, and K. Tape. 2001. Increasing shrub abundance in the Arctic. *Nature* 411 (6837): 546–47.

Tarduno, J. A., D. B. Brinkman, P. R. Renne, R. D. Cottrell, H. Scher, and P. Castillo. 1998. Evidence for extreme climatic warmth from Late Cretaceous Arctic vertebrates. *Science* 282 (5397): 2241–44.

Taylor, K. C., P. A. Mayewski, R. B. Alley, E. J. Brook, A. J. Gow, P. M. Grootes, D. A. Meese, E. S. Saltzman, J. P. Severinghaus, M. S. Twickler, J.W.C. White, S. Whitlow, and G. A. Zielinski. 1997. The Holocene Younger Dryas transition recorded at Summit, Greenland. *Science* 278 (5339): 825–27.

Taylor, Stuart Ross. 1998. *Destiny or Chance: Our Solar System and Its Place in the Cosmos*. Cambridge; New York: Cambridge University Press.

Temeles, E. J., and W. J. Kress. 2003. Adaptation in a plant-hummingbird association. *Science* 300 (5619): 630–33.

Temeles, E. J., I. L. Pan, J. L. Brennan, and J. N. Horwitt. 2000. Evidence for ecological causation of sexual dimorphism in a hummingbird. *Science* 289 (5478): 441–43.

Thewissen, J.G.M., M. J. Cohn, L. S. Stevens, S. Bajpai, J. Heyning, and W. E. Horton. 2006. Developmental basis for hind-limb loss in dolphins and origin of the cetacean body plan. *Proceedings of the National Academy of Sciences of the United States of America* 103 (22): 8414–18.

Thomas, Larry. 2002. *Coal Geology*. Chichester, West Sussex, England: Wiley.

Thompson, John N. 1982. *Interaction and Coevolution*. New York: Wiley.

Timofeeff, M. N., T. K. Lowenstein, S. T. Brennan, R. V. Demicco, H. Zimmermann, J. Horita, and L. E. von Borstel. 2001. Evaluating seawater chemistry from fluid inclusions in halite: Examples from modern marine and nonmarine environments. *Geochimica Et Cosmochimica Acta* 65 (14): 2293–300.

Timofeeff, M. N., T. K. Lowenstein, M. A. da Silva, and N. B. Harris. 2006. Secular variation in the major-ion chemistry of seawater: Evidence from fluid inclusions in Cretaceous halites. *Geochimica Et Cosmochimica Acta* 70 (8): 1977–94.

Turner, J. Scott. 2000. *The Extended Organism: The Physiology of Animal-Built Structures*. Cambridge, MA: Harvard University Press.

Turney, Jon. 2003. *Lovelock and Gaia: Signs of Life*. Revolutions in Science. New York: Columbia University Press.

Tyrrell, T. 1999. The relative influences of nitrogen and phosphorus on oceanic primary production. *Nature* 400 (6744): 525–31.

———. 2001. Redfield ratio. In *Encyclopedia of Ocean Sciences*, edited by John H. Steele, 2377–87. San Diego: Academic Press.

Tyrrell, T., E. Marañón, A. J. Poulton, A. R. Bowie, D. S. Harbour, and E.M.S. Woodward. 2003. Large-scale latitudinal distribution of *Trichodesmium* spp. in the Atlantic Ocean. *Journal of Plankton Research* 25 (4): 405–16.

Tyrrell, T., J. G. Shepherd, and S. Castle. 2007. The long-term legacy of fossil fuels. *Tellus Series B—Chemical and Physical Meteorology* 59 (4): 664–72.

Tyrrell, T., and R. E. Zeebe. 2004. History of carbonate ion concentration over the last 100 million years. *Geochimica Et Cosmochimica Acta* 68 (17): 3521–30.

Velbel, M. A. 1993. Temperature-dependence of silicate weathering in nature: How strong a negative feedback on long-term accumulation of atmospheric CO_2 and global greenhouse warming. *Geology* 21 (12): 1059–62.

Vitousek, P. M., O. A. Chadwick, T. E. Crews, J. H. Fownes, D. M. Hendricks, and D. Herbert. 1997. Soil and ecosystem development across the Hawaiian Islands. *GSA Today* 7: 1–8.

Vitousek, P. M., S. Hattenschwiler, L. Olander, and S. Allison. 2002. Nitrogen and nature. *Ambio* 31 (2): 97–101.

Volk, Tyler. 1998. *Gaia's Body: Toward a Physiology of Earth*. New York: Copernicus.

———. 2002. Toward a future for Gaia theory. *Climatic Change* 52 (4): 423–30.

———. 2003. Natural selection, Gaia, and inadvertent by-products: A reply to Lenton and Wilkinson's response. *Climatic Change* 58 (1–2): 13–19.

———. 2003. Seeing deeper into Gaia theory: A reply to Lovelock's response. *Climatic Change* 57 (1–2): 5–7.

———. 2004. Gaia is life in a wasteworld of by-products. In *Scientists Debate Gaia: The Next Century*, edited by S. H. Schneider, J. R. Miller, E. Crist, and P. J. Boston. Cambridge, MA: MIT Press.

von Borstel, L. E., H. Zimmermann, and H. Ruppert. 2000. Fluid inclusion studies in modern halite from the Inagua solar saltwork. In *Proceedings of the 8th World Salt Symposium*, edited by R. Geertmann: Amsterdam; New York: Elsevier Press.

Wade, M. J. 1976. Group selection among laboratory populations of *Tribolium*. *Proceedings of the National Academy of Sciences of the United States of America* 73 (12): 4604–7.

Walker, Gabrielle. 2003. *Snowball Earth: The Story of the Great Global Catastrophe That Spawned Life as We Know It*. New York: Crown Publishers.

———. 2007. *An Ocean of Air: Why the Wind Blows and Other Mysteries of the Atmosphere*. 1st US ed. Orlando: Harcourt.

Walker, J.C.G., P. B. Hays, and J. F. Kasting. 1981. A negative feedback mechanism for the long-term stabilization of Earth's surface temperature. *Journal of Geophysical Research—Oceans and Atmospheres* 86 (Nc10): 9776–82.

Waltham, D. 2011. Testing anthropic selection: A climate change example. *Astrobiology* 11 (2): 105–14.

Ward, Peter Douglas. 2009. *The Medea Hypothesis: Is Life on Earth Ultimately Self-Destructive?* Science Essentials. Princeton, NJ: Princeton University Press.

Ward, Peter Douglas, and Donald Brownlee. 2000. *Rare Earth: Why Complex Life Is Uncommon in the Universe*. New York: Copernicus.

Warren, John K. 2006. *Evaporites: Sediments, Resources, and Hydrocarbons*. Berlin; New York: Springer.

Watson, A. 2004. Gaia and observer self-selection. In *Scientists Debate Gaia: The New Century*, edited by S. H. Schneider, J. R. Miller, E. Crist and P. J. Boston. Cambridge, MA: MIT Press.

Watson, A. J. 1999. Coevolution of the Earth's environment and life: Goldilocks, Gaia and the anthropic principle. In *James Hutton: Present and Future*, edited by G. Y. Craig and J. H. Hull. London: Geological Society.

Watson, A. J., and J. E. Lovelock. 1983. Biological homeostasis of the global environment: The parable of Daisyworld. *Tellus Series B—Chemical and Physical Meteorology* 35 (4): 284–89.

Weart, Spencer R. 2003. *The Discovery of Global Warming*. New Histories of Science, Technology, and Medicine. Cambridge, MA: Harvard University Press.

Weber, S. L., and J. M. Robinson. 2004. Daisyworld homeostasis and the Earth System. In *Scientists Debate Gaia: The Next Century*, edited by S. H. Schneider. Cambridge, MA: MIT Press.

Wellman, C. H., P. L. Osterloff, and U. Mohiuddin. 2003. Fragments of the earliest land plants. *Nature* 425 (6955): 282–85.

West, A. J., A. Galy, and M. Bickle. 2005. Tectonic and climatic controls on silicate weathering. *Earth and Planetary Science Letters* 235 (1–2): 211–28.

West, G. B., J. H. Brown, and B. J. Enquist. 1997. A general model for the origin of allometric scaling laws in biology. *Science* 276 (5309): 122–26.

Wharton, David A. 2002. *Life at the Limits: Organisms in Extreme Environments*. Cambridge: Cambridge University Press.

White, A. F., A. E. Blum, T. D. Bullen, D. V. Vivit, M. Schulz, and J. Fitzpatrick. 1999. The effect of temperature on experimental and natural chemical weathering rates of granitoid rocks. *Geochimica Et Cosmochimica Acta* 63 (19–20): 3277–91.

White, T.C.R. 1993. *The Inadequate Environment: Nitrogen and the Abundance of Animals*. Berlin; New York: Springer-Verlag.

Whitfield, John. 2006. *In the Beat of a Heart: Life, Energy, and the Unity of Nature*. Washington, DC: Joseph Henry Press.

Wigley, T.M.L. 2005. The climate change commitment. *Science* 307 (5716): 1766–69.

Wilkinson, D. M. 2004. Homeostatic Gaia: An ecologist's perspective on the possibility of regulation. In *Scientists Debate Gaia: The Next Century*, edited by S. H. Schneider, J. R. Miller, E. Crist and P. J. Boston. Cambridge, MA: MIT Press.

Willenbring, J. K., and F. von Blanckenburg. 2010. Long-term stability of global erosion rates and weathering during Late-Cenozoic cooling. *Nature* 465 (7295): 211–14.

Williams, George C. 1966. *Adaptation and Natural Selection: A Critique of Some Current Evolutionary Thought*. Princeton, NJ: Princeton University Press.

———. 1971. *Group Selection*. Chicago: Aldine Atherton.

———. 1996. *Plan and Purpose in Nature: The Limits of Darwinian Evolution*. Science Masters. New York: Phoenix Mass Market.

Williams, G. E. 1993. History of the Earth's obliquity. *Earth-Science Reviews* 34: 1–45.

Williams, H.T.P., and T. M. Lenton. 2008. Environmental regulation in a network of simulated microbial ecosystems. *Proceedings of the National Academy of Sciences of the United States of America* 105 (30): 10432–37.

Williams, R. G., and M. J. Follows. 2011. *Ocean Dynamics and the Carbon Cycle*. Cambridge: Cambridge University Press.

Wilson, R.C.L., S. A. Drury, and J. L. Chapman. 2000. *The Great Ice Age: Climate Change and Life*. London: Routledge and the Open University.

Wilson, R. W., C. M. Wood, R. J. Gonzalez, M. L. Patrick, H. L. Bergman, A. Narahara, and A. L. Val. 1999. Ion and acid-base balance in three species of Amazonian fish during gradual acidification of extremely soft water. *Physiological and Biochemical Zoology* 72 (3): 277–85.

Wood, A. J., G. J. Ackland, J. G. Dyke, H.T.P. Williams, and T. M. Lenton. 2008. Daisyworld: A review. *Reviews of Geophysics* 46 (1): RG1001.

Wright, S., J. Keeling, and L. Gillman. 2006. The road from Santa Rosalia: A faster tempo of evolution in tropical climates. *Proceedings of the National Academy of Sciences of the United States of America* 103 (20): 7718–22.

Wynne-Edwards, Vero Copner. 1962. *Animal Dispersion in Relation to Social Behaviour*. Edinburgh: Oliver and Boyd.

Yamamoto, Y., D. W. Stock, and W. R. Jeffery. 2004. Hedgehog signalling controls eye degeneration in blind cavefish. *Nature* 431 (7010): 844–47.

Yool, A., and T. Tyrrell. 2003. Role of diatoms in regulating the ocean's silicon cycle. *Global Biogeochemical Cycles* 17 (4): 1103.

Zachos, J., M. Pagani, L. Sloan, E. Thomas, and K. Billups. 2001. Trends, rhythms, and aberrations in global climate 65 Ma to present. *Science* 292 (5517): 686–93.

Zachos, J. C., B. N. Opdyke, T. M. Quinn, C. E. Jones, and A. N. Halliday. 1999. Early Cenozoic glaciation, Antarctic weathering, and seawater Sr-87/Sr-86: Is there a link? *Chemical Geology* 161 (1–3): 165–80.

Zeebe, R. E., and K. Caldeira. 2008. Close mass balance of long-term carbon fluxes from ice-core CO_2 and ocean chemistry records. *Nature Geoscience* 1 (5): 312–15.

Zeebe, R. E., and P. Westbroek. 2003. A simple model for the $CaCO_3$ saturation state of the ocean: The "Strangelove," the "Neritan," and the "Cretan" Ocean. *Geochemistry Geophysics Geosystems* 4 (12).

Zettler, L.A.A., F. Gomez, E. Zettler, B. G. Keenan, R. Amils, and M. L. Sogin. 2002. Eukaryotic diversity in Spain's River of Fire: This ancient and hostile ecosystem hosts a surprising variety of microbial organisms. *Nature* 417 (6885): 137.

ACKNOWLEDGMENTS

IT IS A challenge to avoid errors in a work of this scope, spanning many different subject areas. I am extremely grateful to colleagues who have helped me keep these to a minimum, by proofreading chapters or the whole book. I am indebted to Adrian Martin, Andrew Yool, Tyler Volk, Paul Wilson, Rory Howlett, Lynn Rothschild, David Beerling, Emilio Marañón, Andy Gardner, Andy Ridgwell, and Dave Waltham. Any remaining errors are my own. I am grateful to Ingrid Gnerlich for editorial advice and considerable patience, Leslie Grundfest for production, my mother Janet Tyrrell for assistance with readability and quality of writing, and Lizzy Fitzsimmons for assistance with the bibliography and notes. Meghan Kanabay, Dimitri Karetnikov, Draga Gelt, Kate Davis, Barry Marsh, and especially Athena Drakou provided useful help with figures, and the index was generated by Maria denBoer. The book has also benefited considerably from discussions with Norman Kuring, Rod Wilson, Dick Holland, Jim Kasting, David Catling, John Marshall, Lars Hedin, Sven Thatje, Heiko Pälike, Jeffrey Severinghaus, Paul Crutzen, and Lee Kump. I would also like to thank many others not mentioned here for their generosity in answering requests and providing figures.